Encyclopedia of Soybean: Modern Processing Techniques

Volume VI

Encyclopedia of Soybean: Modern Processing Techniques
Volume VI

Edited by **Albert Marinelli and Kiara Woods**

New York

Published by Callisto Reference,
106 Park Avenue, Suite 200,
New York, NY 10016, USA
www.callistoreference.com

Encyclopedia of Soybean: Modern Processing Techniques
Volume VI
Edited by Albert Marinelli and Kiara Woods

International Standard Book Number: 978-1-63239-301-2 (Hardback)

Contents

Preface

Soybean can be employed for various purposes like biodiesel generation, medicinal purposes, textile, human food, etc. In this book, the technologies and novel features for the applicability of soybean and its products are elucidated substantially. It encompasses a variety of topics and consists of contributions which are organized under sections: modern processing technologies and soybean in aspects of food. The contributors of this book have diverse backgrounds and they possess years of experience in this field. This book would be of interest to agro-scientists, cultivators, students and researchers.

All of the data presented henceforth, was collaborated in the wake of recent advancements in the field. The aim of this book is to present the diversified developments from across the globe in a comprehensible manner. The opinions expressed in each chapter belong solely to the contributing authors. Their interpretations of the topics are the integral part of this book, which I have carefully compiled for a better understanding of the readers.

At the end, I would like to thank all those who dedicated their time and efforts for the successful completion of this book. I also wish to convey my gratitude towards my friends and family who supported me at every step.

Editor

Modern Processing Technologies

Soybean Oil De-Acidification as a First Step Towards Biodiesel Production

C. Pirola, D.C. Boffito, G. Carvoli,
A. Di Fronzo, V. Ragaini and C.L. Bianchi
Università degli Studi di Milano – Dipartimento di Chimica
Fisica ed Elettrochimica, Milano
Italy

1. Introduction

According to the predictive studies of the World Energy Outlook 2009, the global demand of energy is expected to increase till 2030 of about 1.5 percentage points per year. Fossil fuels are expected to remain the main energy source in the world, but in the meantime renewable energy sources (wind, solar, geothermal, bioenergy) will be characterized by a rapid growing rate. Their use and development is strongly encouraged by most of the recent regulations. For instance, as reported by the European Environment Agency Transport (EEA), 2009, the European Union required in the same year to achieve by 2020 at least 10% of mixture of hydrocarbons from renewable and conventional sources for what concerns the energy employed for the transports. In addition, the increase in oil price and the growing interest in environmental issues have recently given a considerable impetus to the research for cleaner and renewable energy sources, in order to ensure a sustainable future.

Biodiesel (BD) is a renewable energy source in liquid form that has many advantages over normal diesel, including lower emissions of gases harmful to humans and environment. The UE directive 2003/30/EC, defines the Biodiesel as "a methyl ester produced from vegetable or animal oil, of diesel quality, to be used as biofuel". Moreover, the National Biodiesel Board (NBB), 1996, responsable for biodiesel ASTM standards, define biodiesel as "the mono alkyl esters of long chain fatty acids derived from renewable lipid feedstock's, such as vegetable oils or animal fats, for use in compression ignition (diesel) engines."

The processes for BD production are well known. According to the NBB, 2007, there are three main routes to BD production from oils and fats:

- Base- catalyzed transesterification;
- Direct acid- catalyzed transesterification;
- Conversion of the oil into fatty acids and then into biodiesel.

At the present BD is mainly produced through the base-catalyzed transesterification for many different reasons:

- Mild reaction conditions, i.e. low temperature and pressure may be adopted;
- High conversions (up to 98.5%) are usually achieved in short times with minimization of side reactions;
- The conversion into BD is direct and no intermediate steps are required;

-      No expensive construction materials are required.

According to the base-catalyzed process, BD is produced through the transesterification of triglycerides contained in oils or fats, with methanol and in the presence of an alkaline catalyst, also yielding glycerin as a by-product (Fig. 1).

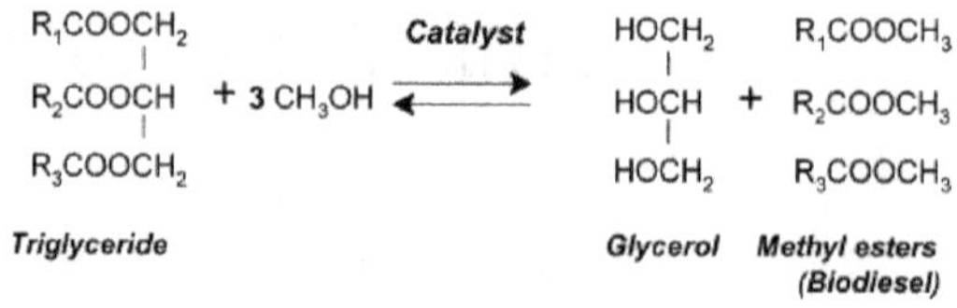

$$R_1COOCH_2 \qquad Catalyst \qquad HOCH_2 \quad R_1COOCH_3$$
$$R_2COOCH \ + 3\,CH_3OH \ \underset{\longleftarrow}{\longrightarrow} \ HOCH \ + \ R_2COOCH_3$$
$$R_3COOCH_2 \qquad\qquad\qquad HOCH_2 \quad R_3COOCH_3$$

Triglyceride   *Glycerol*   *Methyl esters (Biodiesel)*

Fig. 1. Transesterification of a triglyceride for biodiesel production.

Although food-grade oils with low acidity can be employed with few practical problems, their use is strongly discouraged to avoid interference with the human food requirements, besides being not cost-wise competitive with the petroleum-based diesel.

To overcome this problem, waste materials, such as waste cooking oils or animal fat, can be employed.

The use of not refined or waste oils as a feedstock represents a very convenient way in order to lower biodiesel production costs. Crude vegetable oil, waste cooking oils and animal fat are examples of alternative, cheaper, raw materials. The main problem associated with the use of this type of low-cost feedstock lies in its high content of FFA, leading to the formation of soaps during the final transesterification step.

The presence of soaps during the transesterification complicates the reaction resulting in hindering the contact between the reagents and causing difficulties in products separation

Not refined or waste fats require therefore to be standardized by the reduction of the acidity prior to be processed through the transesterification reaction (Bianchi et al., 2010).

However, while BD (pure or mixed) as an alternative fuel to diesel for use in diesel engines is a reality in many states (in France it is usually used in a 5% blend with diesel fuel, in Germany pure, in the USA in the "fleet"), the same cannot be stated for what concerns the use of biofuels as boilers in small, medium or large size plant.

The EU has also published some very restrictive parameters in collaboration with the CEN (*European Committee for Standardization*) to ensure an adequate performance and consequently a higher quality of the BD as biofuel. The required limits for biodiesel properties are listed in the paragraph 4 (European Standard EN 14214).

2. Different technologies for FFA removal from triglycerides (TG)

The starting materials for the biodiesel production are usually vegetable oils and animal fats, indeed constituted mainly by triglycerides (TG) and Free Fatty Acids (FFA), linear carboxylic acids in the range C_{14}-C_{22} with different unsaturation levels. FFA are contained in the oils in their free form as a result of the spontaneous hydrolysis of the starting TG molecules.

Fats have more saturated fatty acids as compositional building blocks than oils. This gives rise to higher melting points and higher viscosities for fats in comparison to oils. The FFA content varies among different lipid sources and also depends on the treatments and storage conditions. In Tab.1 a comparison among soybean oil and other kinds of feedstock is reported, from Lotero et al. (2005).

Fatty acid	Fatty acid composition, wt%							
	Myristic 14:00	Palmitic 16:00	Palmitoleic 16:01	Stearic 18:00	Oleic 18:01	Linoleic 18:02	Linolenic 18:03	Sat. (%)
Soybean oil	0.1	10.6	-	4.8	22.5	52.3	8.2	15.5
Palm oil	1.2	47.9	-	4.2	37	9.1	0.3	53.3
Sunflower oil	-	6.0	-	4.2	18.7	69.3	-	10.2
Lard	1.7	17.3	1.9	15.6	42.5	9.2	0.4	34.6
Yellow grease	2.4	23.2	3.8	13	44.3	7.0	0.7	38.6
Brown grease	1.7	22.8	3.1	12.5	42.4	12.1	0.8	37.0

Table 1.Typical FFA composition of soybean oil and others raw materials for BD production.

As already discussed in the introduction, the chemical transformation of these lipids into biodiesel involves the transesterification of glycerides with alcohols to alkylesters.

Nowadays BD is industrially obtained using alkaline homogeneous catalysts, such as sodium and potassium methoxides and hydroxides. Other possible routes to obtain biodiesel through transesterification and exploiting different catalytic systems are reported in a recent review (Vyas et al., 2010). These include: 1) homogeneous acid catalysis, 2) heterogeneous alkali or acid catalysis; 3) enzymatic catalysis, 4) supercritical conditions without catalyst, 5) microwave or ultrasound assisted reactions. All these methods will be presented in the paragraph 2.2.

Any TG or FFA source (vegetable oil, animal fat or waste grease) may be potentially used as source for biodiesel production through alkali or acid-catalyzed transesterification reaction. In spite of this, a feedstock characterized by a low impurities level and low water and FFA content is required to obtain a valuable, marketable product. In particular, the base-catalyzed transesterification requires high purity reactants (FFA < 0.5wt%, water< 0.1-0.3 wt%), having demonstrated to be very sensitive to the impurities contained in the feedstock (Strayer et al., 1983).

As a matter of fact, the raw material contributes 60-70% to the final manufacture cost of BD obtained from soybean oil. As a consequence, the utilization of expensive raw materials is responsible for the lack of economic competitioness of BD with fossil fuel.

In paragraph 2.1 different methods of performing the transesterification reaction are described, while in paragraph 2.2 various processes to lower the acidity content of the oil are reported.

2.1 Non alkali-catalyzed transesterification for BD production from feedstock with high FFA content

Synthesis of biodiesel via homogeneous acid catalysis: the homogeneous acid-catalyzed reaction rate is reported to be about 4000 time slower than the homogeneous one (Srivastava and Prasad, 2000). Nevertheless, adopting this technology it is possible to perform TG transesterification of not refined oils. Sulphuric acid is reported to be the best performing catalyst. Other homogeneous catalytic systems, such as HCl, BF_3, H_3PO_4 and organic sulphonic acids have also been studied (Liu, 1994). Homogeneous systems require a large molar ratio alcohol to oil (30:1 at least) to reach acceptable reaction rates. On the other hand, by increasing the alcohol amount, the separation costs increase as well.

Reaction rates may also be increased using higher amounts of catalyst. Common catalysts loadings are in the range 1- 5 wt%, while higher catalyst's loadings result in promoting ether formation by alcohol dehydration (Lotero et al., 2005).

The amount of water content in the oil is more critical in the case of the acid-catalyzed transesterification than in the base-catalyzed one (Canakcy and Van Gerpen, 1999). Canakcy reports that esters production can be affected by a water concentration as little as 0.1 wt% and can be almost totally inhibited by water concentrations higher than 5 wt%. This can be explained supposing that water molecules form a sort of shield around the catalyst, preventing its coming in contact with the hydrophobic TG molecules, so inhibiting the reaction. Water can in fact bind acid species in solution more effectively than alcohol. For this reason, in acid catalyzed processes, the water removal step has to be taken into account. The most economical method for water removal from oils is the one acting under gravity separation.

Synthesis of biodiesel via enzymatic transesterification: the enzymatic methods require expensive enzymes such as lipase. On the other hand these methods are affected by water to a less extent than acid-catalyzed process and can tolerate FFA concentration till 30 wt% (Vyas et al, 2010). Besides some advantages such as the mild reaction conditions (50°C for 12-24 h), easy products separation, minimal wastewater treatment and absence of side reactions, there are also some drawbacks such as the contamination of the final product with the residual enzymatic activity and the high costs of this technology (A. Sulaiman, 2007).

Low water contents in the production of BD from soybean oils using lipase as catalyst are reported to lower the enzyme activity (A. Sulaiman, 2007). Nevertheless, an excess of water is not convenient using immobilized lipase (Yuji et al., 1999). Indeed, the water content is a crucial factor, which requires to be optimized basing on the used reaction system.

In any case, the cost of lipase is still the major concern for the industrialization of this technology.

Synthesis of biodiesel via supercritical transesterification: when a fluid or gas is subjected to temperatures and pressures exceeding its critical point, a single fluid phase is present. Solvents containing hydroxyl (OH) groups, as methanol or water, when subjected to supercritical conditions, gain super-acids properties which can be exploited for some kinds of catalysis.

Transesterification reaction of soybean oil in supercritical methanol conditions (350°C; 200 bar) is reported to have been completed in about 25 minutes (Huayang et al., 2007). In supercritical conditions, the use of an excess of alcohol (ratio soybean oil/methanol =1: 40) is also possible, as a single homogeneous phase is present. This results in accelerating the reaction rate as no limitations to the mass transfer due to the presence of interphases occur.

The use of such high temperatures and pressures undoubtedly leads to very huge capital and operating costs and high energy consumption. The scale-up of this process may be therefore very difficult.

Synthesis of biodiesel via Microwave or Ultrasound assisted transesterification: microwave (MW) irradiation activates the smallest degree of variance of polar molecules such as alcohol through the continuous changing of the magnetic field. The production of BD using MW leads to some advantages as short reaction times, low oil/methanol ratio and general reduction of energy consumption (Vyas et al., 2010). MW assisted processes have been studied both in homogeneous and heterogeneous alkali- and acid-catalyzed BD syntheses (Leonelli and Mason, 2010). For this reason MW might be a suitable solution to process feedstock characterized by high initial FFA content. The main problem of the use of MW is the scale-up from the laboratory scale to the industrial plant. The crucial issue is represented by the penetration depth of MW radiation into the absorbing material. Another critical point is the safety aspect concerning the use of this technology, in particular on an industrial scale. Ultrasound (US) is well known as a powerful tool to enhance the reaction rate in a variety of chemical reactions. At high ultrasonic intensities and frequencies between 20 kHz and 100 MHz, a small gas cavity present in the liquid may grow rapidly generating oscillating bubbles; when these bubbles collapse they produce local hot spots of high temperature and pressure able to promote chemical and mechanical effects (Leonelli and Mason, 2010; Colucci et al., 2005). In the biodiesel reacting media, the collapse of these bubbles may be moreover able to disrupt the phase boundary causing emulsification, so impinging one liquid to another as a consequence of the formation of ultrasonic jets (Stavarache et al., 2005). US also introduce turbulence in the system resulting in an improved mechanical mixing: the activation energy required for initiating the reaction can be so easily achieved. Both the advantages and the drawbacks of the US-assisted transesterification are the same described for MW-assisted reaction.

2.2 FFA removal to make oil feedstock suitable to the alkali-catalyzed transesterification

Alkali refining method: in this technology, the removal of FFA is performed adding caustic soda and water to the oil before carrying out the transesterification reaction. In this way the FFA are transformed in fatty acid soaps and then removed by washing. This is a well-established practice in the soybean processing industry (Erikson, 1995). The soybean oil is heated to 70°C and mixed with a caustic solution to form soap and free fatty acids. The amount of FFA measured in the oil determines the flow rate of caustic soda to be added. The washing step is also carried out at 70°C, at a rate of 15% of the crude oil soybean mass flow rate. A certain yield loss occurs as result from the saponification of tryglicerides. The resulting mixture (oil, soap and wash water) is sent to a centrifuge to separate soap and water from the oil. A total quantity of about 1% of oil is lost in the soap and water mixture. The loss of product represents the main drawback of this method. Moreover, this technology gives often rise to problems during the separation phase.

Solvent extraction method: the FFA can be transferred into another phase from the oil one exploiting the difference of solubility in a solvent (e.g. methanol) between the fatty acids and the triglycerides (Ganguli et al., 1998). The oil and the solvent are fed counter-current with a high ratio solvent/oil. After the extraction process, the esterification using H_2SO_4 as acid homogeneous catalyst is performed. The solvent is then separated from the final product, purified and re-used. The main drawbacks of this technology are represented by the high costs lying in the separation and re-use cycle, also due to the presence of emulsions in the extraction reactor.

Hydrolization method: this technology is based on the hydrolyzation of the starting TG into pure FFA and glycerine. This process is typically performed in a counter-current reactor using sulphuric/sulfonic acids and steam. Then, pure FFA undergo to the acid-catalyzed esterification in another counter-current reactor and are converted into methylesters. In this case yields can be higher than 99%. The equipment to be adopted requires being highly acid-resistant.

Glycerolysis: this technique involves the addition of glycerol to the starting TG and the consequent heating to high temperatures (200°C). Zinc chloride is often used as catalyst. This reaction produces mono and dyglicerides, i.e. low FFA oil suitable for the based-catalysed transesterification. A recent patent by Parodi and Marini (WO 2008/007231 A1) deals with this technology and its improvement with a new optimized process design and new kinds of catalysts.

Pre-esterification method: this method will be deeply described in the next paragraphs. The involved technology is based on the esterification of FFA with an alcohol in presence of a homogeneous or heterogeneous acid catalyst. Transesterification is then performed in a second step by using an alkaline homogeneous catalyst.

The not alkali-catalyzed systems for BD production are today used only on the laboratory scale. Moreover, the possibility to maintain and improve the alkali-catalyzed transesterification process as main route to BD production is an important requirement by all the currently working BD plants.

The pre-esterification process is the only method not resulting in a loss of final product, differently from all the other technologies previously described in this paragraph. These last give moreover rise to problems during the separation phase and require therefore high energy exploitation.

The main drawback of the pre-esterification method, if performed with the use of a homogeneous acid catalyst, consists in the necessity of the catalyst's removal from the oil before the transesterification step. This problem, as will be discussed in the next paragraphs, can be solved using a heterogeneous catalyst.

3. Pre-esterification methods by heterogeneous acid catalysis

Nowadays BD synthesis using homogeneous catalysis is considered not advisable. In fact, all processes involving homogeneous catalysis give raise to problems such as product purification and catalyst recovery. In addition, homogeneous acid catalysts are strongly corrosive.

Even a very small amount of residual acid catalyst in the final BD could cause engine problems; hence, an extensive washing with water is required to remove the catalyst residuals from the systems and obtain marketable products.

The use of heterogeneous catalysts prevents neutralization and separation costs, besides being not corrosive, so avoiding the use of expensive construction materials. Another important advantage is that the recovered catalysts can be potentially used for a long time and/or multiple reaction cycles. For all these reasons, the FFA pre-esterification method using heterogeneous acid catalysts is usually preferred to the homogeneously-catalyzed process.

Different solid acid catalysts have been studied in the recent years (Goodwin Jr. et al., 2005). They can be classified into two main categories: inorganic materials and ion-exchange resins functionalized with $-SO_3H$ groups. The main advantage of inorganic materials is represented by their higher thermal stability compared to the resin-based ones. For these last the maximum operating temperatures are in fact around 140° C. Nevertheless, the effectiveness of the $-SO_3H$ active groups for the catalysis of the FFA esterification reaction has been proved even at low temperatures (< 100°C).

Among the different types of inorganic solid materials used for the production of esters, the most popular are the zeolitic compounds. The acid strength of these materials can be modulated changing the Si/Al ratio. In addition, by adopting zeolites as catalysts, it is possible to choose among different pore structures and surface hydrophobicity. Only large-pore zeolites have been used in FFA esterification to avoid limitation to the mass transfer of both reactants and products inside the catalyst's pores. Their use along with high operating temperatures may lead to the formation of undesired by-products (Corma and Garcia, 1997). Silica molecular sieves with amorphous pore walls, as MCM-41, are not sufficiently acid to catalyze the esterification process. The introduction of aluminum, zirconium or titanium into the silica matrix to improve the acid strength is not advisable, due to the easy deactivation to which these materials are usually subjected when water is present in the reaction system (Goodwin Jr. et al., 2005). Another possibility is represented by sulfated zirconia (SO_4^{2-}/ZrO_2), which has already been experimented in other kinds of esterification reactions (Bianchi et al., 2003), both in monophasic and biphasic systems. The main drawback of this type of catalyst lies in its fast deactivation due to the sulphate groups leaching, which may be favored by the presence of water in the system. Others similar materials that can be employed in the FFA esterification reaction are: sulfated tin oxide (SO_4^{2-}/SnO_2), prepared from meta-stanic acid, which is characterized by higher acidity compared to sulfated zirconia, and tungstaned zirconia, characterized by lower acidity but higher resistance to deactivation (Di Serio et al, 2008).

Recently, sulfonated carbons (the so called "sugar catalysts", derived by incomplete carbonization of simple cheap sugar), were reported to have a good performance in the FFA esterification (Takagaki et al., 2006). These carbon-based acids are thermally stable up to 230°C, and are characterized by very low surface area (1-2 m^2 g^{-1}) and amorphous structure. Their high acid strength, due to the electron-withdrawing capacity of the polycyclic aromatic rings, besides to the surface hydrophobicity, makes these catalysts highly suitable for FFA esterification in oils (Goodwin Jr. et al., 2005).

Mixed zinc and aluminium oxide (Bournay et al., 2005) is an inorganic material industrially adopted in the Hepsterfip-H technology, developed by the Institute Français du Petrol and used in a plant producing 160000 t/y started up in 2006 (Santacesaria et al., 2008). In this case the range of the operating temperature is 200-250 °C.

The ion-exchange resins are characterized by a gel structure of microsphere that forms a macroporous polymer (generally copolymers of divinylbenzene and styrene) with sulfonic Brønsted acid groups as active sites. Due to their polymeric matrix, such materials have limited thermal stability (< 140°C) and low structural integrity at high pressure. Their swelling capacity controls substrate accessibility to the acid sites and for some kinds of reactor the effective operating volume of the catalytic bed. Once swelled in a polar medium, such as methanol, the resins pores are able to become macropores, so contributing to reduce the diffusive limitations in the working conditions. Recent studies dealing with the use of acid ion exchange resins demonstrated the possibility to obtain excellent results in FFA esterification in mild temperature and pressure conditions, as reported in the following papers: (Santacesaria et. al., 2005; Pirola et al., 2010) (T= 85°C) and (Bianchi et al., 2010) (T = 65°C). The total pressure inside the system is given by the methanol vapour pressure at the reaction temperature.

Several kinds of ion-exchange resins are commercially available from various producers and differ to each other for what concerns acidity strength, surface area, porosity, swelling, characteristics and disposal of acid groups. In Table 2, some features of a series of Amberlysts by Dow Chemical ® and D5081 resin by Purolite are reported.

A distinguishing feature of A46 and D5081 is represented by the location of the active acid sites: these catalysts are in fact sulphonated only on their surface and not inside the pores. Consequently, A46 and D5081 are characterized by a smaller number of acid sites per gram if compared to other Amberlysts®, which are also internally sulphonated.

Catalyst	A15d	A36d	A39w	A40w	A46w	A70w	D5081
Surface area (m²/g)	53	33	32	33	75	36	n.d.
Average pore diameter (Å)	300	240	230	170	235	220	n.d.
Total pore volume (cc/g)	0.40	0.20	0.20	0.15	0.15	0.20	n.d.
Acidity (meq H+/g)	4.7	5.4	5.0	2.2	0.43	2.55	1.00
Max. operating temperature (K)	393	423	403	413	393	463	403

[a]: Nitrogen BET; [b]: Dry weight

Table 2. Characteristics of some ion-exchange resins (Amberlyst® - Dow Chemical).

The main advantage represented by the use of these catalysts lies in the possibility of adopting very mild reaction conditions. In particular, working at temperatures lower than the methanol boiling point (64,7°C), FFA esterification can be performed without overpressure. In this way no expensive and complex plants are required, making this technology adaptable also for little biodiesel manufacturers. Another interesting aspect of these catalysts is their small deactivation even after long operating periods. In fact, if no particular critical conditions are present in the system during the process (e.g. mechanical fragmentation of the catalyst (Pirola et al., 2010) or presence of metallic ions as Fe^{3+} in the starting TG (Tesser et al., 2010)), no remarkable diminution of the catalytic performance is observed for several operating hours.

Different types of reactors exploiting these ion-exchange resins have been proposed for FFA esterification (Santacesaria et al., 2007; Pirola et al., 2010). The most studied system is a slurry configuration reactor.

The main drawback of the slurry system lies in the fragmentation of the catalyst's particles due to their collision one against the other and against the inner reactor's walls (Pirola et al., 2010).

Alternatives to the slurry reactor are the PFR (Plug Flow) reactor, the Carberry-type reactor, the chromatographic reactor or spray tower loop reactor.

4. Experimental part

4.1 Oil characterization

Oil characterization before proceeding with the standardization of the raw material is a very important issue. Some properties remain in fact unchanged from the starting material to the finished biodiesel, or they are anyway predetermined. It is so important to check that the values of such chemical and physical oil properties are in range with those required by the standard regulations (see Table 3).

The experimental procedures to get the values of such properties are also standardized and are indicated in the regulations. The following are parameters for starting oil that can affect the quality of the final biodiesel.

• Sulfur and Phosphorous content:

High sulphur fuels cause greater engine wear and in particular shorten the life of the catalyst. Biodiesel derived from soybean oil, as well as from pure rapeseed oil, is known to contain virtually no sulphur (Radich, 2004; Zhiyuan et al., 2008).

The phosphorus content of the vegetable oil depends mainly on the grade of refined oil and arises mainly from phospholipids within the starting material. Measurement of the SO_2 from sulphur is accomplished by ultraviolet fluorescence (ASTM D5453, 2002), whereas the analytical method to determine phosphorous requires an Inductively Coupled Plasma Atomic Emission Spectrometry (ASTM International, 2002).

• Linoleic acid methyl ester and polyunsaturated methyl esters

Soy, sunflower, cottonseed and maize oils contain a high proportion of linoleic fatty acids, so affecting the properties of the derived ester with a low melting point and cetane number. Quantitative determination of linoleic acid methyl ester is accomplished by gas chromatography with the use of an internal standard after the substrate has been transesterificated and allows also the quantification of the other acid methyl esters (Environment Australia, 2003).

A typical fatty acid methyl esters composition of soybean oil and other feedstock oils is given in Tab. 1, paragraph 2.

• Iodine Value

The iodine value (IV) is an index of the number of double bonds in biodiesel, and therefore is a parameter that quantifies the degree of unsaturation of biodiesel. Both EN and ASTM standard methods measure the IV by addition of an iodine/chlorine reagent.

Soybean oil is reported to have an IV ranging from about 117 to 143 (Knothe, 1997), having quite the same unsaturation level of sunflower oil.

• Cold Filter Plugging Point

The cold filter plugging point (CFPP) is the temperature at which wax crystals precipitate out of the fuel and plug equipment filters. At temperatures above this point, the fuel should give trouble free flow. These limits are to be decided by each EU member state according to its climate conditions, whereas the US ASTM D 6751 does not set any limit.

The test requires that the sample is cooled and, at intervals of fixed temperature, is drawn through a standard filter so determining the temperature at which the fuel is no longer filterable within a specified time limit.

The CFPP of soybean oil is reported to be around -5°C (Georgianni et al., 2007; Ramos et al., 2009), i.e. accomplishing only a part of the EU members countries (Meher et al., 2006).

• Cetane Number

The cetane number (CN) measures the readiness of a fuel to auto-ignite when injected into the engine. It is also an indication of the smoothness of combustion. The CN of biodiesel depends on the distribution of fatty acids in the original oil. The CN determination is accomplished with the use a diesel engine called *Cooperative Fuel Research* (CFR) engine, under standard test conditions. The CFPP of soybean oil is reported to be higher than 50 (Ramos et al., 2009), so matching in the most cases the limit required by both EN and ASTM biodiesel standards.

4.2 Oil standardization: the esterification reaction

As already remarked in paragraph 3, pre-esterification of FFA in oils assumes great importance to obtain a feedstock suitable to be processed in the transesterification reaction.

In the recent years the authors have deepened the study of the pre-esterification process investigating the effect of the use of different kinds of oils, different types of reactors and catalysts and different operating conditions.

In the following paragraphs, the most relevant aspects of the experimental work and the results obtained by the authors for what concerns the pre-esterification process are reported.

Specification	Units	limits		Method
		Min	Max	
Ester content	% (m/m)	96.5		EN 14103
Density 15°C	kg/m^3	860	900	EN ISO 3675 EN ISO 12185
Viscosity 40°C	mm^2/s	3.50	5.00	EN ISO 3104
Sulphur	mg/kg	-	10.0	preEN ISO 20846 preEN ISO 20884
Carbon residue (10% dist.residue)	% (m/m)	-	0.30	EN ISO 10370
Cetane number		51.0		EN ISO 5165
Sulphated ash	% (m/m)	-	0.02	ISO 3987
Water	mg/kg	-	500	EN ISO 12937
Total contamination	mg/kg	-	24	EN 12662
Cu corrosion max		-		EN ISO 2160
Oxidation stability, 110°C	h (hours)	6.0		EN 14112
Acid value	mg KOH/g	-	0.5	EN 14104
Iodine value	gr I$_2$/100 gr	-	120	EN 14111
Linoleic acid ME	% (m/m)	-	12.0	EN 14103
Methanol	% (m/m)	-	0.20	EN 14110
Monoglyceride	% (m/m)	-	0.80	EN 14105
Diglyceride	% (m/m)	-	0.20	EN 14105
Triglyceride	% (m/m)	-	0.20	EN 14105
Free glycerol	% (m/m)	-	0.02	EN 14105
Total glycerol	% (m/m)	-	0.25	EN 14105
GpI metals (Na+K)	mg/kg	-	5.0	EN 14108 EN14109
Gp II metals (Ca+Mg)	mg/kg	-	5.0	EN14538
Phosphorous	mg/kg	-	5.0	EN 14538

Table 3. Standard specifications for biodiesel (automotive fuels).

In the following table (Table 4) the IV obtained by the authors using the standard procedure are listed for different kinds of not refined feedstock.

Oilseed	Iodine Value (g I$_2$/100 g fat)
Brassica juncea (Indian mustard)	111
Brassica napus (Rapeseed)	115
Cartamus tinctorius (Safflower)	109
Heliantus annus (Sanflower)	143
Nicotiana tabacum (Tobacco)	137
Waste Cooking Oil	54.0

Table 4. Iodine values of some potential feedstock for biodiesel production.

4.2.1 General reaction conditions

A remarkable aspect of the proposed process is represented by the mild operative conditions, i.e. low temperature (between 303 and 338 K) and atmospheric pressure.

Each single reaction has been carried out for six hours withdrawing samples from the reactor at pre-established times and analysing them through titration with KOH 0.1 M. The percentage of FFA content per weight was calculated as otherwise reported (Marchetti & Errazu, 2007, Pirola et al. 2010).

Unless otherwise specified, all the esterification experiments have been conducted using a slurry reactor as the one represented in Fig. 2a. A slurry reactor is the simplest type of catalytic reactor, in which the catalyst is suspended in the mass of the regents thanks to the agitation. In Fig. 2b a typical kinetic curve for the esterification reaction performed with soybean oil is displayed.

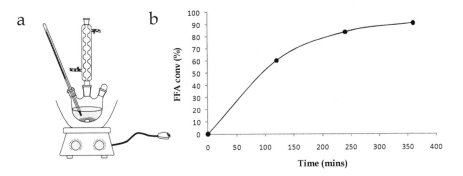

Fig. 2. a) Scheme of the slurry reactor; b) Example of kinetic curve of crude soybean oil with initial acidity=5% (wt.): FFA conversion (%) vs. time, slurry reactor, T=338K, catalyst: Amberlyst® 46, weight ratio alcohol/oil= 16:100, weight ratio catalyst/oil=1:10.

Much attention has been paid by the authors to the use of acid ion exchange resins. Amberlysts ®, i.e. a commercial product by Dow Advanced Materials, and D5081, a catalyst at the laboratory development stage by Purolite® have been successfully applied in this reaction. The main features of the employed catalysts are reported in Tab. 2 and described in paragraph 3.

4.2.2 Effect of the use of different kinds of oil

In Fig. 3 the results from the esterification reaction of different starting oils are shown.

From the graph it can be noted that the lowest final acidity values are obtained with the refined materials, in spite of their initial acidities are the highest due to the addition of pure oleic acid. Refined oils are undoubtedly more easily processable with the esterification in comparison to crude oils, probably due to their lower viscosity which does not result in limitations to the mass transfer of the reagents towards catalysts.

This result has been confirmed by the addition of rapeseed oil, less viscous, to the waste cooking oil in different ratios: increasing the ratio of rapeseed oil to waste oil, the FFA conversion after 6 hours increases.

The differences in the acidic composition seem not to affect the yield of the reaction; in fact, similar values of FFA conversions are obtained for both the soybean oil and the animal fat, in spite of their different acidic compositions. Indicative compositions of some oils used in the experimentation are given in Tab.1

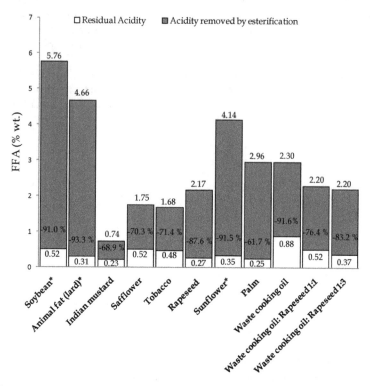

Fig. 3. Acidity removed by esterification (6 h) and residual acidity of different oils used as raw material: slurry reactor, T=338K, catalyst: Amberlyst® 46 weight ratio alcohol/oil= 16:100, weight ratio catalyst/oil=1:10; *commercial, refined fats with the addition of pure oleic acid.

4.2.3 Comparison among different catalysts at different loadings and temperatures

A comparison among the different kinds of Amberlyst at different temperatures has been performed by the authors in a recently published paper dealing with the de-acidification of animal fat (Bianchi et al., 2010). In Fig. 4 the results of this study are summarized.

Under the applied conditions, all the catalysts perform quite well in the esterification reaction, with the exception of A40. Its unsatisfactory performance can be explained taking into account its lower specific surface area and a lower acid site concentration if compared to other Amberlysts. Being these two parameters directly connected to catalytic activity, their simultaneous deficiency is clearly the cause of the unsatisfactory performance.

The catalytic performances of the sample A46 appear to be remarkable, in spite of the low concentration of active acid sites. This result can be explained considering the particular configuration of the catalytic particles, where acid sites are located only on its surface, thus being immediately and easily available for the reaction.

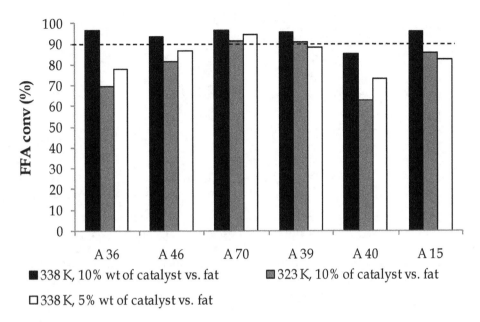

Fig. 4. FFA conversion (%) after 6 h. Comparison of Amberlysts in different operative conditions, initial acidity=5%, slurry reactor, T=338K, weight ratio alcohol/oil= 16:100, weight ratio catalyst/oil=1:10. The dotted line represents the value of FFA conversion necessary to obtain a feedstock with FFA content 0.5% per weight, i.e. suitable for industrial applications.

A70 shows the best performance in all the operative conditions. For this reason, it was further tested to evaluate its catalytic activity in milder operating conditions, i.e. lower temperatures and lower catalyst/fat ratio. The results thus obtained are displayed in Figs. 5a and b, showing that the activity of the catalyst decreases as the reaction temperature or its concentration decrease. However it is worth remarking that even at room temperature.
Catalysts A46 and D5081, have been compared at different temperatures and catalyst's loadings. The results of this study are summarized in the Figs. 6 a and b.
As can be seen from the graphs, catalyst D5081 shows better results than A46 in milder operative conditions. This can be easily explained by the higher number of acid sites located on its surface (compared with Tab. 2). In particular, the use of a ratio of 10% of catalyst D5081 vs. oil allows reaching the maximum conversion in 2 hours. The outcome of this study suggested that a fixed amount of acid active sites per gram of FFA was required to reach the maximum of conversion in 4 hours. Based on the experimental data, this amount was found to be equal to 1,2 meq of H^+.
To verify this hypothesis, different batches of sunflower oil with different initial acidity were prepared and then de-acidified by loading a quantity of A46 corresponding to 1,2 meq of H^+ per gram of FFA. The obtained results are shown in Fig. 7 and confirm the hypothesis set out above. In fact, a complete conversion, corresponding to a FFA concentration lower than 0,5%, is reached after 4 hours regardless of the initial FFA amount.

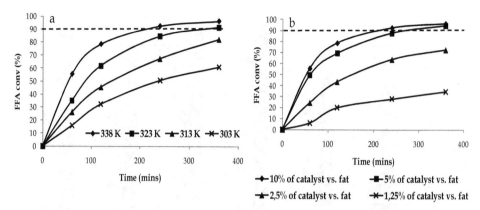

Fig. 5. a) FFA conversion (%) vs reaction time in presence of A70 for different reaction temperatures, initial acidity=5%, slurry reactor, 10 % per weight (wt) of catalyst vs. fat, weight ratio alcohol/oil= 16:100; b) FFA conversion (%) vs. time in presence of A70, slurry reactor, weight ratio alcohol/oil= 16:100, T = 338K with different amounts of the catalysts. The dotted line represents the value of FFA conversion necessary to obtain a feedstock with a FFA content 0.5% per weight, i.e. suitable for industrial applications.

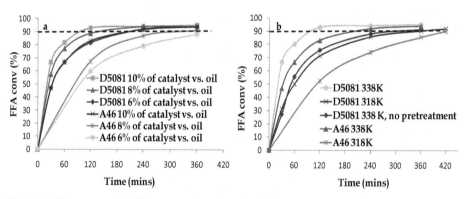

Fig. 6. a) FFA conversion (%) vs reaction time for different amounts of catalysts A46 and D5081, rapeseed oil with initial acidity=5%, slurry reactor, weight ratio alcohol/oil= 16:100, T=338K; b) FFA conversion (%) vs. time at different temperatures, slurry reactor, 10 % per weight (wt) of catalyst vs. fat, weight ratio alcohol/oil= 16:100, T = 338K. The dotted line represents the value of FFA conversion necessary to obtain a feedstock with a FFA content 0.5% per weight, i.e. suitable for industrial applications.

4.2.4 Study of catalysts' lifetime

A crucial parameter for the industrial application is the catalyst lifetime; this parameter has been evaluated by the authors in a recent work (Pirola et al., 2010) by performing ninety consecutive batch de-acidification runs, each lasting 6 hours, were conducted using crude palm oil or soybean oil as a feedstock and Amberlyst® 46 as a catalyst. The final FFA

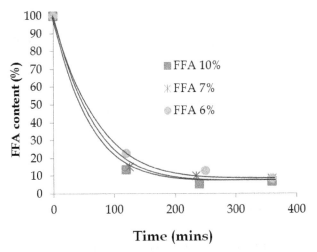

Fig. 7. FFA content (%) vs. reaction time: kinetic curves of the FFA esterification in sunflower oil with different initial acidities using a fixed catalyst/FFA ratio, slurry reactor, weight ratio alcohol/oil= 16:100, T = 338K.

conversions, measured at the end of each of the 6–hour reactions, are reported in Fig. 8 as a function of the run number. At the end of the recycles, a decrease of activity of about 25% was observed, to be probably ascribed to some fragmentation of catalyst's particles. For further details, (Pirola et. al., 2010).

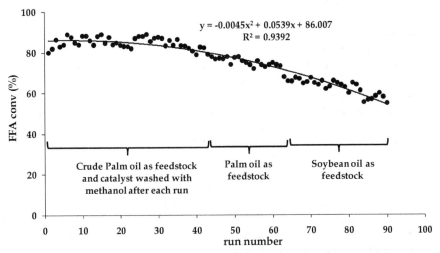

Fig. 8. FFA conversion % after 6 h during 90 successive runs with the same Amberlyst 46® sample. Slurry reactor, T= 338 K, weight ratio alcohol/oil= 16:100, weight ratio catalyst/oil=1:10.

The recycle of the use of catalyst was also performed for D5081 in the FFA esterification of rapeseed oil. The obtained results are outlined in the following graphs (Figs. 9a and b).

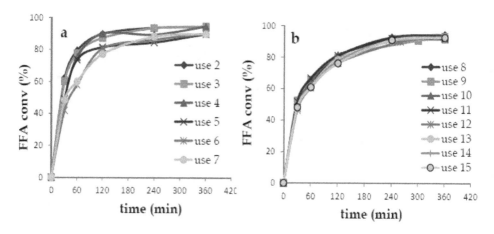

Fig. 9. FFA conversion % vs. reaction time of a) recycles 2 to 7 and b) 8 to 15, rapeseed oil, initial acidity=5% (oleic acid), slurry reactor, T = 338 K, weight ratio alcohol/oil= 16:100, weight ratio catalyst/oil=1:10.

In the initial stage (use 2÷7) the resin does not results in well defined kinetic curves. The reason of slight diminution of FFA conversion with time is probably ascribable to catalyst's settling in the system as it has to adapt to the ambient of reaction before giving a stable performance. From the 8th recycle of catalyst's use on, the curves of FFA conversion overlap: the conversion reached after pre-established times is the same for different runs with the use of the same batch of catalyst.

4.2.5 Reactors

The experimental results discussed in the previous paragraph suggested that a packed-bed reactor, where the catalyst particles are immobilized inside it, could eliminate the mechanical stress of the catalyst particles typical of a slurry reactor. On the other hand, a packed-bed reactor makes the contact between the organic phase (oil/FFA mixture) and methanol less effective. For this reason, in the employed experimental setup a mixing chamber was located just before the catalytic reactor.

The reaction in both continuous and semicontinuous modes was conducted using the experimental setup shown in Fig. 10.

The methanol/oil mixture is taken from the vessel (a), where it is continuously mechanically stirred (b), and then it is admitted into the mixing chamber (d) by a pump (c). This chamber (0.2 L) is located just before the catalytic reactor in order to obtain the maximum contact between oil and methanol (not fully soluble) inside the catalytic bed (e). The catalytic reactor (0.5 L) contains a packed bed of Amberlyst 46® (7 g). The pump flow is maintained at 10 mL min⁻¹, so obtaining a contact time in the catalytic bed equal to 1 min.

In the semi-continuous experiments the reaction stream, leaving the catalytic reactor (e) was returned to the vessel (a); in continuous experiments the reaction stream from the catalytic reactor (e) was continuously discharged from the system.

In Fig. 11, the results obtained for both crude palm oil and soybean oil are reported using the experimental setup shown in Fig. 11 as a semicontinuous reactor.

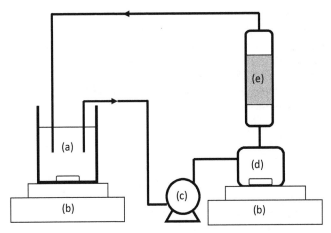

Fig. 10. Continuous or semi-continuous experimental set-up: (a) feeding vessel, (b) mechanic stirrer, (c) pump, (d) mixing chamber, (e) catalytic packed bed reactor.

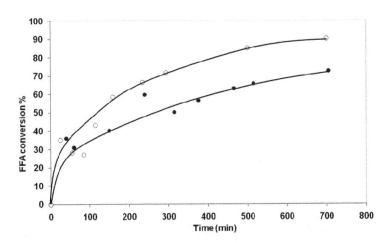

Fig. 11. Semi-continuous experiments with 7 g of catalyst; FFA conversion % vs. time using crude palm oil (o) or soybean oil (•). Pump flow = 10 mL min^{-1}; T = 338 K, molar ratio FFA/alcohol = 1: 6, weight ratio catalyst/ oil = 1: 10.

The FFA conversion increases with time using both oils, but it is higher using the crude palm oil (FFA conversion after 700 min: about 90% and 70% for palm and soybean oil, respectively). This result can be explained considering the difference between the two different raw materials, which concerns both the composition of the substrate (different unsaturation levels between the two oils) and the composition of the FFA (only oleic acid for soybean oil and a mixture of C_{12}-C_{18} acids for crude palm oil). These differences obviously affect both the lifetime of the oil/methanol emulsion and the diffusional aspects along the catalytic bed.

To improve the stability of the methanol/oil emulsion, the authors substituted the classical mixing chamber with an emulsificator based on five co-axial rotating ring gears which are able to break the biphasic mixture into very tiny drops. Using this device and starting from two entirely separated liquid phases (oil and methanol) it was possible to obtain a much more stable emulsion. In Fig. 12 a comparison between FFA conversion obtained with the classical mixing chamber and the emulsificator is reported: the better results reached using the emulsificator are evident.

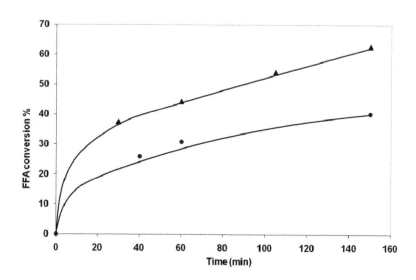

Fig. 12. Semi-continuous experiment: FFA conversion % vs. time using crude palm oil as feedstock and having before the catalytic bed: classical mechanic stirring reactor (•) or emulsificator at 500 rpm (▲). Pump flow = 10 mL min^{-1}; T = 338 K, molar ratio FFA: alcohol = 1: 6, weight ratio catalyst: oil = 1: 10.

The methanol/oil emulsion lifetime without the emulsificator device, measured in the same way at the exit of the mixing chamber, is of about 15 seconds. Referring to our system, the contact time between the catalyst and the reactant is 1 min for 7 g of catalyst (pump flow = 10 mL min^{-1}): for this reason without the emulsificator device a significant part of the catalytic bed (about 75%) does not work.

4.3 Oil transformation: the transesterification reaction and biodiesel characterization

The transesterification reaction has been performed by the authors on the raw materials de-acidified with the esterification process described in the previous paragraph.

Sodium methoxide (MeONa) was employed as catalyst. MeONa is known to be the most active catalyst for triglycerides transesterification reaction, but it requires the total absence of water (Schuchardt, 1996). For this reason, the unreacted methanol and the reaction water were evaporated from the de-acidified oils before processing them with the transesterification reaction.

The employed experimental setup was the same as displayed in Fig. 2.

Being the transesterification an equilibrium reaction, it was performed in two steps, removing the formed glycerine after the first step. The adopted conditions were the following:

- 1st step: weight ratio methanol/oil=20:100, weight ratio MeONa/oil=1:100, 233 K, 1,5 h
- 2nd step: weight ratio methanol/oil=5:100, weight ratio MeONa/oil=0.5:100, 233 K, 1 h

The total ester content is a measure of the completeness of the transesterification reaction. Many are the factors that affect ester yield in the transesterification reaction: molar ratios of glycerides to alcohol, type of catalyst(s) used, reaction conditions, water content, FFA concentration, etc. (Environment Australia, 2003).

The European preEN14214 biodiesel standard sets a minimum limit for ester content of >96.5% mass, whereas the US ASTM D 6751 biodiesel standard does not set a specification for ester content.

Mono- and di-glycerides as well as tri-glycerides can remain in the final product in small quantities. Most are generally reacted or concentrated in the glycerine phase and separated from the ester.

The analyses of methyl esters and unreacted mono-, di- and triglycerides are accomplished through gas chromatography.

The detailed requirements for biodiesel according to both EN 14214 and US ASTM D 6751 are listed in paragraph 1.

5. Simulation

In order to develop a process simulation of the FFA esterification, able to predict the reaction progress, a thermodynamic and kinetic analysis was performed.

5.1 Thermodynamic aspects

The considered reaction system turns out to be a highly non-ideal system, being formed by a mixture of oil, methylester, methanol, FFA and water. The interactions among these molecules are absolutely not ideal, in fact they are only partially soluble and a two phase system is formed if the quantity of methanol is greater than 6-8 wt%.

Indeed, the activity coefficients are used not only for the phase and chemical equilibria calculations, but also for the kinetic expressions. Modified UNIFAC model was used adopting the parameters available in literature and published by Gmehling et al. (2002).

5.2 Reaction kinetics

The considered reaction is the following (Fig.13).

$$ FFA + methanol \rightleftharpoons methylester + water $$

Fig. 13. FFA esterification reaction and hydrolysis (indirect reaction) considered in the process simulation.

The oil is considered as non-reacting solvent, being present in large quantity.

A pseudohomogeneous model was used for describing the kinetic behaviour of the reaction (Pöpken et al., 2000). The adopted model is displayed in the following equation:

$$r = \frac{1}{m_{cat}} \frac{1}{\upsilon_i} \frac{dn_i}{dt} = k_1 a_{FFA} a_{methanol} - k_{-1} a_{methylester} a_{water}$$

where:

r= reaction rate

m_{cat}= dry mass of catalyst, gr

υ_i= stoichiometric coefficients of component i

n_i= moles of component i

t = reaction time

k_1= kinetic constant of direct reaction

k_{-1}= kinetic constant of indirect reaction

a_i= activity of component i

The temperature dependence of the rate constant is expressed by the Arrhenius law:

$$k_i = k_i^0 \exp\left(\frac{-E_{a,i}}{RT}\right)$$

where k_i^0 and $E_{a,i}$ are the pre-exponential factor and the activation energy of the reaction i, respectively (i=1 for the direct reaction, i=-1 for the indirect reaction), T is the absolute temperature and R the Universal Gas Constant.

The adopted parameters set is the same reported by Steinigeweg (Steinigeweg & Gmehling, 2003). The absolute values of pre-exponential factors were corrected as reported in Table 5, so to take into account the presence of both a second liquid phase and a different type of catalyst.

Reaction (i)	E_A (kJ mol^{-1})
Esterification (1)	68.71
Hydrolisis (-1)	64.66

Table 5. Kinetic Parameters for the adopted pseudohomogeneous Kinetic Model.

The ratio of pre-exponential factors is: $K_{eq}^T = k_1^0/k_{-1}^0 = 60.7841$.

All the simulations were carried using Batch Reactor of PRO II by Simsci – Esscor.

In the next picture (Fig. 14) the comparison between the experimental and calculated FFA conversion is shown for the esterification reaction of an acid rapeseed oil.

The initial composition (% by weight) entered in the simulation program is: oil (83.2%), FFA (3.47 %), methanol (13.3%).

The model turned out to be able to reproduce qualitatively the behaviour of different systems, characterized by different starting acidities values, at different temperatures and with an high impurities content.

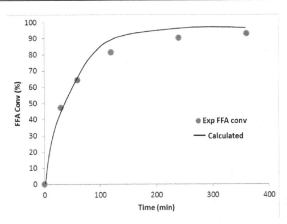

Fig. 14. Experimental and calculated (pseudohomogeneous model) FFA conversion % vs. reaction time, rapeseed oil, initial acidity= 3.47% (oleic acid), slurry reactor, T= 338 K, weight ratio alcohol/oil=16:100, weight ratio catalyst/oil=1:10.

6. Conclusions

The use of not refined or waste oils as a feedstock represents a very convenient way in order to lower biodiesel production costs. The main problem associated with the use of this type of low-cost feedstock lies in its high content of FFA, leading to the formation of soaps during the final transesterification step. These materials require therefore to be standardized by the reduction of their acidity and different de-acidification methods have been described in this chapter. Among them, a new technology based on an esterification reaction heterogeneously catalyzed and performed at mild operative conditions, i.e. low temperature (between 303 and 338 K) and atmospheric pressure has been proposed and described. Several kinds of ion-exchange resins, commercially available, have been used as heterogeneous catalysts, different one form the other for what concerns acidity strength, surface area, porosity, swelling, characteristics and disposal of acid groups.

The experimental tests were performed using different reactors (CSTR or PFR), starting oils (in comparison with the results obtained for soybean oil), catalyst/oil ratio and working temperature. All these experimental parameters have been optimized in order to obtain, at the end of the reaction, a concentration of FFA suitable for the transesterification reaction for the biodiesel production (FFA < 0.5 wt%). A crucial parameter for the industrial application is the catalyst lifetime and this parameter has been evaluated by performing ninety consecutive batch de-acidification runs, each lasting 6 hours, with the same catalyst sample (Amberlyst 46®) in a slurry reactor. At the end of the recycles, a decrease of activity of about 25% was observed, to be ascribed to some fragmentation of catalyst's particles, that collide against one another and against the reactor walls. To overcome this problem a packed bed configuration have been adopted and optimized. At last, a process simulation of the FFA esterification, able to predict the reaction progress, through a thermodynamic and kinetic analysis, was successfully performed. A pseudohomogeneous model was used for describing the kinetic behaviour of the reaction, using a modified UNIFAC model for the calculation of the activity coefficients (used not only for the phase and chemical equilibria calculations, but also for the kinetic expressions).

7. Acknowledgements

The authors gratefully acknowledge the financial support by the Italian Ministero delle Politiche Agricole, Alimentari e Forestali (project SUSBIOFUEL – D.M. 27800/7303/09).

8. References

Allen, C.A.W. Watts, K.C., Ackman, R.G. & Pegg M.J. (1999). Predicting the Viscosity of Biodiesel Fuels from their Fatty Aid Ester Composition. *Fuel*, Vol. 78, (September 1999), pp. 1319-1326

ASTM International (2002). Standard Specification for Biodiesel Fuel - Blend Stock (B100) for Distillate Fuels - ASTM D 6751-02, in *Annual Book of ASTM Standards*, Vol. 05.04

ASTM International (2002). Determination of Total Sulfur in Light Hydrocarbons, Motor Fuels and Oils by Ultraviolet Fluorescence Method D5453 – 00 (2002), in *Annual Book of ASTM Standards*, Vol. 05.03

Bianchi, C.L., Boffito, D.C., Pirola, C., Ragaini, V. (2010). Low temperature de-acidification process of animal fat as a pre-step to biodiesel production. *Catalysis Letters*, Vol. 134, (November, 2009), pp. 179-183.

Bianchi, C.L., Ragaini, V., Pirola, C., Carvoli, G. (2003). A new method to clean industrial water from acetic acid via esterification, *Applied Catalysis B Environmental*, Vol. 40, (June 2003), pp. 93-99.

Bournay, L., Casanave, D., Delfort, B., Hillion, G., Chodorge, J.A., (2005). New heterogeneous process for biodiesel production: a way to improve the quality and the value of the crude glycerin produced by biodiesel plants. *Catalysis Today*, Vol. 106, (2005). pp. 190-192.

Canakci, M., Van Gerpen, J. (1999). Biodiesel production via acid catalysis. *Trans ASAE*, Vol. 42, (1999), pp. 1203-1210.

Colucci, J.A., Borrero, E., Alape, F. (2005). Biodiesel from an alkaline transesterification reaction of soybean oil using uyltrasonic mixing. *Journal of the American Oil Chemists' Society,*, Vol. 82, (2005), pp. 525-530.

Corma, A., Garcia, H. (1997). Organic reactions catalyzed over solid acids. *Catalysis Today*, Vol. 38, (April 1997), pp. 257-308. ISNN 0920-5861

Di Serio, M., Tesser, R., Pengmei, L., Santacesaria, E. (2008). Heterogeneous catalysis for biodiesel production. *Energy & Fuel*, Vol. 22. (September 2008), pp. 207-217.

Environment Australia (2003). National Standards for Biodiesel – Discussion Paper, In *Setting National Fuels Quality Standards*, 0 642 54908 7 pp. 1-196, Retrieved from <www.ea.gov.au/atmosphere/transport/biodiesel/index.html>

Erickson, D.R., Alkaline Refining. (1995). *Practical Handbook of Soybean Processing and Utilization*, Erickson, D.R., ed, AOCS Press, Champlain, Illinois, 1995.

Ganquli, K.L. ,van Immersel, A.R., Michaelides, G.C., van Putte, K.P., Turksma, H., (1998), U. S Pat. Nos. 1,371,342

Georgianni, K.G. Kontominas, M.G. Tegou, E. Avlonitis, D. & Gergis V. (2007). Biodiesel Production: Reaction and Process Parameters of Alkali-Catalyzed Transesterification of Waste Frying Oils. *Energy Fuels*, Vol 21, No. 5, pp. 3023-3027

Gmehling, J. Wittig, R. Lohmann, J. Joh, R. (2002). A Modified UNIFAC (Dortmund) Model. 4. Revision and Extension. *Industrial & Engineering Chemistry Research*, Vol. 41, pp. 1678-1688.

Huayang, H., Tao, W., Shenlin, Z. (2007). Continuous production of biodiesel fuel from vegetable oil using supercritical methanol process. *Fuel*, Vol. 86, (2007), pp. 442-447.

Knothe, G. Dunn, R.O. & Bagby M.O. (1997). Biodiesel: the Use of Vegetable Oils and their Derivates as Aternative Fuels, In *Fuels and Chemicals from Biomass*, Badal, C. Saha, J., pp. 172-208, American Chemical Society, , ACS Symposium Series, Retrieved from <http://pubs.acs.org/doi/abs/10.1021/bk-1997-0666.ch010>

Leonelli, C., Mason, T.J. (2010). Microwave and ultrasonic processing : Now a realistic option for industry. *Chemical Engineering and Processing*, Vol. 49, (June 2010), pp. 885-900.

Liu, K.S. (1994). Preparation of Fatty-Acid Methyl Esters for Gas-Chromatographic Analysis of Lipids in biological materials. *J. Am. Oil Chem Soc.*, Vol. 71 (1994), pp. 1179-1187.

Lotero, E., Liu, Y., Lopez, D.E., Suwannakarn, K., Bruce, D.A., Goodwin Jr., J.G. (2005). Synthesis of Biodiesel via acid Catalysis. *Industrial & Engineering Chemistry Research*, Vol. 44, (January 2005), pp. 5353-5363.

Marchetti, J.M. & Errazu, A.F. (2008) Comparison of Different Heterogeneous Catalysts and Different Alcohols for the Esterification Reaction of Oleic Acid. *Fuel*, Vol. 87, (June 2008), pp. 3477-3480, ISSN 0016-2361

Meher, L.C., Vidaya Sagar, D. & Naik S.N. (2006). Technical Aspects of Biodiesel Production by transesterification – a Review. *Renewable and Sustainable Energy Reviews*, Vol. 10, (2006), pp. 248-268

Moser, B.R. Haas, M.J. Jackson M.A. Erhan S.Z. & List G.R. (2007). Evaluation of Partially Hydrogenated Methyl Esters of Soybean Oil as Biodiesel. *European Journal of Lipid Science and Technology*, Vol. 109, (2007), pp. 17-24

Öezbay, N. Oktar, N. &. Tapan, N. (2008). Esterification of Free Fatty Acids in Waste Cooking Oils (WCO): Role of Ion-exchange Resins. *Fuel*, Vol. 87, No. 10-11, pp. 1789-1798

Pirola, C., Bianchi, C.L., Boffito, D.C., Carvoli, G., Ragaini, V. (2010). Vegetable oil deacidification by Amberlyst : study of catalyst lifetime and a suitable reactor configuration. *Industrial & Engineering Chemistry Research*, Vol. 49 (2010), pp. 4601-4606.

Pöpken, T. Götze, L. Gmehling, J. (2000). Reaction Kinetics and Chemical Equilibrium of Homogeneously and Heterogeneously Catalyzed Acetic Acid Esterification with Methanol and Methyl Acetate Hydrolysis. *Industrial & Engineering Chemistry Research*, Vol. 39, pp. 2601-2611.

Prankl H & Worgetter M. (1996). Influence of the Iodine Number of Biodiesel to the Engine Performance. *Proceedings of the Alternative Energy Conference*, American Society of Agricultural Engineers, 1996

Radich, A. (1998). Biodiesel Performance, Costs, and Use. Energy Information Administration Retrieved from <http://www.eia.doe.gov/oiaf/analysispaper/biodiesel/pdf/biodiesel.pdf>

Ramos, M.J. Fernández, C.M. Casas, A. Rodríguez, L. & Pérez, Á. (2009). Influence of Fatty Acid Composition of Raw Materials on Biodiesel Properties. *Bioresource Technology*, Vol. 100, pp. (261-268), ISSN 0960-8524

Santacesaria, E., Tesser, R., Di Serio, M., Guida, M., Gaetano, D., Garcia Agrda, A., Cammarota, F. (2007). Comparison of different reactor configurations for the reduction of free acidity in raw materials for biodiesel productio. *Industrial & Engineering Chemistry Research*, Vol. 46 (March, 2007), pp. 8355-8362.

Schäfer A. (1991). Pflanzenölfettsäure-Methyl-Ester als Dieselkraftstoffe / Mercedes-Benz, *Proceedings of Symposium Kraftstoffe aus Pflanzenöl für Dieselmotoren*, Technische Akademie, Esslingen, Germany, 1991

Schuchardt, U. Vargas, R.M. & Gelbard, G. (1996). Transesterification of soybean oil catalyzed by alkylguanidines heterogenized on different substitude polystyrenes. *Journal of Molecular Catalysis A: Chemical A*, Vol. 109, pp. 1381-1169

Srivastava, A., Prasad, R. (2000). Tryglicerides-based diesel fuel. *Renewable Sustainable Energy Reviews*, Vol. 4, (2000), pp. 111-133.

Stavarache, C., Vinatoru, M., Nishimura, M., Maeda, Y. (2005). Fatty Acids methyl esters from vegetable oil by means of ultrasonic energy. *Ultrasonics Sonochemistry*, Vol. 12, (2005), pp. 525-530.

Steinigeweg, S. & Gmehling, J. (2003). Esterification of a Fatty Acid by Reactive Distillation. *Industrial & Engineering Chemistry Research*, Vol. 42, pp. 3612-3619

Steinigeweg, S. & Gmehling, J. (2003). Esterification of a Fatty Acid by Reactive Distillation. *Industrial & Engineering Chemistry Research*, Vol. 42, pp. 3612-3619.

Strayer, R. C., Blake, J. A., Craig, W. K. (1983). Canola and high erucic rapeseed oil as substitutes for diesel fuel: Preliminary tests. *Journal of the American Oil Chemists Society*, Vol. 60, (1983),pp. 1587-1592.

Sulaiman, A. (2007). Production of biodiesel: possibilities and challenges. *Biofuel Bioprod. Bioref.* Vol. 1, (2007), pp. 57-66.

Takagaki, A., Toda, M., Okamura, M., Kondo, J. N., Hayashi, S., Domen, K., Hara, M. (2006). Esterification of higher fatty acids by a novel strong solid acid. *Catalysis Today*, Vol. 116 (2006), pp. 157-161.

Tate, R.e. Watts, K.C. Allen, C.A.W. & Wilkie K.I. (2006). The Viscosities of Three Biodiesel Fuels at Temperatures up to 300°C. *Fuel*, Vol. 85, pp. 1010-1015, ISSN 0016-2361

Tesser, R., Di Serio, M, Casale, L., Sannino, L., Ledda, M., Santacesaria, E. (2010). Acid exchange resins deactivation in the esterification of free fatty acids. *Chemical Engineering Journal*, Vol. 161, (April, 2010), pp. 212-222.

Tesser, R., Di Serio, M., Guida, M., Nastasi, M., Santacesaria, E. (2005). Kinetics of oleic acid esterification with methanol in the presence of triglycerides. *Industrial & Engineering Chemistry Research*, Vol. 44 (2005), pp. 7978-7982.

Vyas, A.P., Verma, J.L., Subrahmanyam, N. (2010). A review on FAME production processes. *Fuel*, Vol. 89, (August 2009), pp. 1-9

Yuji, S., Yomi, W., Taichi, S., Akio, S., Hideo, N., Hideki, F., Yoshio, T. (1999). Conversion of vegetable oil to biodiesel using immobilized Candida Antarctica Lipase. *Journal of the American Oil Chemists' Society*, Vol. 76, (1999), pp. 789-793.

Zheng, S. Kates, M. Dubé M.A. & McLean D.D. (2006) Acid-catalyzed Production of Biodiesel from Wastre Frying Oil. *Biomass and Bioenergy*, Vol. 30, No 3, pp. 267-272

Zhiyuan, H. Piqiang, T. Xiaoyu & Y. Diming, L. (2008). Life cycle energy, environment and economic assessment of soybean-based biodiesel as an alternative automotive fuel in China. *Energy*, Vol. 33, pp. 1654-1658, ISSN 0360-5442

2

Processing of Soybean Oil into Fuels

Joanna McFarlane
Oak Ridge National Laboratory[1]
USA

1. Introduction

1.1 Rationale for processing soybean oil into a fuel

Abundant and easily refined, petroleum has provided high energy density liquid fuels for a century. However, recent price fluctuations, shortages, and concerns over the long term supply and greenhouse gas emissions have encouraged the development of alternatives to petroleum for liquid transportation fuels (Van Gerpen, Shanks et al. 2004). Plant-based fuels include short chain alcohols, now blended with gasoline, and biodiesels, commonly derived from seed oils. Of plant-derived diesel feedstocks, soybeans yield the most of oil by weight, up to 20% (Mushrush, Willauer et al. 2009), and so have become the primary source of biomass-derived diesel in the United States and Brazil (Lin, Cunshan et al. 2011). Worldwide ester biodiesel production reached over 11,000,000 tons per year in 2008 (Emerging Markets 2008). However, soybean oil cannot be burned directly in modern compression ignition vehicle engines as a direct replacement for diesel fuel because of its physical properties that can lead to clogging of the engine fuel line and problems in the fuel injectors, such as: high viscosity, high flash point, high pour point, high cloud point (where the fuel begins to gel), and high density (Peterson, Cook et al. 2001).

Industrial production of biodiesel from oil of low fatty-acid content often follows homogeneous base-catalyzed transesterification, a sequential reaction of the parent triglyceride with an alcohol, usually methanol, into methyl ester and glycerol products. The conversion of the triglyceride to esterified fatty acids improves the characteristics of the fuel, allowing its introduction into a standard compression engine without giving rise to serious issues with flow or combustion. Commercially available biodiesel, a product of the transesterification of fats and oils, can also be blended with standard diesel fuel up to a maximum of 20 vol.%. In the laboratory, the fuel characteristics of unreacted soybean oil have also been improved by dilution with petroleum based fuels, or by aerating and formation of microemulsions. However, it is the chemical conversion of the oil to fuel that has been the area of most interest. The topic has been reviewed extensively (Van Gerpen, Shanks et al. 2004), so this aspect will be the focus in this chapter. Important aspects of the chemistry of conversion of oil into diesel fuel remain the same no matter the composition of

---

[1] This manuscript has been authored by UT-Battelle, LLC, under Contract No. DE-AC05-00OR22725 with the U.S. Department of Energy. The United States Government retains and the publisher, by accepting the article for publication, acknowledges that the United States Government retains a non-exclusive, paid-up, irrevocable, world-wide license to publish or reproduce the published form of this manuscript, or allow others to do so, for United States Government purposes.

the triglyceride. Hence, although the focus in this book is on soybean oil, studies on other plant based oils and simulated oils have occasional mention in this chapter. Valuable data can be taken on systems that are simpler than soybean based oils, with fewer or shorter chain components. Sometimes the triglycerides will behave differently under reaction conditions, and when relevant, these have been noted in the text.

1.2 Transesterification and homogeneous base catalysis

Processing of soybean oil into a diesel compatible fuel through transesterification has received much recent attention as the most likely route to large-scale adoption of bio-based diesel. To improve flow characteristics, the triglyceride that constitutes the soybean oil has to be broken apart into smaller molecules. Fragmentation of the triglyceride takes place through a transesterification mechanism, a three step process that yields a molecule of esterified fatty acid at each step, shown below in Reaction (1) (Freedman, Pryde et al. 1984). Initially, the soybean oil reacts with a molecule of methanol, in the form of a reactive methylate in the case of base catalysis, to cleave a long-chain fatty acid fragment from the glycerine backbone that becomes a methyl ester, depicted as R^1 in the reaction below. The residual chains (R^2,R^3) attached to the backbone comprise a diglyceride after the first step, a monoglyceride after the second step, before the final decomposition to glycerine, or 1,2,3-propanetriol, and an ester at the last step. Commercially, a base such as sodium hydroxide or methylate is used to catalyze the transesterification process, promoting the reaction between the alcohol and the oil.

$$C_3H_5(CO_2R^1)(CO_2R^2)(CO_2R^3) + CH_3OH \rightarrow CH_3O_2R^1 + C_3H_5(OH)(CO_2R^2)(CO_2R^3)$$
$$C_3H_5(OH)(CO_2R^2)(CO_2R^3) + CH_3OH \rightarrow CH_3O_2R^2 + C_3H_5(OH)_2(CO_2R^3) \qquad (1)$$
$$C_3H_5(OH)_2(CO_2R^3) + CH_3OH \rightarrow CH_3O_2R^3 + C_3H_8O_3$$

Triglyceride + 3 Methanol → 3 Methyl Esters + Glycerine

In commercial parlance, the glycerine that is produced by transesterification is termed free glycerine, and the unreacted tri-, di-, and monoglycerides are called bound glycerine, usually expressed as wt.%. The base catalyst is usually introduced as anhydrous sodium methylate, to minimize the amount of water in the system as this leads to saponification, Reaction (2). The amount of base catalyst typically used is only slightly over 1 vol.% of the methanol, again to reduce formation of soapy emulsions. While the stoichiometry of the process demands a mole ratio of methanol to oil of 3, commercially the ratio is doubled to push the reaction to completion. In the US, biodiesel must have a bound glycerine content of less than 0.24 wt.% and a free glycerine content of less than 0.3 wt.% to be sold commercially. Standards for biodiesel purity are based either on the removal of contaminants before the oil feedstock is esterified or on the separation of unwanted by-products (ASTM 2007; ASTM 2008).

$$CH_3O_2R + NaOH \xrightarrow{H_2O} NaO_2R + CH_3OH \qquad (2)$$

Methanol and base catalyst (in the form of NaOH or sodium methylate) are the reagents of choice in industrial production because of their being less expensive than other reagents. Potassium hydroxide has the advantage of a lower rate of saponification. Other alcohols can be used, primarily ethanol. Longer chain alcohols have better miscibility with the oil and hence higher yields, however, the product esters also become more difficult to separate from glycerine.

1.3 Esterification and homogeneous acid catalysis

The conversion of fatty acids to esters can also be catalyzed by acids in the esterification reaction scheme shown in Equation (3). The mineral acids commonly used as catalysts include sulfuric or hydrochloric acid. This chemical route is less popular when working with good quality soybean oil as a feedstock, because of the low free fatty acid content of the oil. For degraded or lower quality feedstocks, however, the advantages of avoiding large amounts of bound and free glycerine production as happens during transesterification can be desirable. Before triglycerides can be subjected to esterification, they must be saponified using base, such as NaOH, to strip apart the acylglyceride chains. Treatment with acid follows to protonate and form fatty acids. In the expression (3) below, the acid allows a complex to form between the triglyceride and the alcohol, which then falls apart to give a methyl ester and the diglyceride. Similarly to transesterification, the reaction progresses sequentially through a number of steps, not all shown in (3).

$$C_3H_5(CO_2R)_2(CO_2R) + CH_3OH \underset{}{\overset{H^+}{\rightleftharpoons}} C_3H_5(CO_2R)_2(COROH(CH_3OH))$$

$$C_3H_5(CO_2R)_2(COROH(CH_3OH)) \rightarrow C_3H_5(CO_2R)_2OH + CH_3O_2R$$

(3)

1.4 Conventional processing of soybean oil into methyl esters

Vegetable oils, including soybean oil, have complex compositions, which include a variety of fatty acid chain lengths. Soybean oil consists primarily of palmitic, oleic, linoleic, and linolenic acid chains, with a typical mixture given in Table 1 (Holčapek, Jandera et al. 2003). The actual composition depends on the source of the oil and can vary from one variety to another (Mello, Pousa et al. 2011). In addition to the variation in the fatty acid chains linked to the glyceryl backbone, processing of soybean oil will induce some degradation in a fraction of the triglyceride molecules to yield free fatty acid fragments. These compounds will not undergo base-catalyzed transesterification and must be esterified under acidic catalytic conditions.

Separation of the free fatty acids from the intact triglyceride molecules prior to conversion to esters is one of the challenges of processing soybean oil to biodiesel. Food grade soy oil can have very low free fatty acid content, less than 4%, in comparison with other oils, such as olive oil with up to 20%. However, lower quality feedstocks being considered for fuel production have higher free fatty acid content, with the highest concentration being present in waste oil that has usually been subjected to repeated heating cycles before being salvaged for biodiesel production. To process waste soybean oil, a combination of transesterification and esterification can be used, shown schematically in Figure 1. The waste oil passes through a centrifugal separator to remove water and suspended solids. The oil then moves to a tank of acid catalyst, H_2SO_4, and methanol. Upon esterification, three phases will form and separate: a rag layer containing acid, water, and methanol, a layer of unreacted oil, and the esterified products on the bottom of the tank. The lower two layers go through to the transesterification reactor, a reaction environment that does not degrade the already-formed methyl esters.

Triglyceride Chains	Normalized Mole Fraction	Ln (Linolenic) C18:3	L (Linoleic) C18:2	O (Oleic) C18:1	P (Palmitic) C16:0
LLLn	0.18	1	2	0	0
LLL	0.34	0	3	0	0
OLL	0.27	0	2	1	0
LLP	0.21	0	2	0	1
Mole fraction		0.083	0.751	0.083	0.083
Molecular weight, g	278	278	280	282	256

Table 1. Triglyceride Composition of Soybean Oil.

Recent expansion of biodiesel manufacture has resulted in increased interest among commercial enterprises to minimize the cost of feedstock materials and waste production and to maximize the efficiency of production. Hence, the technical issues limiting the feasibility of biodiesel production have received a lot of attention in the last decade. The next section discusses new approaches to converting soy oil to biodiesel highlighting the advantages that new technologies give over standard homogeneous base or acid-catalysis. Some ideas for improvement focus on gains in chemical kinetics or mass transfer, and others seek to reduce the amount of reagent methanol or simplify separations in pretreatment or posttreatment (preparation for sale). The next sections also present some of the drivers for advances in conversion technologies, along with recently published discoveries in making fuel from soybeans.

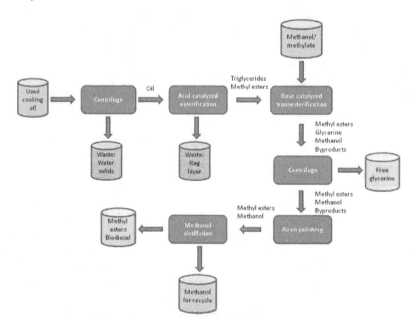

Fig. 1. A simplified flowsheet for the conversion of waste oil into methyl esters.

2. Advances in conversion technologies

Although homogeneous catalysts are used to promote the kinetics of the conversion of soybean oil to biodiesel, speeding up the process has inherent limitations. The reagent oil and alcohol, usually methanol, have limited miscibility, and so the reaction occurs primarily at the interface between the two liquid phases. Mass transport to the interface can be increased by rapidly mixing or forcing methanol into solution using higher pressures. Commercial processes are carried out at about 80°C, a temperature above the normal boiling point of methanol, to ensure conversion to the required ASTM specification. Even under pressurized conditions, the conversion takes at least half an hour in a batch reactor. At lower temperatures conversion typically take several hours. Several studies have investigated the use of supercritical methanol in the transesterification of soybean oil. Process intensification methods have been applied such as the use of rapid mixing and separation of products in a centrifugal contactor.

In addition to the issues with mass transfer, the kinetics of the three step transesterification can limit the overall conversion of the triglycerides to the esters. As the concentrations of intermediates increase, the rate of the back reactions can become significant with respect to the rates of the forward reactions. The online removal of glycerine to drive the process to completion has been attempted with some success; however, conversion to ASTM specification still takes minutes to complete. Addition of catalyst gives rise to saponification of the esters, resulting in phase separation and foaming, causing difficulties in processing. The reaction kinetics of transesterification has been modeled successfully with a three step forward and backward mechanism (Freedman, Butterfield et al. 1986; Noureddini & Zhu 1997). Although rate constants do vary with the type of oil and the processing conditions, the success with the model suggests that the constraints on reaction rate can be predicted and mitigated in a developing a optimized process flowsheet for soybean oil conversion to biodiesel.

Also problematic are multiple separation steps required during the conversion of soy oil to biodiesel. These include: pretreatment and removal of contaminants, separation of free-fatty acids from triglycerides, removal of free glycerine, washing to remove base catalyst, polishing of product in a resin bed, capture and recycle of unreacted methanol. In conventional processing, these steps can take several hours. However, the goal of some of the newer technologies being investigated in the laboratory is to minimize separation requirements.

A variety of new conversion technologies are being investigated to facilitate the conversion of soybean oil into biodiesel. Many of these ideas have been captured in recent reviews (i.e., (Lin, Cunshan et al. 2011), and a summary is given in the next section. Table 2 gives a synopsis of some of the new technologies, listing advantages and disadvantages for development on a commercial scale.

2.1 Advances in catalysis
2.1.1 High temperature cracking and heterogeneous catalysis

Thermal cracking of triglycerides, as opposed to transesterification discussed earlier, has been carried out for over 100 years, with a recent focus on converting fats and oils to liquid fuels (Maher & Bressler 2007). The cracking process takes place at high temperatures, 300-500°C, and atmospheric pressure producing alkanes, alkenes, aromatics and carboxylic acids, that can be separated by distillation (Lima, Soares et al. 2004). The resulting mixture has a lower viscosity than the parent oil. Yields tend to be low in comparison with transesterification, although up to 77% conversion of soybean oil has been observed with the use of a high quality, edible oil as the starting material. Although pyrolysis has been tested successfully on used cooking oil, fatty acid salts, and soaps, the low yields and the wide variety of chemicals produced in pyrolysis have made this process uneconomical. Difficulties with the pyrolytic method include the formation of char and cokes, as well as oxygenated compounds that need to be removed if the products are to be used as diesel substitutes. However, if the complex chemistry can be understood, and the decomposition pathways leading to aromatics and olefins as well as to more desirable alkanes identified, better control of reactor conditions to give desired products should be possible.

Processing Technology	Conditions & Reagents	Advantages	Disadvantages
Homogeneous catalysis	80°C. Mole ratio[2] ~6 (more needed for acid catalysis).	Used on commercial scale. Base catalysis has high yields	Batch process. Many separations needed. Corrosion.
Heterogeneous catalysis	160°C. Mole ratio ~12.	Flash separation of H_2O & excess alcohol. Catalyst recyclable. H_2O tolerant.	Small batch process. Lower yields <1 h. 80-90% yield at 3 h.
Supercritical alcohol	250°C. 200 bar. Mole ratio 30-80.	Complete conversion. Processing time < 1 h.	High mole ratio. Small scale only. High temperature. High pressure.
Enzymatic catalysis	50°C. 2 h reaction time. Mole ratio is 3.	Low temperature. Low mole ratio. Catalyst regeneration. Tolerant to H_2O & free fatty acids.	Long reaction time. Rate depends on enzyme loading. Scaling difficult. Low yields. Expensive catalyst.
Hydrodeoxygenation	300-350°C. 45-70 bar H_2. Mix with petroleum oil.	Diesel fraction linear alkanes produced. 100% conversion.	Long reaction time (hundreds of hours). High temperature. Mid pressure H_2.
Ultrasonic	45°C. 1 h reaction time. Mole ratio is 6.	Low temperature. Continuous process. Scalable.	Longer reaction time/multiple passes (bound glycerine after one pass is 80-90 wt%).
Centrifugal reactor/separator	80°C. 2.6 bar. Mole ratio is 4.8.	90% conversion after 2 min. Staged approach to get ASTM spec.	Pressurized system.
Metal foam	100°C. 5 bar. Mole ratio is 10. 1 wt.% catalyst.	95% after 3.3 min. Low power (<1% conventional). 0.9 L/h throughput.	Pressurized system.

Table 2. Summary of soybean oil conversion technologies to biodiesel.

---

[2] Methanol-to-oil or alcohol-to-oil mole ratio unless otherwise specified.

Better selectivity may be achieved through the use of catalysts in the pyrolysis process (Maher & Bressler 2007). For instance, molecular sieve materials, being porous with high surface area, exhibit high catalytic reactivity. The tetrahedral structures of zeolites, or crystalline aluminosilicate AlO_4-SiO_4 materials, show localized areas of high reactivity associated with the cations in the structure. Heterogeneous catalysis for pyrolysis of oils carried out at temperatures of 300-500°C over zeolites produces paraffins, olefins, carboxylic acids and aldehydes (Lima, Soares et al. 2004). Other studies using a protonated zeolite H-ZSM5 (85kPa He) have shown relatively more olefins and aromatics being produced from a variety of lipid starting materials and very little formation of oxygenated species. The reaction only generates a small amount of alkanes, and what is produced comes in the form of gases such as propane, and so is not appropriate for diesel fuel.

Metal catalysts have also been used for deoxygenation, Pt and Pd on activated carbon, at 300°C under nitrogen. Detailed analysis of the chemistry show that fragmentation of the triglyceride occurs more quickly than decarboxylation of the fatty acid chains. The fatty acid chains eventually form alkanes, with lighter hydrocarbons coming from β-fission at the double bonds. Lighter alkanes, CO_2, and CO come from the glycerol backbone. These studies demonstrate alkane production in the gasoline and diesel fraction range, with yields as high as 54% at 92% conversion for soybean oil (Morgan, Grubb et al. 2010).

A drawback to using catalyzed heterogeneous pyrolysis has been coking of the catalyst, requiring frequent cycles of oxidative regeneration (Milne, Evans et al. 1990). Fractionating batch reactors may allow the selective removal of alkanes, increasing their relative abundance, but yields are still relatively low (62 wt%) with high coke production (38 wt%) (Dandik & Aksoy 1999). Using mesoporous MCM-41 (1.93 nm pore size) as a catalyst showed lower gas production than H-ZSM5, with the best results being observed for palm oil, with 97.72% being converted overall and a yield of linear hydrocarbons C13-C17 in the diesel range of 42.52 wt.% (Twaiq, Zabidi et al. 2003). Palm oil differs from soybean oil with a higher fraction of shorter chain triglycerides, 50% C12 and 16% C14 and so these results may not relate directly to the conversion of soybean oil. In the same study, the authors show that experimentation with an oil of higher average molecular weight, in this case palm olein oil, showed a lower conversion and higher coke formation.

2.1.2 Low-temperature heterogeneous catalysis

Heterogeneous catalysis at low temperatures promises advantages over conventional processing in phase separation and avoidance of the use of strong caustic or acidic reagents. Some catalytic systems have proven to be more robust to fatty acids and to water than heterogeneous base catalysis (Zeng, Deng et al. 2009). The catalysts commonly used include transition metals and inorganic oxide systems that promote esterification and transesterification, besides the molecular sieves that are used in pyrolysis (discussed in Section 2.1.1).

A recent review gives details on supported solid metal oxides that have been used both for the transesterification and the esterification of oils to biodiesel (Zabeti, Daud et al. 2009). The transition metal oxides (alumina, tin, and zinc) form Lewis acids with the metal atoms acting as electron acceptors. Alkaline earth oxides (magnesium, calcium, and strontium) form Brønsted bases through the oxygen atoms in the structure. Because of the colocation of acidic and basic sites, the activity of the catalyst is often described in both of these terms (Yan, DiMaggio et al. 2010). In a series of steps, Figure 2, the metal atom coordinates with both the oxygen of the carbonyl group in the acylglyceride or fatty acid and the alcohol, liberating a water molecule. The basic site can stabilize transfer of a proton from a fatty acid to water. The product ester forms within the supported complex or transition state, which decomposes regenerating the active metal oxide. Oxides such as alumina or silica can exhibit catalytic activity at acidic sites, dehydrating and decarboxylating fatty acids and triglycerides (Boz, Degirmenbasi et al. 2009). Acid-base catalysts can also be used in high temperature pyrolysis as well as for transesterification reactions.

Yields tend to be lower with heterogeneous catalysis in comparison with homogeneous catalysis (Section 1) because of reduced interfacial contact, not only between the oil and alcohol phases but also with the catalytic surfaces. To mitigate this limitation, methanol-to-oil ratios are usually high, 12 or greater; several wt% catalyst is often used; and reactions continue for a number of hours to drive the conversion to completion. Hence, studies are carried out in batch microreactors or autoclaves where extreme conditions can be controlled. Co-solvents have been used to improve the miscibility of the reagents (Yang & Xie 2007). Another way of improving reaction rate to get higher yields is to use high surface area catalysts and catalyst supports. For instance, nanoscale MgO has been used to achieve a 99% yield of methyl ester at 523°C and 24 MPa (Wang & Yang 2007). In autoclave studies of esterification at 160°C, a mass ratio of methanol: fatty acid: catalyst of 4: 10: 0.1 generated yields of up to 74% after only 1 h of residence time. The lowest yield, 32%, occurred in systems without the hetereogeneous catalyst, with methanol: fatty acid: catalyst 4: 10: 0.0, showing that the catalyst had a significant effect on reaction rate (Mello, Pousa et al. 2011). The same group showed that higher yields could be achieved after 3 h of reaction time. They also demonstrated that the catalyst could be regenerated at least ten times using centrifugation and cleaning in solvent without an observable loss in performance.

Heterogeneous catalysis continues to generate much interest in the research community. Surface area and morphology appear to have a greater influence over catalyst activity than the chemistry of the catalysts. Although some of the conversions show promise, the extreme temperatures or pressures currently required for effective heterogeneous catalysis, as well as the relatively low yields in comparison with homogeneous catalysis, preclude them from being used on a large or commercial scale. However, many of the catalysts being considered appear quite robust, and although subject to coking and other deactivation processes, can be regenerated many times.

2.1.3 Enzymatic catalysis

Lipases, naturally occurring enzymes, have been used to catalyze the transesterification of triglycerides. The mechanism is thought to be a two step process, where the lipase reacts with one substrate to form a product and an intermediate enzyme, followed by reaction with another substrate to give a final product and the regenerated enzyme (Varma, Deshpande et al. 2010). The advantages of enzymatic processing are high yields of methyl esters, milder reaction conditions, high tolerance of water contamination, and easy separation of free glycerine. The lipase process can be done in a number of different solvents, including supercritical CO_2. In the case of enzymatic catalysis, the loading of the enzyme has a profound effect on the initial rate of the reaction, and loadings of 5-10% w/w were found to be optimal. Enzymatic catalysis can be used for both esterification and transesterification, and a variety of oils and alcohols as feedstocks; however, processing conditions can be different depending on the starting material and desired product. Processing can take hours to reach equilibrium, typically achieved at when the reaction reaches about 50-70% conversion. Yields have been limited by inhibition of the catalyst by the alcohol, although the enzyme can be regenerated by driving off the alcohol to regain its activity.

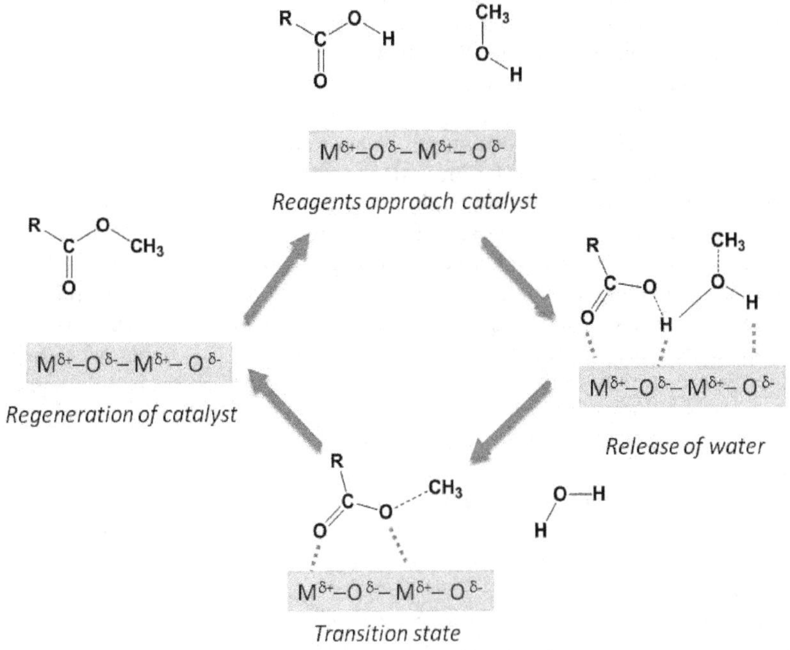

Fig. 2. Heterogeneous catalytic formation of a methyl ester from a fatty acid precursor.

2.2 Supercritical alcohols

A way of driving the transesterification reaction to completion without requiring catalyst is to perform the reaction under supercritical conditions. Many types of oils have been esterified in this way, including soy oil (Zhou, Wang et al. 2010). Both methanol and ethanol have been used as reagents (Rathore and Madras 2007). Pressures and temperatures are high for these processes, so that the conditions in the reactor exceed the critical point of the alcohols involved in the reaction. Pressures greater than 200 bar and temperatures exceeding 300°C are typical, although conversions of soybean oil have been successful at temperatures as low as 250°C. Because of the extreme conditions, these processes have only been demonstrated in the laboratory at bench scale. With a large excess of alcohol, the transesterification process can be described as a pseudo-first order reaction, and rate constants have been measured for a number of different alcohols reacting with a variety of oils (Varma, Deshpande et al. 2010). Rates of conversion in ethanol are greater than in methanol because of the greater miscibility of ethanol and the oil reagent. The rates also depend on the fatty acid content of the oil, being inversely proportional to the saturated fatty acid content.

2.3 Continuous and intensified processing

Conventional batch processing of soybean oil to biodiesel can take several hours, especially when post-conversion separation and polishing steps are included, see Figure 1. Each batch has to be tested against ASTM specifications before sale and a batch that has been compromised must be recycled back into the feed loop, adding cost. Properties of the fuel product can change because variations in the feedstock or changes in process condition. If implemented, a continuous process has the advantage of allowing online control of reagent flows, temperature and pressure conditions, to achieve good conversion, reducing the need for recycling of impaired product.

To achieve high conversion in a continuous process, however, issues such as the non-miscibility of reagents and mass transfer limitations in the transesterification process have to be overcome. Process intensification, an engineering concept that gained attention through investigations in the 1970s at the University of Newcastle (Stankiewicz & Moulijin 2002), is a way of enhancing mass transfer, thus reducing the capital cost of a chemical plant through a smaller plant size and reagent inventory and reducing operating costs through decreased energy consumption and feedstock required per unit mass of the product. Centrifugal phase contact and separation is an example of an intensified technique that enhances mass transfer at high throughput and minimizes the inventory of solvents (Tsouris & Porcelli 2003). Another example is to use bubble formation to increase the interfacial area of immiscible fluids, which can be induced by introducing energy to the system through acoustic coupling (Cintas, Mantegna et al. 2010).

Process intensification methodology has been adapted to enhance the pretreatment of biodiesel feedstocks, the conversion reactions, or the posttreatment separation of reaction products. A cavitation reactor was used in the process intensification of the homogeneous acid (H_2SO_4) catalyzed esterification of simulant fatty acids (Kelkar, Gogate et al. 2008). High throughput ultrasonic irradiation at 21.5 kHz coupled with a stirred tank was used to make a fine emulsion of oil and methanol, thereby increasing the interfacial area. The reactor achieved a yield of >80% methyl esters from soybean oil (Cintas, Mantegna et al. 2010). In this apparatus, temperatures were kept low, ~45°C, to prevent boiling of methanol in the microwave reactor. A sonochemical reactor has also been used to enhance the base-catalyzed transesterification of lightly used cooking oil as well as food grade vegetable oil (Hingu, Gogate et al. 2010).

Centrifugal mixing has been applied to biodiesel production (Peterson, Cook et al. 2001), because of its ease of operation, rapid attainment of steady state, high mass transfer, phase separation efficiencies, and compact size (Leonard, Bernstein et al. 1980). The high shear force and turbulent mixing achieved in a contactor minimize the effect of diffusion on the reaction rate of transesterification, pushing it to be limited only by the reaction kinetics. The contactor has been used as a low-throughput homogenizer, employing very low flow rates to increase residence times to tens of minutes (Kraai, van Zwol et al. 2008; Kraai, Schuur et al. 2009).

At ORNL, we have combined the reaction of oil and methoxide with the online separation of biodiesel and glycerol into one processing step, using a modified centrifugal contactor. Two distinct phases enter the reactor (reagents: methanol and base catalyst; and vegetable oil), and two distinct phases leave the reactor/separator (products: glycerol and methyl ester), thus demonstrating process intensification in high-throughput biofuel production. The ORNL reactor separator was modified from a commercial unit, Figure 3a, to increase the residence time from a few seconds to a few minutes by achieving hold-up in the mixing zone, Figure 3b. (Birdwell, Jennings et al. 2009). In the ORNL tests, base-catalyzed transesterification of soybean oil was carried out at continuous flow conditions at 60°C and in static pressurized tests at 80°C (McFarlane, Tsouris et al. 2010).

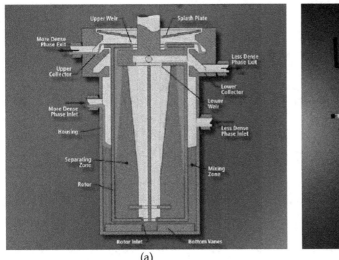 

(a)   (b)

Fig. 3. Reactor-separator housing: a) commercial unit schematic, b) modified contactor housing.

Besides bubble formation and stirring, another way of achieving high turbulence and good mass transfer for the production of biodiesel is through the use of reactors involving tortuous flow pathways. These concepts were first tested on microreactors, involving zigzag channels (Wen, Yu et al. 2009). Although high conversions were achieved, 99.5% at 28s residence time, scaling the reactor up from microliter s^{-1} flow rates has not been possible. More recently, turbulence has been achieved by passing the reagents through porous metal foam, which can be made to have a high pore density (50 pores per inch) and a relatively low pressure drop (0.6 MPa). At 100°C and with a methanol-to-oil mole ratio of 6, a conversion of 90.5% was observed (Yu, Wen et al. 2010). With the foam, the arithmetic mean drop size of the disperse phase was about 3 μm. By balancing the effect of smaller, high surface area bubbles at high flow rates, with the lower residence time, conversions were pushed to 95 mol% with a flow rate of 0.9 L·h^{-1}. While high for a microreactor, this flow rate is much lower than for competing continuous technologies.

In all continuous processes, the conversion of soybean oil to esters is limited by residence time in the reactor. Producers and investigators have focused on the kinetics of transesterification to determine if conversions to methyl ester are limited by mass transfer effects or by slow kinetics (Darnoko & Cheryan 2000; Karmee, Mahesh et al. 2004). In the transesterification reaction, mass transfer limitations early in the process become superseded by kinetic limitations when trying to achieve high yields of methyl esters. In the case of the Oak Ridge experiments, although 90% conversion was achieved in 2 min, a 22 min residence time at 80°C was needed to achieve ASTM specification grade fuel, ~98% conversion, Figure 4. Hence, in both the centrifugal processing and the ultrasonic reaction, multiple stages were found to shorten reaction time and reduce energy consumption. The online

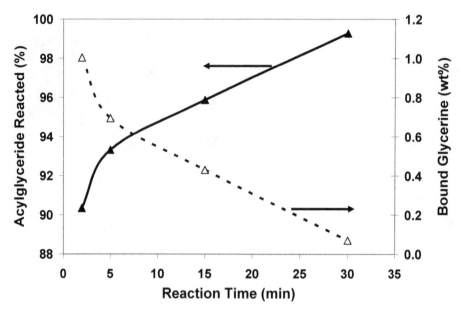

Fig. 4. Yield of batch transesterification reaction in continuous contactor in terms of the weight percent of triglyceride reacted (▲) and remaining total bound glycerine (∆) as a function of reaction time (80°C, above ambient pressure to 2.6 bar, 3600 rpm rotor speed). The arrows indicate the conversion goal of <0.24 wt% bound glycerine, or 97.8% conversion of acylglyceride.

separation of free glycerine removes a sink for the base catalyst (Cintas, Mantegna et al. 2010), as well as reduces back reactions to form bound glycerine species (McFarlane, Tsouris et al. 2010). The accelerated reaction achieved with online separation also prevents thermal degradation of the methyl esters, arising from beta scission adjacent to the carbonyl group and cleavage of the unsaturated bonds in the fatty acid chains (Nawar & Dubravcic 1968; Osmont, Catoire et al. 2010).

3. Generation of fungible fuels from plant oils and new technologies for de-oxygenation

Even after esterification, the product biodiesel can be substituted directly for standard diesel fuel only to a limited percentage and is normally restricted from portions of the United States common carrier distribution system[3]. Although biodiesel has a similar cetane number to hexadecane, the higher oxygen content causes changes in the combustion profile and can enhance corrosion of engine seals (Haseeb, Fazal et al. 2011). The higher oxygen content also means that the heating value of methyl esters is slightly lower than standard diesel, although the reduction is not nearly as large as is when comparing ethanol to gasoline. The

---

[3] ASTM specifications allow 5 vol.% fatty acid methyl esters (FAME) in commercial diesel fuel.

lower volatility and higher oxygen content of biodiesel change the injection profile in a compression engine, and hence the ignition timing and production of pollutants, for instance decreasing soot and increasing the NO_x in the exhaust (Ra, Rietz et al. 2008; Toulson, Allen et al. 2011). This active area of study has an impact on high efficiency clean combustion engines, the vanguard of advanced diesel engine design. In most standard vehicles, biodiesel concentrations are limited to a blend of 20% to mitigate the effects of its physical properties being different from those of standard diesel fuel (Mushrush, Willauer et al. 2009), such as poor cold flow. In addition, biodiesel has a limited shelf life and can form precipitates and go rancid in storage, causing problems in distribution.

One method of producing deoxygenated products from soybean oil is to use a high temperature (350-450°C) hydrogenation process rather than transesterification to make fuels. This hydroprocessing, carried out over supported catalysts, is different than the pyrolytic schemes described in some detail in Section 2.1.1 because hydrogen is introduced directly into the reactor. Heavier paraffinic fragments are produced rather than the small gaseous alkanes made in pyrolysis. The process, as applied to triglycerides, has been reviewed by Donnis and colleagues (Donnis, Egeberg et al. 2009). Hydrotreating experiments on triglycerides have used the same conventional catalysts used in hydroprocessing oil, such as sulfided NiMo or CoMo on alumina under relatively low pressures of H_2S/H_2 mixtures (Huber, O'Connor et al. 2007). The process includes several chemical steps to give alkanes as a final product, including: hydrogenation of C=C bonds; decarboxylation (removal of CO_2); decarbonylation (removal of CO); and dehydration (hydrodeoxygenation (HDO) to convert COOH to H_2O). The glycerin backbone may react to form methane or propane (Donnis, Egeberg et al. 2009). By carefully controlling temperature and reaction time the yield of the paraffinic diesel-fraction, or straight chain C15-C18, can be maximized. Although some studies show that catalyzed hydroprocessing over nickel generates too many aromatics and cyclic compounds, tailoring of HDO products by additional isomerization steps has been suggested to produce branched alkanes (Jakkula, Niemi et al. 2004). This would give a biorefinery the ability to produce the desired fuel properties for vehicular use without the need for blending, giving a product similar to Fischer-Tropsch diesel fuel from natural gas.

Huber and colleagues have also shown that the bio-derived oils can be hydrotreated along with petroleum oils, suggesting that a processing can take place within an existing refinery to lower the capital cost. Issues with hydroprocessing vegetable oils rather than petroleum include: the high oxygen content of biomass can increase heat load in the reactor and cause leaching of sulfur from the catalyst; water and CO_2 generated during the hydrotreatment can reduce catalyst lifetime and must be removed from the product; and also the large triglyceride molecules can clog catalysts with pore sizes of less than 2 nm (Tiwari, Rana et al. 2011). Mesoporous molecular sieves, such as MCM-41, or alumina can have the advantage of a high surface area and activity, but also have much larger pore diameters than zeolites (Kubicka, Simacek et al. 2009), and so may be useful in a combined bio-petro refinery.

Another route to achieving a hydrocarbon rich fuel from soybean oil is through deoxygenation of the esters after the transesterification process has taken place. In this case the biodiesel produced from soybean oil is further reacted to form a hydrocarbon fuel. The processing involves deoxygenation to remove the ester moiety from the hydrocarbon chain. With this step, the product becomes completely miscible with standard diesel fuel and can be introduced at any step in the supply chain, either at the refinery or at the filling station. Note that if blending is done at the terminal or filling station, the product has to meet completely ASTM specifications. Some of these processes involve hydrogen and some do not.

The hydrogenation of methyl octanoate, as a simulant for methyl esters from biodiesel, has been carried out over an N-ZSM5 zeolite catalyst under atmospheric pressure H_2 (Danuthai, Jongpatiwut et al. 2009). The experiments were run over a few hours at temperatures up to 500°C, and showed 99.7% conversion of the ester to C1-C7 alkanes – a third comprising ethane, and small aromatics (C6-C9). Residual oxygenated species comprised only 2.8%. The group also found that the aromatic fraction increased with the time in the reactor, and that H_2O promoted the catalytic activity of the zeolite by enhancing production of an acid byproduct, obviously undesirable as a fuel component. Tests with methyl octanoate, a smaller molecule than methyl esters derived from soybean oil, showed conversion to alkanes and aromatics through formation of a high molecular weight ketone intermediate. The patent literature suggests that similar results have been achieved with longer fatty acid chain methyl esters from soybean and other oils (Craig 1991). Reaction 4 shows the overall conversion process of a methylester to a linear alkane by hydrodeoxygenation: step A) removing the oxygen as CO_2 or methanol followed by formation of the enol, and step B) involving hydrogenation and dehydration of the enol to the linear alkane (Donnis, Egeberg et al. 2009).

A)

$$CH_3OCOC_nH_{2n+1} \rightarrow CH_2{=}C_{n-1}H_{2(n-1)}+CO_2+CH_4 \xrightarrow{H_2} C_nH_{2n+2}$$

$$CH_3OCOC_nH_{2n+1} \xrightarrow{H_2} CHOCH_2C_{n-1}H_{2(n-1)+1}+CH_3OH$$

$$CHOCH_2C_{n-1}H_{2(n-1)+1} \underset{\text{rearrangment}}{\overset{\text{enol}}{\rightleftharpoons}} CH(OH){=}CHC_{n-1}H_{2(n-1)+1} \qquad (4)$$

B)

$$CH(OH){=}CHC_{n-1}H_{2(n-1)+1} \xrightarrow{H_2} CH_2(OH)C_nH_{2n+1} \xrightarrow[-H_2O]{+H_2} C_{n+1}H_{2(n+1)+2}$$

$$CH(OH){=}CHC_{n-1}H_{2(n-1)+1} \xrightarrow[-H_2O]{+H_2} CH_2{=}CHC_{n-1}H_{2(n-1)+1} \xrightarrow{H_2} C_{n+1}H_{2(n+1)+2}$$

As discussed in Section 2.1.1, non-hydrogenated direct catalytic cracking of triglycerides can lead to products with greater oxygen content than desirable for fuels. Better control of the cracking process can be engineered when starting with an esterified feedstock. A recent example is the use of supported platinum and bimetallic platinum-tin catalysts in the deoxygenation of methyl octanoate, methyl dodecanate, and soybean oil by reactive distillation at 320 to 350°C. By manipulating the residence time and the catalyst properties, selectivity for paraffins of 80% was achieved. Overall yields were low, suggesting this process requires more investigation before commercialization (Do, Chiappero et al. 2009; Chiappero, Do et al. 2011).

4. Feasibility of using plant oils for fuels in comparison with petroleum, ethanol, and lignocellulosic feedstocks

The use of soybean oil in production of biodiesel has been primarily limited by economic factors, in particular the cost of the feedstock. Less expensive fuel can be made from degraded starting material such as waste oil. Energy crop alternatives to seed oils have also been proposed (Vinokurov, Barkov et al. 2010). However, the processing of feedstock with higher free fatty acid content adds complexity to the manufacturing process, particularly because of the variability in composition and treatment prior to conversion. The solution to tightening of petroleum supply will likely involve liquid fuel generation from a variety of sources. As should have been apparent from the previous discussion, the processing of biomass-derived oils into burnable esters depends on the chemical composition of the feedstock: the relative concentration of free fatty acids, the saturated versus unsaturated fatty acid chains, impurities and water content.

An additional cost is associated with the alcohol used to convert the seed oil to biodiesel, typically used in amounts well above stoichiometric to push the reaction to completion. An analysis was recently done at ORNL where the cost of a three stage biodiesel manufacturing process was assessed based on the reactor-separator reactor discussed in Section 2.3 (Ashby & McFarlane 2010). In order to optimize the process, environmental conditions such as temperature, pressure, and the starting proportion of methanol–to–oil were all varied individually. Each of these aspects of the production affected the residence time and the fraction of soybean oil converted during the reaction, hence the economics of the process. The analysis gave the projected capital cost for a new plant and its projected profit in the first five years. These analyses revealed that reactions run at higher temperatures needed less time to convert a larger fraction of triglyceride. Reactions with a greater proportion of methanol–to–oil had a higher yield at a residence time of 600s than those with a lower ratio. Figure 5 shows the effects of adding additional methanol at various stages of a five stage process.

An economic analysis shows that production of biodiesel should be more profitable in a three contactor series than a single reactor given similar process conditions, i.e., temperature and ratio of methanol–to–oil, in spite of the costs associated with the reactor and pumps for each additional stage. In the long term, the feedstock soybean oil comprised the highest fraction of the operating expenses, ranging from 70-80% of the total. The cost of the alcohol was also found to be significant, but could be minimized through recycling, thereby also reducing the carbon footprint of the process. In this analysis, the production of biodiesel from soybean oil could only become profitable if the product could be sold at about 1.5 times the cost of the soybean oil feedstock (assuming a 300,000 gal/year operation amortized over 5 years). Based on simulation of the chemical kinetics of soybean oil transesterification, the highest yield of methyl esters in the shortest time arose from using three reactor-separators in series, each with a 200s residence time, recycling of all excess methanol, a 4.5–to–1 initial proportion of methanol–to–oil, and an operating temperature of 100°C. Similar analyses have been done for other reactor configurations, feedstocks, alcohols, and catalysts, to assess the viability of these process designs for commercial production of biodiesel (Peterson, Cook et al. 2001).

In the case of soybean utilization, the feedstock costs appear to dominate the potential use of biomass conversion to supplant petroleum-derived diesel in any of the reactor configurations being considered (Lin, Cunshan et al. 2011). However, the economics of biodiesel production can be improved if value added products can be developed from the byproduct glycerine. Janaun and Ellis give many of these in their review: catalytic conversion to oxidized products such as propylene glycol; biological conversion to lipids and citric acid; fuel oxygenates; gasification to H_2 and syngas; remediation of acid mine drainage; and in agriculture as animal feed (Janaun & Ellis 2010).

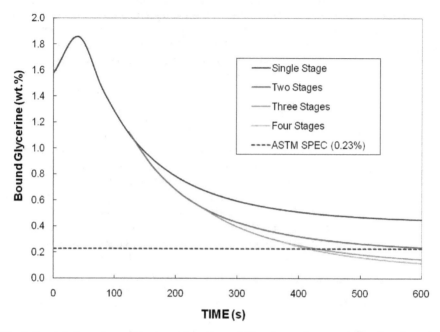

Fig. 5. Bound glycerine content as a function of time for various options of methanol addition, at 80°C: Single stage involves no additional input of methanol over 600s reaction time; Two stages involves adding 10% original methanol charge at 2nd stage; Three stages involves adding an extra 5% original methanol charge at 3rd stage; Four stages has no additional methanol as the effect is minimal by this point.

Another aspect worth consideration is that unless specifically designed to do so, compression engines are not constructed to handle the higher oxygen content of biofuels such as biodiesel or ethanol. Hence, many alternative fuels under consideration are blended to give the properties needed for engine performance and fuel stability, 10% ethanol in gasoline being a common example. However, fuel from different sources may not be compatible. Biodiesel, with its high oxygen content, mixes well with standard diesel, but not with purely paraffinic Fischer-Tropsch fuel. The aromatics in standard diesel solubilize the olefinic chains and electron-rich esters, where as tertiary carbons in the Fischer-Tropsch paraffins appear to form stable hydroperoxides with degradation products in the biodiesel (Mushrush, Willauer et al. 2009). If the biodiesel contains unreacted free-fatty acids, phase separation and precipitates are likely to form. One possibility is to hydrogenate the biodiesel to create a fully hydrocarbon fuel, as discussed earlier in Section 3. Another is to exploit the properties of other biomass-derived fuels to produce a blend with properties that meet the requirements for compression ignition engines. For instance, lignin has the potential to become a biofuel feedstock can be broken down into appropriately sized aromatic fragments, which can be used as additives to diesel fuel or to biodiesel methyl esters (Gluckstein, Hu et al. 2010). The properties of the blend will have the high cetane number and the high lubricity of the biodiesel methyl esters, but with the reduced viscosity and low cloud point of the aromatics. Hence, while a pure biofuel may have some undesirable characteristics, mixtures of alternative fuels may be compatible with standard diesel engines. An assessment of mixtures of diesel compatible formulations has been performed by the Fuels for Advanced Combustion Engines (FACE) Project and target properties are presented in Table 3. The average properties of marketed diesel fuel are shown in brackets (Gallant, Franz et al. 2009).

Property	Range for FACE project	Standard Diesel Fuel	Soybean Methyl Esters
Cetane number	30–55	43–51	51
Aromatics (%)	20–45	32	0
T90 Distillation (°C)	270–340	320	Not applicable
Specific gravity (g cm^{-3})	0.803–0.869	0.82–0.86	0.884
Heat of combustion (kJ/kg)	7790-7980	7850	Reduced by 9-13%
Kinematic viscosity (mm^2 s^{-1})	1.319–3.218	1.90–4.1	4.08
Cloud Point (°C)	-19.5--55.5	-18--30	-0.5
Flash Point (°C)	53–74	55	131
Pour point (°C)		-25	-4
Constraints	< 15 ppm sulfur, <4% olefins Smooth distillation curve		

Table 3. Fuel Formulation Property Targets for Compression Ignition Engines.

Biofuel production in the US and Brazil is dominated by ethanol, where as biodiesel has greater importance in Europe (Rusco & Walls 2009). In some respects the issues with ethanol and biodiesel are similar, competition for agricultural resources with food, oxygen content and lower heating value, and distributed production (Kalnes, Marker et al. 2007). Varying fuel standards can further complicate distribution, leading to lower pipeline capacity and increased storage requirements. For instance, ethanol, even blended with gasoline, currently is not transported through pipelines because of its high affinity for water resulting in corrosiveness and phase separation. However, ethanol is a simple molecule that has the same composition no matter the source, and its impact on petroleum refining can be assessed on a large scale. This is not the case for plant-based biodiesels, from which a variety of fuels can be produced depending on the plant variety and processing conditions. Depending on the regulatory environment and governing standards, this may further break up the markets for biodiesel production and distribution. For example, southern regions will better be able to tolerate higher cloud points than northern, both for pipeline, truck and rail transport, as well as for combustion in passenger vehicles. The cost of the adoption of biofuels needs to be assessed along with benefits, such as reduction in greenhouse gas emissions, energy security, or support of US agriculture (Rusco & Walls 2008).

5. Conclusions

Although the price of diesel fuel has increased, economical production of biodiesel is a challenge because of (1) the increasing price of soybean oil feedstocks and reagent methanol, (2) a distributed supply of feedstocks that reduces the potential for economies of scale, (3) processing conditions that include pressures and temperatures above ambient, and (4) multiple processing steps needed to reduce contaminant levels to ASTM specification D6751 limits (Vasudevan & Briggs 2008). Much of the cost of biodiesel production is related to the conversion of the oil to the methyl ester and so there has been an emphasis to research improved methods of converting soybean oil to biodiesel. However, most of these studies have taken place at the bench scale, and have not demonstrated a marked improvement in yield or reduced oil-to-methanol ratio in comparison with standard base-catalyzed transesterification.

One aspect that has a short term chance of implementation is the improvement of the conversion process by the use of a continuous rather than batch process, with energy savings generated by combined reaction and separation, online analysis, and reagent methanol added by titration as needed to produce ASTM specification grade fuel. By adapting process intensification methods, recycled sources of soybean oil may also be used for diesel production, taking advantage of a lower priced feedstock material.

Even if the economics of production are feasible, biodiesel distribution is complicated by thermal stability and degradation over time, and the physical properties of methyl esters make them undesirable for standard compression ignition engines in concentrations greater than 20% in a blend with diesel fuel. Generation of truly fungible fuel from biomass is now being investigated through a variety of routes. However, it is too early to judge which will become the most viable.

The promise of soybean-generated biodiesel is that of a truly fungible, thermodynamically and economically viable technology providing a biomass replacement for a petroleum product. The use of biodiesel has the potential to reduce the amount of CO_2 released to the atmosphere by the transportation sector; to provide an additional source of liquid fuel that can be produced in small distributed operations; and to allow the processing of waste oil-to-energy that can result in enhanced lifecycle efficiencies as well as reduced environmental footprint.

6. Acknowledgments

Research on biodiesel manufacture at Oak Ridge National Laboratory was sponsored by the Laboratory Directed Research and Development Program; the United States Department of Energy Office of Efficiency and Renewable Energy Technology Commercialization and Deployment Program's Technology Commercialization Fund; and by Nu-Energie, LLC, under CRADA #01377. Oak Ridge National Laboratory is managed by UT-Battelle, LLC, for the U.S. Department of Energy. The author would also like to thank Dr. Bruce Bunting of the National Transportation Research Center at ORNL for his helpful comments.

7. References

Ashby, E. & McFarlane, J. (2010). *Optimization of Biodiesel Production in a Centrifugal Contactor*, Poster Presentation, May 5, 2010. Oak Ridge National Laboratory, Oak Ridge.

ASTM. (2007). *Standard Test Method for the Determination of Free and Total Glycerin in B-100 Biodiesel Methyl Esters by Gas Chromatography*, ASTM International. ASTM Method D6854.

ASTM. (2008). *Standard Specification for Biodiesel Fuel Blend Stock (B100) for Middle Distillate Fuels*, ASTM International. D6751-07b.

Birdwell, J. F. J., Jennings, H. L., McFarlane, J. & Tsouris, C. (2009). *Integrated reactor and centrifugal separator and uses thereof*, UT-Battelle. US Patent Application WO/145954.

Boz, N., Degirmenbasi, N. & Kalyon, D. M. (2009). Conversion of biomass to fuel: Transesterification of vegetable oil to biodiesel using KF loaded nano-gamma-Al_2O_3 as catalyst. *Applied Catalysis B-Environmental*, Vol. 89, No. 3-4, pp. (590-596), 0926-3373.

Chiappero, M., Do, P. T. M., Crossley, S., Lobban, L. L. & Resasco, D. E. (2011). Direct conversion of triglycerides to olefins and paraffins over noble metal supported catalysts. *Fuel*, Vol. 90, No. 3, pp. (1155-1165), 0016-2361.

Cintas, P., Mantegna, S., Gaudino, E. C. & Cravotto, G. (2010). A new pilot flow reactor for high-intensity ultrasound irradiation. Application to the synthesis of biodiesel. *Ultrasonics Sonochemistry*, Vol. 17, No. 6, pp. (985-989), 1350-4177.

Craig, W. (1991). *Production of hydrocarbons with a relatively high cetane raing*, US Patent 4,992,605, USA.

Dandik, L. & Aksoy, H. A. (1999). Effect of catalyst on the pyrolysis of used oil carried out in a fractionating pyrolysis reactor. *Renewable Energy*, Vol. 16, No. 1-4, pp. (1007-1010), 0960-1481.

Danuthai, T., Jongpatiwut, S., Rirksomboon, T., Osuwan, S. & Resasco, D. E. (2009). Conversion of methylesters to hydrocarbons over an H-ZSM5 zeolite catalyst. *Applied Catalysis a-General*, Vol. 361, No. 1-2, pp. (99-105), 0926-860X.

Darnoko, D. & Cheryan, M. (2000). Kinetics of palm oil transesterification in a batch reactor. *Journal of the American Oil Chemists Society*, Vol. 77, No. 12, pp. (1263-1267), 0003-021X.

Do, P. T., Chiappero, M., Lobban, L. L. & Resasco, D. E. (2009). Catalytic deoxygenation of methyl-octanoate and methyl-stearate on Pt/Al_2O_3. *Catalysis Letters*, Vol. 130, No. 1-2, pp. (9-18), 1011-372X.

Donnis, B., Egeberg, R. G., Blom, P. & Knudsen, K. G. (2009). Hydroprocessing of bio-oils and oxygenates to hydrocarbons. Understanding the reaction routes. *Topics in Catalysis*, Vol. 52, No. 3, pp. (229-240), 1022-5528.

Emerging Markets (2008). Biodiesel 2020, In: *Emerging Markets Online* 2nd Ed. Date of access March 24, 2011, Available from: <www.emerging-markets.com>

Freedman, B., Butterfield, R. O. & Pryde, E. H. (1986). Transesterification kinetics of soybean oil. *Journal of the American Oil Chemists Society*, Vol. 63, No. (1375-1380), 0003-021X.

Freedman, B., Pryde, E. H. & Mounts, T. L. (1984). Variables affecting the yields of fatty esters from transesterified vegetable oils. *Journal of the American Oil Chemists Society*, Vol. 61, No. 10, pp. (1638-1643), 0003-021X.

Gallant, T., Franz, J. A., Alnajjar, M. S., Storey, J. M. E., Lewis, S. A., Sluder, C. S., Cannella, W. C., Fairbridge, C., Hager, D., Dettman, H., Luecke, J., Ratcliff, M. A. & Zigler, B. T. (2009). Fuels for advanced combustion engines research diesel fuels: Analysis of chemical and physical properties. *SAE International Journal of Fuels and Lubricants*, Vol. 2, No. 2, pp. (262-272), 978-0-7680-1800-4 (proceedings), 1946-3952 (journal).

Gluckstein, J. A., Hu, M. Z., Kidder, M., McFarlane, J., Narula, C. K. & Sturgeon, M. R. (2010). *Final report: Investigation of catalytic pathways for lignin breakdown into monomers and fuels*, (ORNL/TM-2010/281), Oak Ridge National Laboratory.

Haseeb, A. S. M. A., Fazal, M. A., Jahirul, M. I. & Masjuki, H. H. (2011). Compatibility of automotive materials in biodiesel: A review. *Fuel*, Vol. 90, No. 3, pp. (922-931), 0016-2361

Hingu, S. M., Gogate, P. R. & Rathod, V. K. (2010). Synthesis of biodiesel from waste cooking oil using sonochemical reactors. *Ultrasonics Sonochemistry*, Vol. 17, No. 5, pp. (827-832), 1350-4177.

Holčapek, M., Jandera, P., Zderadička, P. & Hrubá, L. (2003). Characterization of triacylglycerol and diacylglycerol composition of plant oils using high-performance liquid chromatography - Atmospheric pressure chemical ionization mass spectrometry. *Journal of Chromatography A*, Vol. 1010, No. (195-215), 0021-9673.

Huber, G. W., O'Connor, P. & Corma, A. (2007). Processing biomass in conventional oil refineries: Production of high quality diesel by hydrotreating vegetable oils in heavy vacuum oil mixtures. *Applied Catalysis A-General*, Vol. 329, No. (120-129), 0926-860X.

Janaun, J. & Ellis, N. (2010). Perspectives on biodiesel as a sustainable fuel. *Renewable & Sustainable Energy Reviews*, Vol. 14, No. 4, pp. (1312-1320), 1364-0321.

Kalnes, T., Marker, T. & Shonnard, D. R. (2007). Green diesel: A second generation biofuel. *International Journal of Chemical Reactor Engineering*, Vol. 5, No. (A48), 1542-6580.

Karmee, S. K., Mahesh, P., Ravi, R. & Chadha, A. (2004). Kinetic study of the base-catalyzed transesterification of monoglycerides from pongamia oil. *Journal of the American Oil Chemists Society*, Vol. 81, No. 5, pp. (425-430), 0003-021X.

Kelkar, M. A., Gogate, P. R. & Pandit, A. B. (2008). Intensification of esterification of acids for synthesis of biodiesel using acoustic and hydrodynamic cavitation. *Ultrasonics Sonochemistry*, Vol. 15, No. 3, pp. (188-194), 1350-4177.

Kraai, G. N., Schuur, B., van Zwol, F., Haak, R. M., Minnaard, A. J., Feringa, B. L., Heeres, H. J. & de Vries, J. G. (2009). Process intensification. Continuous two-phase catalytic reactions in a table-top centrifugal contact separator. In: *Catalysis of Organic Reactions*, 123, M. L. Prunier, pp. (39-49), 978-1-4200-7076-7, CRC Press-Taylor & Francis Group, Boca Raton.

Kraai, G. N., van Zwol, F., Schuur, B., Heeres, H. J. & de Vries, J. G. (2008). Two-phase (bio)catalytic reactions in a table-top centrifugal contact separator. *Angewandte Chemie-International Edition*, Vol. 47, No. 21, pp. (3905-3908), 1433-7851.

Kubicka, D., Simacek, P. & Zilkova, N. (2009). Transformation of vegetable oils into hydrocarbons over mesoporous-alumina-supported CoMo catalysts. *Topics in Catalysis*, Vol. 52, No. 1-2, pp. (161-168), 1022-5528.

Leonard, R. A., Bernstein, G. J., Ziegler, A. A. & Pelto, R. H. (1980). Annular centrifugal contactors for solvent-extraction. *Separation Science and Technology*, Vol. 15, No. 4, pp. (925-943), 1520-5754.

Lima, D. G., Soares, V. C. D., Ribeiro, E. B., Carvalho, D. A., Cardoso, E. C. V., Rassi, F. C., K.C., M. & Suarez, P. A. Z. (2004). Diesel-like fuel obtained by pyrolysis of vegetable oils. *Journal of Analytical and Applied Pyrolysis*, Vol. 71, No. 2, pp. (987-996), 0165-2370.

Lin, L., Cunshan, Z., Vittayapadung, S., Xiangquian, S. & Mingdong, D. (2011). Opportunities and challenges for biodiesel fuel. *Applied Energy*, Vol. 88, No. 4, pp. (1020-1031), 0306-2619.

Maher, K. D. & Bressler, D. C. (2007). Pyrolysis of triglyceride materials for the production of renewable fuels and chemicals. *Bioresource Technology*, Vol. 98, No. (2351-2368), 0960-8524.

McFarlane, J., Tsouris, C., Birdwell, J. F. J., Schuh, D. L., Jennings, H. L., Pahmer Boitrago, A. M. & Terpstra, S. M. (2010). Production of biodiesel at kinetic limit achieved in a centrifugal reactor/separator. *Industrial and Engineering Chemistry Research*, Vol. 49, No. 7, pp. (3160-3169), 0888-5885.

Mello, V. M., Pousa, G., Pereira, M. S. C., Dias, I. M. & Suarez, P. A. Z. (2011). Metal oxides as heterogeneous catalysts for esterification of fatty acids obtained from soybean oil. *Fuel Processing Technology*, Vol. 92, No. 1, pp. (53-57), 0378-3820.

Milne, T. A., Evans, R. J. & Nagle, N. (1990). Catalytic conversion of microalgae and vegetable oils to premium gasoline, with shape-selective zeolites. *Biomass*, Vol. 21, No. 3, pp. (219-232), 0144-4565.

Morgan, T., Grubb, D., Santillan-Jimenez, E. & Crocker, M. (2010). Conversion of triglycerides to hydrocarbons over supported metal catalysts. *Topics in Catalysis*, Vol. 53, No. 11-12, pp. (820-829), 1022-5528.

Mushrush, G. W., Willauer, H. D., Bausermann, J. W. & Williams, F. W. (2009). Incompatibility of fischer-tropsch diesel with petroleum and soybean biodiesel blends. *Industrial and Engineering Chemistry Research*, Vol. 48, No. 15, pp. (7364-7367), 0888-5885.

Nawar, W. W. & Dubravcic, M. F. (1968). Thermal decomposition of methyl oleate. *Journal of the American Oil Chemists Society*, Vol. 45, No. 2, pp. (100-102), 0003-021X.

Noureddini, H. & Zhu, D. (1997). Kinetics of transesterification of soybean oil. *Journal of the American Oil Chemists Society*, Vol. 74, No. 11, pp. (1457-1463), 0003-021X.

Osmont, A., Catoire, L. & Dagaut, P. (2010). Thermodynamic data for the modeling of the thermal decomposition of biodiesel. 1. Saturated and monounsaturated FAMES. *Journal of Physical Chemistry A*, Vol. 114, No. 11, pp. (3788-3795), 1089-5639.

Peterson, C. L., Cook, J. L., Thompson, J. C. & Taberski, J. S. (2001). Continuous flow biodiesel production. *Applied Engineering in Agriculture*, Vol. 18, No. 1, pp. (5-11), 0883-8542.

Ra, Y., Rietz, R. D., McFarlane, J. & Daw, S. (2008). Effects of fuel physical properties on diesel engine combustion using diesel and biodiesel fuels. *SAE International Journal of Fuels and Lubricants*, Vol. 1, No. 1, pp. (703-718), 1946-3952.

Rusco, F. W. & Walls, W. D. (2008). *Biofuels, Petroleum Refining, and the Shipping of Motor Fuels*, University of Calgary, TP-08005, Calgary AB.

Rusco, F. W. & Walls, W. D. (2009). *Biofuels and the Fungibility of Motor Fuels*, International Association for Energy Economics. Third Quarter: 49-52.

Stankiewicz, A. & Moulijin, J. A. (2002). Process intensification. *Industrial & Engineering Chemistry Research*, Vol. 41, No. 8, pp. (1920-1924), 0888-5885.

Tiwari, R., Rana, B. S., Kumar, R., Verma, D., Kumar, R., Joshi, R. K., Garg, M. O. & Sinha, A. K. (2011). Hydrotreating and hydrocracking catalysts for processing of waste soya-oil and refinery-oil mixtures. *Catalysis Communications*, Vol. 12, No. 6, pp. (559-562), 1566-7367.

Toulson, E., Allen, C. M., Miller, D. J., McFarlane, J., Schock, H. J. & Lee, T. (2011). Modeling the autoignition of fuel blends with a multistep model. *Energy & Fuels*, Vol. 25, No. 2, pp. (632-639), 0887-0624.

Tsouris, C. & Porcelli, J. V. (2003). Process intensification—Has its time finally come? *Chemical Engineering Progress*, Vol. 99, No. 10, pp. (50-55), 0360-7275.

Twaiq, F. A., Zabidi, N. A. M., Mohamed, A. R. & Bhatnagar, A. K. (2003). Catalytic conversion of palm oil over mesoporous aluminosilicate MCM-41 for the production of liquid hydrocarbon fuels. *Fuel Processing Technology*, Vol. 84, No. 1-3, pp. (105-120), 0378-3820.

Van Gerpen, J., Shanks, B., Pruszko, R., Clements, D. & Knothe, G. (2004). *Biodiesel Production Technology*, National Renewable Energy Laboratory, NREL/SR-510-36244, Golden CO.

Varma, M. N., Deshpande, P. A. & Madras, G. (2010). Synthesis of biodiesel in supercritical alcohols and supercritical carbon dioxide. *Fuel*, Vol. 89, No. 7, pp. (1641-1646), 0016-2361.

Vasudevan, P. T. & Briggs, M. (2008). Biodiesel production – current state of the art and challenges. *Journal of Industrial Microbiology and Biotechnology*, Vol. 35, No. 5, pp. (421-430), 1367-5435.

Vinokurov, V. A., Barkov, A. V., Krasnopol'skaya, L. M. & Mortikov, E. S. (2010). New methods of manufacturing alternative fuels from renewable feedstock sources. *Chemistry and Technology of Fuels and Oils*, Vol. 46, No. 2, pp. (75-78), 0009-3092.

Wang, L. & Yang, J. (2007). Transesterification of soybean oil with nano-MgO or not in supercritical and subcritical methanol. *Fuel*, Vol. 86, No. 3, pp. (328–333), 0016-2361.

Wen, Z., Yu, S.-T., Yan, J. & Dahlquist, E. (2009). Intensification of biodiesel synthesis using zigzag micro-channel reactors. *Bioresource Technology*, Vol. 100, No. 12, pp. (3504-3560), 0960-8524.

Yan, S. L., DiMaggio, C., Mohan, S., Kim, M., Salley, S. O. & Ng, K. Y. S. (2010). Advancements in heterogeneous catalysis for biodiesel synthesis. *Topics in Catalysis*, Vol. 53, No. 11-12, pp. (721-736), 1022-5528.

Yang, Z. & Xie, W. (2007). Soybean oil transesterification over zinc oxide modified with alkali earth metals. *Fuel Processing Technology*, Vol. 88, No. 6, pp. (631–638), 0378-3820.

Yu, X. H., Wen, Z. Z., Lin, Y., Tu, S. T., Wang, Z. D. & Yan, J. Y. (2010). Intensification of biodiesel synthesis using metal foam reactors. *Fuel*, Vol. 89, No. 11, pp. (3450-3456), 0016-2361.

Zabeti, M., Daud, W. M. A. W. & Aroua, M. K. (2009). Activity of solid acid catalysts for biodiesel production: A review. *Fuel Processing Technology*, Vol. 90, No. 6, pp. (770-777), 0378-3820.

Zeng, H. Y., Deng, X., Wang, Y. J. & Liao, K. B. (2009). Preparation of Mg-Al hydrotalcite by urea method and its catalytic activity for transesterification. *AIChE Journal*, Vol. 55, No. 5, pp. (1229-1235), 0001-1541.

Zhou, C., Wang, C. W., Wang, W. G., Wu, Y. X., Yu, F. Q., Chi, R. A. & Zhang, J. F. (2010). Continuous production of biodiesel from soybean oil using supercritical methanol in a vertical tubular reactor: I. Phase holdup and distribution of intermediate product along the axial direction. *Chinese Journal of Chemical Engineering*, Vol. 18, No. 4, pp. (626-629), 1004-9541.

3

Soybean Biodiesel and Metrology

Vanderléa de Souza, Marcos Paulo Vicentim, Lenise V. Gonçalves,
Maurício Guimarães da Fonseca and Viviane Fernandes da Silva
INMETRO- National Institute of Metrology, Standardization and Industrial Quality;
Directorate of Industrial and Scientific Metrology; Division of Chemical Metrology
Brazil

1. Introduction

Biodiesel is a renewable fuel defined as a monoalkyl ester derived from vegetable oils, animal fats or microbial oils (algae, bacteria and fungi). The conversion of the fats or oils from these raw materials into biodiesel is possible through enzymatic or chemical reactions, which the most widely employed and studied is the transesterification reaction, involving alcohol and a catalyst. Such process converts triacylglycerols into esters of fatty acids molecules, which present physical-chemical properties and cetane number similar to diesel (Krawczyk, 1996; Ma & Hanna, 1999; Li *et al.*, 2008; ASTM D6751, 2008; Moser, 2009; Knothe *et al.*, 2005; Knothe & Steidley, 2005).

Vegetable oils were first tried for combustion in engines since the early creation of Diesel engines, in the end of 19th century. At that age, the higher cost and lower availability of these oils compared to the just developed petroleum derivates, associated to the higher homogeneity and efficiency gain up to 35% utilizing diesel, led to the complete abandonment of vegetable oils for combustion in engines. However, in the last century, the supply stability of petroleum by some countries has changed, causing drastic petroleum price raise. Thus, worldwide discussions concerning petroleum dependence were retaken, and since the second half of 90's utilization of fuels derived from renewable sources, including biodiesel, has increased in Brazil, Europe, USA and Asia (Costa *et al.*, 2003). In Brazil, social factors, such as new job opportunities, also stimulated biodiesel production.

The direct use of vegetable oils as fuel in compression ignition engines could be considered, but they are problematic due to their high viscosity (about 11-17 times greater than diesel fuel) and low volatility. These oil types do not burn completely and form carbon deposits in the fuel injectors of diesel engines. The viscosity of vegetable oils can be better improved with transesterification reaction, a process which seems to insure very good outcomes in terms of lowering viscosity and enhancing other physicochemical properties. Transesterification is a chemical reaction which proceeds under heat and involves triacylglycerols and an alcohol of lower molecular weights (typically methanol, ethanol, isopropanol or butanol) using homogeneous or heterogeneous substances as catalyst, which typically is an acid or a base, to yield biodiesel and glycerol (Ferella *et al.*, 2010), as presented in Figure 1.

Almost all biodiesel is produced from virgin vegetable oils using the base-catalyzed technique as it is the most economical process for treating virgin vegetable oils, requiring only low temperatures and pressures and producing over 98% conversion yield (provided the starting oil is low in moisture and free fatty acids). However, biodiesel produced from other sources or by other methods may require acid catalysis which is much slower (Ataya et al., 2007).

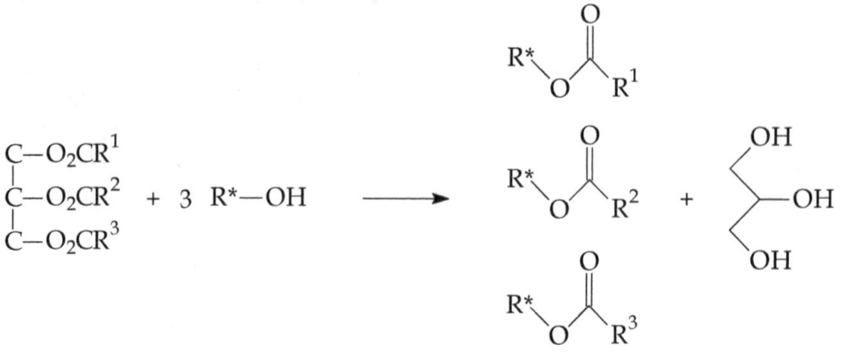

$$
\begin{array}{c}
C-O_2CR^1 \\
C-O_2CR^2 \\
C-O_2CR^3
\end{array}
\quad + \quad 3 \ R^*-OH \quad \longrightarrow
$$

1   **2**   **3**   **4**

Fig. 1. A general representation of transesterification reaction between a triacylglycerols (1) and an alcohol (2) to give alkyl esters of fatty acids (3) and glycerol (4).

The purification of biodiesel is a crucial step for production of a high-quality product, and choosing of the appropriate techniques is important for this biofuel to become economically viable. Biodiesel and glycerol are typically mutually soluble, but a notable difference in density between biodiesel (880 kg/m3) and glycerol (1050 kg/m3, or more) phases is a property that allows the employment of simple separation techniques such as gravitational settling or centrifugation. Washing can be also applied to remove free glycerol, soap, excess alcohol, and residual catalyst. But in this case, drying of alkyl ester is needed to achieve the stringent limits of biodiesel specification on the amount of water (Atadashi et al., 2011).

Biodiesel presents physical-chemical properties and cetane number similar to diesel, but this biofuel has several advantages over the fossil fuel (petrodiesel). Biodiesel is biodegradable, its sources are renewable, it respects the carbon cycle, and it presents lower toxicity, essentially no sulfurous and no aromatic compounds. The substitution of conventional diesel by biodiesel would reduce sulfur emissions by 20%, carbonic anhydride by 9.8%, non-burned hydrocarbonates by 14.2%, particulate material by 26.8%, and nitrogen oxide by 4.6%, thus reducing most regulated exhaust emissions. Biodiesel presents superior lubricity, higher flash point and positive energy balance (Albuquerque, 2006).

Biodiesel has been utilized blended to petrodiesel for internal combustion engines. Most countries utilize a system known as "factor B" to indicate the volumetric concentration of biodiesel in the blends. So, B100 indicates that a sample is pure biodiesel, while B20 or B5, for example, indicates that the blend has 20% (v/v) or 5% (v/v), respectively, of biodiesel. Some authors report that mixtures containing up to 20% biodiesel can generally be employed in diesel engines without modifications, but most of the authors and the vehicle produces do not recommend employing mixtures containing more than 5% biodiesel (Biodieselbr.com, 2011; Fueleconomy.gov, 2011; Biodiesel.org, 2009). Nowadays, biodiesel sold in Brazil and Europe is a B5 fuel (Biodieselbr.com, 2011; Biopowerlondon.co.uk, 2011).

However, a primary disadvantage of biodiesel is inferior oxidative and storage stability versus petrodiesel, lower volumetric energy content, reduced low temperature operability, susceptibility to hydrolysis and microbial degradation, as well as higher nitrogen oxide emissions (Albuquerque, 2006; Moser, 2009; Knothe et al., 2005; Knothe & Steidley, 2005). Also, the esters which biodiesel is composed can attach to water attributing to this biofuel the hygroscopic property. Some water content also comes from the extraction and transesterification processes. The presence of water in biodiesel reduces the calorific value and enhances engine corrosion. Moreover, water promotes the growth of microorganisms and increases the probability that oxidation products are formed during long-term storage. These oxidation products can cause disturbances in the injection system and in the engine itself (Schlink & Faas, 2009).

The most significant vegetable oils produced worldwide during 2009 were palm (45.13 MMT), soybean (37.69 MMT), rapeseed/canola (21.93 MMT), and sunflower (11.45 MMT) oils (United States Department of Agriculture, 2010). Generally, the most abundant oils or fats in a region are most commonly used as feedstocks for biodiesel production. Thus, for production of biodiesel, rapeseed/canola and sunflower oils are principally used in Europe, palm oil predominates in tropical countries, and soybean oil and animal fats are most commonly used in the USA and Brazil (Moser, 2009; Knothe et al., 2005).

As globalization increases, there is a need for harmonization of technical parameters regarding several products and services (especially commodities) provided around the world. In order to achieve greater transparency, reliability and suitability among these products and services, International Standards have been developed and must be followed by those countries and companies that take part in this worldwide trade. For biodiesel it is not different. Biodiesel must be certified as compliant with accepted fuel standards before combustion in diesel engines. In Europe, specifications for this biofuel are described by the European Committee for Standardization (CEN) through standard EN 14214:2008 (EN 14214:2008, 2009); in the United States the specifications must be according to ASTM D6751-08a (ASTM D6751, 2008); and in Brazil, fuels and biofuels are regulated by National Agency of Petroleum, Gas and Biofuels (ANP), through Resolution ANP no. 7 from March 19, 2008 (ANP Resolution n°7, 2008).

However, international standards are not conclusive, and many times they also are not suitable or accurate for many parameters of many products and services. Several parameters have to be exhaustively studied for the convergence of some rules. In the case of biodiesel, standards that have been applied for this biofuel were originally developed for diesel analysis and adapted for biodiesel. Just a few standards have been developed specifically for biodiesel up this time. Standards have to be constantly attending to the modernization of products, markets and methodologies. In some cases, technical methods employed to analyze some kind of product is simply adapted to analyze another similar one. An example for Biodiesel concerns the water determination method. The water content in biodiesel is ruled by EN 14214:2008 (EN 14214:2008, 2009), ASTM D6751-08a (ASTM D6751, 2008) and in Brazil by Resolution ANP no. 7 (ANP Resolution N°7, 2008). All of them settle the maximum water content as 0.05% (w/w). These standards require Karl Fischer titration for water determination, as described in ISO 12937:2000 (ISO 12937:2000, 2000). Otherwise, this ISO standard was created considering petrodiesel analyses and specifications, but now it has been adapted for water content determination in biodiesel. These adaptations of methods cause errors or at least low accuracy in analyses. Furthermore, many standards, like this one, do not describe a method setting exactly the parameters to be employed by the apparatus.

Determining such low water content in non-aqueous substances with high accuracy is not an easy task and just a few works have been published regarding moisture in biodiesel. Assessments for high accuracy determination of water in biodiesel have been performed by the Laboratory of Organic Analyses from INMETRO (Brazilian Institute of Metrology), which optimized some parameters for commercial soybean biodiesel utilizing Coulometric Karl Fischer Titration coupled to Auto-sampler Oven (Vicentim et al., 2010). Experiments ongoing by this group are still verifying the necessity of further optimizations for biodiesel samples produced form another sources (data unpublished yet).

Complete discussions regarding the need for International Standards, their applications, the mechanisms for biodiesel obtainment, its chemical and physicochemical properties, and considerations about the importance of metrology and its influence on biodiesel quality will be presented in the next sessions of this chapter.

2. Biodiesel synthesis

All over the world there are many research groups searching for fuels from renewable sources due to the imminent serious depletion of fossil resources and also due to an increasing societal ecological environmental awareness. Many types of alternative energy sources have been studied, as solar, wind, water, nuclear (through the cleavage of radioisotopes) and plant biomass. However, nowadays, the only ready-to-use technologies for automotive renewable energy supply, and that has presented excellent results, are the production and utilization of the so-called biofuels, like the bioethanol from sugar cane, corn starch, sugar beet and the biodiesel, especially that one produced from oily crops.

Biodiesel can be defined as mono alkyl esters of fatty acids derived from animal fat and vegetable oils (researches are ongoing for utilization of microbial oils), and obtained mainly through the transesterification reaction. In a general way, this reaction involves triacylglycerols (which are esters) reacting with a small chain aliphatic alcohol, generally methanol, ethanol, isopropanol or butanol, producing a new ester and an alcohol, as shown in Figure 1 (Pinto et al, 2005).

Biodiesel can be derived from the following processes: pyrolysis, cracking, alcoholysis, esterification and transesterification of fats and oils which is the most commonly used process.

Processes like pyrolysis and cracking produce many side products, the reactions are not very selective and the processes require many steps, like removing ash and solid products for example. Pyrolysis, strictly defined, is the conversion of one substance into another by means of heat or by heat with the aid of a catalyst. It involves heating in the absence of air or oxygen and cleavage of chemical bonds to yield small molecules. Pyrolytic chemistry is difficult to characterize because of the variety of reaction paths and the variety of reaction products. The pyrolyzed material can be vegetable oils, animal fats, natural fatty acids and methyl esters of fatty acids. The first pyrolysis of vegetable oil was conducted in an attempt to synthesize petroleum from vegetable oil. Since World War I, many investigators have studied the pyrolysis of vegetable oils to obtain products suitable for fuel. Catalysts have been used in many studies, largely metallic salts, to obtain paraffines and olefins similar to those present in petroleum sources. Soybean oil was thermally decomposed and distilled in air and nitrogen sparged with a standard ASTM distillation apparatus. The total identified hydrocarbons obtained from the distillation of soybean and high oleic safflower oils were 73-77 and 80-88% respectively. The main components were alkanes and alkenes, which accounted for approximately 60% of the total weight. Carboxylic acids accounted for another 9.6-16.1% (Fangrui & Milford, 1999).

Esterification is a process that consists in two main steps. In the first one the oil is saponified with sodium hydroxide followed by acidification, washing and drying, obtaining a mix of fatty acids. In the final steps the fatty acids are esterified with a small chain alcohol, like methanol, ethanol or isopropyl alcohol.

2.1 Transesterification using catalysts

In a general way transesterification reaction occur catalyzed by an acid (Gerpen, 2005), alkali (Rinaldi et al, 2007), enzyme (Mendes et al, 2011 & Watanabe et al, 2002) or employing heterogeneous catalysis (Mell et al, 2011). The main heterogeneous catalysts are zeolites (Suppes et al, 2004), clays (Jaimasith et al, 2007), ion-exchange resins (Honda et al, 2007) and oxides.

The most used way of catalysis is employing an alkali. The reaction mechanism under alkaline condition occurs in two steps: In the first step sodium hydroxide reacts with methanol, in an acid-base reaction producing a strong base, sodium methoxide and water. In the second step sodium methoxide reacts as a nucleophile and attacks the three carbonyl carbons from the triacylglycerol. A very unstable tetrahedral intermediate is obtained. As a result, the cracking of the triacylglycerol occurs, obtaining three methyl esters (biodiesel) and glycerol.

The most employed transesterifying agent is methanol. Other alcohols may also be used in the preparation of biodiesel, such as ethanol, propanol, isopropanol, and butanol. Ethanol is of particular interest primarily because it is less expensive than methanol in some regions of the world, and biodiesel prepared from bioethanol is completely bio-based. Butanol may also be obtained from biological materials, thus yielding completely bio-based biodiesel as well. Methanol, propanol, and isopropanol are normally produced from petrochemical materials such as methane obtained from natural gas in the case of methanol. Some conditions utilized in these reactions are described below:

Methanolysis: The classic reaction conditions for the methanolysis of vegetable oils or animal fats are 6:1 molar ratio of methanol to oil, 0.5 wt.% alkali catalyst (with respect to TAG), 600 rpm, 60°C reaction temperature, and 1 h reaction time to produce FAME and glycerol.

Ethanolysis: The classic conditions for ethanolysis of vegetable oils or animal fats are 6:1 molar ratio of ethanol to oil, 0.5 wt.% catalyst (with respect to TAG), 600 rpm, 75°C reaction temperature, and 1 h reaction time to produce fatty acid ethyl esters (FAEE) and glycerol.

Butanolysis: The classic conditions for butanolysis of vegetable oils or animal fats are 6:1 molar ratio of butanol to oil, 0.5 wt.% catalyst (with respect to TAG), 600 rpm, 114°C reaction temperature, and 1 h reaction time to produce fatty acid butyl esters and glycerol. Butanol is completely miscible with vegetable oils and animal fats because it is significantly less polar than methanol and ethanol. Consequently, transesterification reactions employing butanol are monophasic throughout. The monophasic nature of butanolysis reactions also complicates purification of the resultant butyl esters (Moser, 2009).

2.1.1 Homogeneous catalysts

Conventional processes include the use of homogeneous alkaline catalysts — NaOH, KOH, NaOMe and KOMe — under mild temperatures (60-80 °C) and atmospheric pressure. There are two main factors that affect the cost of traditional biodiesel production: the cost of raw materials and the cost of processing (multiple steps), though the commercialization of resultant glycerol can share the production costs with biodiesel, improving the overall process profitability. In order to reduce the costs associated with feedstock, waste cooking oils, animal fats or non-edible oils could be used. However, the use of homogeneous alkaline catalysts in the transesterification of such fats and oils involves several troubles due to the presence of large amounts of free fatty acids (FFAs). Of course, alkaline catalysts can be used to process these raw materials, but a large consumption of catalyst as well as methanol is compulsory to achieve biodiesel of standard specifications. Thus, FFA concentration in the oil inlet stream is usually controlled below 0.5% (w/w), avoiding the formation of high soap concentrations as a consequence of the reaction of FFAs with the basic catalyst. The soap causes processing problems downstream in the product separation because of emulsion formation. Usually, this problem is overcome through a previous esterification step where FFAs are firstly esterified to FAMEs using a homogeneous acid catalyst, and then, once the acid homogeneous catalyst has been removed, transesterification of triacylglycerols is performed as usual by means of an alkaline catalyst. Likewise, homogeneous acid catalysts (H_2SO_4, HCl, BF_3, H_3PO_4) have been proposed to promote simultaneous esterification of FFAs and transesterification of triacylglycerols in a single catalytic step, thus avoiding the pre-conditioning step when using low cost feedstock with high FFA content. However, these catalysts are less active for transesterification than alkaline catalysts and therefore higher pressure and temperature, methanol to oil molar ratio and catalyst concentration are required to yield adequate transesterification reaction rates. Hence, despite its insensitivity to free fatty acids in the feedstock, acid-catalyzed transesterification has been largely ignored mainly due to its relatively slower reaction rate. (Melero et. al, 2009).

2.1.2 Heterogeneous catalysts
The use of heterogeneous catalysts (Wang et al, 2007 and Leclercq et al, 2001) has as main advantage the reaction work-up, the post reaction treatment, the purification steps and the separation steps. These catalysts can be easily removed from the reaction medium and even can be reused. Another interesting factor is based in the fact that these catalysts avoid the formation of undesirable side products, like the saponification products (Botts et al, 2001; Thomasevic & Marincovic, 2003). The biggest difficulty at this type of reaction is the diffusion between the systems oil/catalyst/solvent (Gryglewics, 1999). For the soybean biodiesel production, the catalysts that have been commonly employed are tin, zinc and aluminum, as Al_2O_3, ZnO and $(Al_2O_3)_8(SnO_2)$, for example (Mello et al, 2011). Other processes have used an heterogeneous catalyst of a spinel mixed oxide of two (non noble) metals, which eliminate several neutralization and washing steps needed for process using heterogeneous catalysts (Helwani et. al, 2009).

2.1.3 Alkaline catalysts
Alkaline catalysis (Zhou et al, 2003) is a procedure that generally uses sodium and potassium alkoxides, and some times sodium and potassium hydroxides or carbonates. Among these three groups, alkoxides have the advantage of performing reactions at mild temperatures, they provide high yields of esters derived from fatty acids and they are not corrosives like the acid catalysts. On the other hand, these catalysts are hygroscopic, more expensive and usually result in side products, such as the saponification ones.

2.1.4 Acid catalysts

Sulfuric and hydrochloric acid compounds are the main catalysts. This catalysis (Mohamad and Ali, 2002) has the advantage of avoiding the formation of side products and obtaining high yield formation of alkyl esters. However, reactions in acid media are highly corrosive and the work up is more difficult, seen it needs a special treatment to neutralize the reaction medium

2.1.5 Enzymes catalysts

A fourth class of catalysts employed for biodiesel production is enzyme (Fukuda et al, 2001). The enzymes allow the use of mild temperature reactions, between 20 and 60 °C, excess of alcohol is dispensed, the reactions can be performed with or without a solvent and the catalyst can be reused several times (Shimada et al, 2002). Other great advantages are the easy work-up, dispensing neutralization and deodorization of the reaction medium (Bielecki et al, 2009). The disadvantages are related to the fact that enzymes are very specifics, expensive and very sensible to alcohols, causing their deactivation (Manduzzi et al, 2008). The literature (Modi et al, 2007) shows that this problem can be solved by the use of small amounts of water (Kaieda et al, 2001 and Ban at al, 1999). Another research group showed that the use of organic solvents (Narasisham et al, 2008) can activate the enzymes, in special the use of dioxane and petroleum ether (Dennis et al, 2008), for example. Watanabe and his research group has developed a methodology to produce biodiesel from soybean degummed oil by the use of the lipase (an enzyme specific for hydrolysis of lipids, like triacyglycerides) from *Candida antarctica* in a free solvent system (Watanabe et al, 2002). Another procedure was performed by Liu et al. (2005) studying the acyl group migration with immobilized lipozyme TL catalyzing the production of biodiesel from soybean oil (Noureddini et al, 2005).

2.2 New process for biodiesel obtainment
2.2.1 Microwave and Ultrasound

Many researches seek for the improvement of catalysts in biodiesel production. Reactions employing ultrasound (Santos et al, 2009) and microwave (Leadbeater & Stencel, 2006) techniques represent a great advance. Ultrasound (Chand et al, 2010) and microwaves (Barnard et al, 2007) as auxiliary techniques facilitate the interaction between methoxide ions and reagents, increasing the process efficiency, obtaining higher yields in a shorter reaction time. Reaction employing these techniques can be performed at mild temperatures due to a higher kinetic energy in the reaction medium, facilitating also the miscibility among the reactants (Fukuda, 2001).

2.2.2 Transesterification using supercritical fluids

This is a non catalytic method to produce biodiesel, which has the several advantages. One of them concerns the shorter reaction time than the traditional catalyzed transesterification. This is possible because the initial reaction lag time is overcome due to the reaction is proceeded in a single homogeneous phase since the supercritical methanol is fully miscible with the vegetable oils. Moreover, the reaction rate is very high and the subsequent purification is much simpler than that of the conventional process. The supercritical route is also characterized by high yield because of simultaneous transesterification of triacylglycerols and esterification of fatty acids. This process is environmentally friendly seen that waste water containing alkali or acid catalysts is not produced. The disadvantages of this process regard the high costs, the necessity of a high pressure system (200-400 bar), high temperatures (350-400°C) and high methanol/oil rates (Balat, M.H., 2008; Melero et. al, 2009).

2.2.3 Hydrotreating

Hydrotreating is a process that produces biodiesel through a hydrotreatment of triacylglycerols. The hydrocarbons are produced by two reaction pathways: hydrodeoxygenation (HDO) and hydrodecarboxylation (HDC). n-Alkanes originating from HDO have the same carbon number as the original fatty acid chain, i.e., even carbon number, typically 16 or 18. Water and propane are the main reaction by-products of this route (Snare et al, 2007).

2.3 Biodiesel purification process

Separation and purification of biodiesel is a critical task. Normally, the crude biodiesel produced by homogeneous catalysis can be separated from glycerol by simple gravitational settling or centrifugation, due to their notable difference in phase density (biodiesel 880 kg/m3 and glycerol 1050 kg/m3 or more). Washing ester phase with water or an acid mineral or base solution to remove base/acid catalyst residues, for example, can be also applied to remove free glycerol, soap, excess alcohol, and residual catalyst. Finally the biodiesel is dried after neutralization. For methanol recycle vacuum distillation can be used prior to glycerin purification. When the biodiesel is obtained by a heterogeneous catalysis, this one is removed by a filtration process.

However, these conventional technologies and other ones like decantation, washing with ether and the use of absorbents have proven to be inefficient, time and energy consumptive, and less cost effective. On the other hand, the involvement of membrane reactor and separative membrane shows great promise for the separation and purification of biodiesel. Membrane technology needs to be explored and exploited to overcome the difficulties usually encountered in the separation and purification of biodiesel (Zhang et al, 2003; Atadashi et al., 2011).

3. Biodiesel properties and their influence on engine performance

3.1 The fatty acid composition of feedstocks and the influence on the properties of biodiesel fuel

The most common feedstocks for biodiesel production are commodities such as vegetable oils derived from soybean, palm and sunflower seed. These materials possess fatty acid profiles consisting primarily of five fatty acids with carbon chains containing 16 to 18 carbon atoms (C16 to C18) namely palmitic acid (hexadecanoic-C16: 0), stearic acid (octadecanoic, C18: 0), oleic acid (9 (Z)-octadecenoic - C18: 1), linoleic (9 (Z), 12 (Z)-octadecadienoic acid - C18: 2), linolenic (9 (Z), 12 (Z), 15 (Z) , octadecatrienoic acid- C18: 3). The proportions of different fatty acids in feedstocks influence the properties of biodiesel. Some of the most relevant properties to be considered for a biodiesel candidate to be used as a substitute for diesel fuel (or blended with the same) are cetane number, viscosity, cold flow properties and oxidative stability. Lubricity is another important parameter for a fuel but it is independent on the fatty acid composition.

Two major problems to be overcome in biodiesel are the poor properties at low temperatures and low oxidative stability. In most cases these two problems occur with the same sample. They result from physical and chemical properties of fatty esters, the major components of biodiesel and minor constituents that arise during the transesterification reaction or are from raw materials.

The profile of methyl esters found in greater proportion in soybean is about 11% C16: 0, 4% C18: 0, 21-24% C18: 1, 49-53% C18:2, 7-8% C18: 3 which provides cetane number in the range of 48-52, kinematic viscosity at 40 °C equal to 4.10 to 4.15 mm²s⁻¹ and cloud point approximately equal to 0 ºC (Knothe et al., 2005, Mittelbach and Remschmidt, 2004). Rapessed (canola) methyl esters have a fatty acid profile approximately 4% C16:0, 2% C18:0, 58-62% C18:1, 21-24% C18:2, 10-11% C18:3 and present cetane number in the range of 51-55, kinematic viscosity at 40 °C around 4,5 mm²s⁻¹ and cloud point of approximately -3 °C (Knothe et al., 2005, Mittelbach and Remschmidt, 2004). Thus the difference in fatty acid profile, more specifically concerning C18:1 and C18:2 contents, which had their values almost reversed in the case presented, causes a noticeable change in fuel properties.

Many researches have focused on resolving or at least reducing problems related to low oxidative stability and cold flow properties of biodiesel. Some trials in this way involves the addition of additives and changes in the composition of fatty esters, that can be reached varying either the reactive alcohol or the oil fatty acid profile. Changing the fatty acid profile can be achieved by physical methods, genetic modification of feedstock or use of alternative feedstocks with different fatty acid profiles.

Important features regarding the use of neat biodiesel or its blends with diesel fuel include reduced emissions, with the exception of nitrogen oxides, compared to petrodiesel (petroleum-derived diesel fuel), biodegradability, absence of sulfur, inherent lubricity, positive energy balance, higher flash point, compatibility with existing infrastructure for distribution of fuel, to be renewable and a domestic source. The American ASTM D6751-08a, the European EN 14214:2008 and the Brazilian ANP nº 7 standards deal with the technical specifications for biodiesel to be used in internal combustion cycle diesel engine taking into account the advantage of utilizing the existing infrastructure for distribution of diesel ensuring fuel quality for the final consumer. Table 1 shows the specifications recommended by American, European and Brazilian standards aiming biodiesel utilization as fuel.

3.2 The influence of cetane number on combustion and atmospheric emissions

The cetane number (CN) is a dimensionless parameter related to the ignition delay time after fuel injection into the combustion chamber of a diesel engine. A higher cetane number results in a shorter ignition delay time and vice versa. A cetane scale was established, being hexadecane commonly used as reference compound, with CN = 100, and 2,2,4,4,6,8,8-heptamethylnonane, a highly branched compound with poor ignition quality in a diesel engine, with CN =15.

The cetane scale explains why the triacylglycerols, such as those found in vegetable oils, animal fats and their derivatives, are suitable alternatives to diesel fuel. The reason is the long chain, linear and unbranched fatty acids, chemically similar to those in n-alkanes of conventional diesel fuels with good quality.

The cetane number of fatty esters increases with the increase of saturation and carbon chain. Thus, the CN of methyl palmitate and methyl stearate (C16: 0 and C18: 0) is greater than 80 (Knothe et al., 2003), the CN of methyl oleate (C18: 1) is in the range of 55-58, the methyl linolenate is (C18: 2) around 40 and the methyl linolenato is (C18: 3) around 25. Esters derived from branched alcohols such as isopropanol have CN values comparable to methyl esters or other ester with alkylic chain (Knothe et al., 2003, Zhang & Gerpen, 1996) linear, although the cost of production once isopropyl alcohol is more expensive than methanol and ethanol costs.

In general biodiesel does not require additives to improve cetane number, because its cetane number generally reaches the minimum values established in the international technical specifications. An exception may be the methyl esters of soybean that did not reach the minimum of 51 set by EN 14214:2008 (EN 14214:2008, 2009) but usually reach the minimum set of 47 recommended in ASTM D6751 (ASTM D6751-08a, 2008), as shown in Table 1.

Cetane number may influence both the quality of combustion and vehicle emissions. Several international agencies like the EPA (Environmental Protection Agency - USA) and the CONAMA (National Environment Council - Brazil) set limits and goals for reducing pollutants automotive emissions. In diesel cycle engines, the main pollutants are hydrocarbons, carbon monoxide, nitrogen oxides (NOx) and particulate matter. Reducing these emissions requires improving the combustion process, the treatment of exhaust gases from existing engines and technical fuels specifications. A low cetane number leads to difficulties in cold starting, increases emissions and noise level of combustion. If the cetane number is high may occur an increase in particulate emissions but NOx emissions decrease. Samples of biodiesel with low level of triacylglicerols, especially those with polyunsaturated fatty acids of C18:3, should show low levels of NOx emissions. Linear correlation was obtained between the level of unsaturation of biodiesel indicated by iodine number, the density of biodiesel and NOx emissions (McCormick et al., 2001). Thus little amounts of unsaturated fatty acids may reduce the density and the NOx emissions. An important property of biodiesel is its ability to reduce total particulate emissions of the engine and also carbon monoxide and hydrocarbons contents of exhaust gases. However biodiesel causes an increase in NOx emissions. Increasing CN to a certain level (around 60) implies in the reduction of NOx emissions (Landommatos et al., 1996).

An experiment was conducted with the OM 611 diesel engine light load of Damler Benz with ultra low sulfur content diesel (ULSD), conventional diesel and B20 blend of pure methylic soybean biodiesel and ULSD. The results obtained with the B20 blend showed no differences in NOx content compared to the two reference diesel fuels. Reductions of particulate matter by 32% and 14%, respectively, compared to conventional diesel fuel and USLD were observed with B20 blend (Sirman, et al., 2000).

The causes for the increase of NOx associated with biodiesel for fuel injection systems are related to a small displacement in the range of fuel injection which is caused by differences in mechanical properties of biodiesel compared to conventional diesel (Tat & van Gerpen, 2003; Monyem et al., 2001). Due to the higher modulus of compressibility (or sound speed) of biodiesel, there is a faster transfer of the pressure wave of the injection pump to the injector needle resulting in anticipation of lifting the needle and the production of a small advance in the injection interval. It was observed that samples of B100 derived from soybeans produces an increase of one degree in the injection interval, which was accompanied by a four degree at the start of combustion (Sybist & Boehman, 2003). Strategies that can be used to reduce NOx emissions to a level equivalent to that of conventional diesel involve increase of cetane number by use of additives.

Specification	Standards					
	ASTM D6751-08a		EN 14214:2008		ANP n º7	
	Test method	Limit	Test method	Limit	Test method	Limit
Cetane number	ASTM D613; D6890	47 minimum	EN ISO 5165	51 minimum	ASTM D613; D6890 EN ISO 5165	Report
Kinematic viscosity	ASTM D445	1.9-6.0 mm²s⁻¹	EN ISO 3104	3.5-5.0 mm²s⁻¹	ASTM D445; EN ISO 3104; ABNT NBR10441	3.0-6.0 mm²s⁻¹
Oxidative stability	EN 14112	3h minimum	EN 14112	6h minimum	EN 14112	6h minimum
Cloud Point	ASTM D2500	Report	-	-	-	-
Cold filter plugging point	-	-	EN 116	Depending on time of year and location	ASTM 6371; EN 116; ABNT NBR14747	19 ºC
Cold soak/Filterability	Annex to D6751	Report	-	-	EN 12662	24 mg/Kg (max.)

ASTM - American Society for Testing and Materials; ISO - International Standards Organization; ANP - National Agency of Oil, Gas and Biofuels ; NBR - Brazilian Standard; ABNT - Brazilian Association of Technical Standards

Table 1. Specifications of biodiesel standards that affect the properties of alkyl esters as fuel in diesel cycle engines [a].

It was observed that B20 blends of soy diesel respond well to conventional peroxide di-t-butyl, a cetane improver, when tested on DDC Series 60 engines of 1991 (McCormick , et al., 2001). The biodiesel NOx was reduced by 6.2% without the contribution of 9.1% in reducing emissions of particulate matter to be compromised and B20 blend produced no noticeable increase in NOx of this engine. The peroxide, di-t-butyl nitrate and 2-ethylhexyl were tested in a similar engine (Sharp, 1994) and the reduced levels of NOx in exhaust emissions were confirmed. Notice the economy of this procedure if necessary high levels of additives.

3.3 The importance of viscosity in the use of biodiesel as fuel

Viscosity is one of the properties that most affect the use of biodiesel as a fuel since the atomization process, the initial stage of combustion in a diesel engine, is significantly affected by the viscosity of the fuel. The viscosity of the transesterified oils, ie, biodiesel is less than their vegetable oil sources, which explains the failure to use pure vegetable oils as alternative fuels to diesel. The high viscosity of untransesterified oils leads to operational problems in diesel engine for example increased engine deposits. Viscosity in the form of Kinematic viscosity is specified in quality standards of biodiesel, which exhibit a range with minimum and maximum values for this parameter. Although there are standardized methods for determining the kinematic viscosity as shown in Table 1, several studies have been conducted to predict the viscosity of biodiesel from its composition of fatty ester.

Assuming a soy biodiesel made from soybean oil containing 0.1% C14: 0, 10.3% C16: 0, 4.7% C18: 0, 22.5% C18: 1, 54 1% of C18: 2 and 8.3% of C18: 3, Allen et al., 1999 using the equation of Grunberg & Nissan, 1949, modified, predicted the value of 3.79 mm^{2}s^{-1} for the viscosity of methylic soybean biodiesel. The viscosity of fatty esters increases with the chain length and with increasing degree of saturation (Kern and Van Nostrand, 1949).

This rule also applies to alcohol used in the reaction, since the viscosity of ethyl esters is slightly higher than that of methyl esters. The configuration of double bonds also influences the viscosity. If there are only double bonds in cis configuration is observed remarkable reduction in viscosity, as well as esters with double bonds in the trans configuration have viscosities similar to the corresponding saturated esters (Kern & Van Nostrand, 1949).

3.4 Cold flow properties of biodiesel and its blends with diesel
3.4.1 Biodiesel and cold flow properties

The mixtures such as biodiesel do not possess defined melting points, but melting ranges. This fact reflects in the specifications used in biodiesel standards.

The cloud point (CP) is the temperature at which the first solids appear, but the fuel can still flow, although these solids can lead to fuel filter plugging (Dunn and Bagby, 1995).

The pour point, usually a few degrees below the cloud point, is the temperature at which the fuel can no longer be freely poured.

Several other methods exist for determining the low-temperature properties of biodiesel. These are the cold filter plugging point (CFPP) and low-temperature flow test (LTFT) (Dunn and Bagby, 1995). The CP and CFPP are included in biodiesel standards without severity since in ASTM D6751 for the value of CP only a report is required and the CFPP value in EN 14214 can vary with time of year and geographic location. The low-temperature properties of biodiesel are also influenced by the properties of individual components. The melting point of fatty esters generally increase with chain length (although chains with odd numbers of carbon have slightly lower melting points that the preceding even-number chain) and increasing saturation (Knothe, 2009).

Intending to provide the industry with an independently generated set of cold flow information on a variety of fuels in the market of the United States in 2009 with the new Ultra Low Sulfur Diesel Fuel (ULSD), cloud point, cold filter plugging point (CFPP) and low temperature flow test (LTFT) methods were used to assess the cold-flow properties for seven different biodiesel fuels blended with four different ULSD fuels representing the span of the market in 2009 (Heck, Thaeler, Howell and Hayes, 2009). The neat fuels were tested in addition to biodiesel blends with ratios of 2% biodiesel (B2), 5% biodiesel (B5), 11% biodiesel (B11), 20% biodiesel (B20) and 50% biodiesel (B50) for cloud point, CFPP, and LTFT. The pour point of the neat biodiesel and B50 blends were also analyzed. Three petrodiesel fuels with cloud points of -47,5 ºC, -16 ºC and -11 ºC were used to produce petrodiesel having target cloud points of -40 ºC, -34,4 ºC, -26,1 ºC, and -12,2 ºC. Seven biodiesel (B100) samples were selected. Three of them with low cloud point, vegetable oil base, in this case soybean oil from various manufacturing processes (distilled biodiesel, non distilled biodiesel from hexane extracted oil, non distilled biodiesel from extruder-expeller oil) were collected from commercial biodiesel producers. Four additional biodiesels of mid to mid-high to high cloud points from commercial biodiesel producers representing mixed saturation levels, blends of soybean, animal, and recycled oil based biodiesel. The cloud point values of the various B100 samples were -2,5 ℃, -2,0 ℃, -1,5 ℃, 1,0 ℃, 7,0 ℃ , 8,0 ℃ , and +12,0 ℃. In some cases the impact on cold flow properties of blending biodiesel with petrodiesel appeared to be mostly linear, while in others the impact was curvilinear. In all cases except a few with low blends of biodiesel where the CFPP of the blend was slightly below that of the petrodiesel, the blended fuel values fell between the pure petrodiesel and pure biodiesel values for all three cold flow measurements.

Biodiesel blends primarily B20 of soybean biodiesel (SME) and diesel n º2 have also been used in a variety of climates including some of the coldest weather on record without cold flow problems. A study to determine the CFPP of blends containing up 20% (SME) and Number 2 (nº 2) diesel was conducted. The University of Missouri prepared the samples that were analyzed in the Cleveland Technical Center in Kansas City, United States. The characteristics of the nº 2 Diesel and the SME are shown in Table 2. The results suggest that the blends with the highest content of biodiesel begin to gel first. Higher concentrations of biodiesel, eg, above 20% may not be suitable for use in cold climates without mixing large quantities of kerosene in combination with cold flow proven enhancers specific to the conventional diesel (National Biodiesel Board- 2007/2008).

Blends's components	Cloud Point (℃)	Pour Point (℃)	CFPP (℃)
n º2 Diesel	-15,6 ℃	-34,4 ℃	-17,2 ℃
Soy Methyl Esters (SME)	0 ℃	-3,9 ℃	-5,6 ℃

Table 2. Characteristics of the components of the blends.

Literature data show that ethyl esters and specially isopropyl esters improve low-temperature properties of biodiesel compared to methyl esters. Isopropyl and isobutyl esters of common soybean oil exhibited crystallization temperatures 7-11 and 12-14 ℃ lower than the corresponding methyl esters (Lee at all., 1995). The data suggest that the fuel blend that begins to gel first contains the highest concentration of biodiesel. Higher concentrations of biodiesel, eg, above 20% may not be suitable for use in cold climates without mixing large quantities of kerosene in combination with proven enhancers of cold-flow properties specific to the conventional diesel.

3.4.2 Diesel fuel background information relevant to biodiesel

The cold weather operability of diesel fuel is defined as the lowest temperature which a vehicle will operate without loss of power due to waxing of the fuel delivery system. Diesel fuels composition and cold flow properties vary greatly across the United States. Cold flow characteristics of diesel fuels are influenced by the source of the crude oil they are made from, how they are refined and if they are blended to improve performance during cold weather. The cold temperature properties of diesel fuel vary across the country depending on the time of year the fuel is produced and the climate. Generally, diesel fuels used in cold climates have better cold flow characteristics than diesel fuels used in warmer regions. Both of these statements have a direct impact on the operability of biodiesel blends in cold weather.

The refining process separates the crude oil into mixtures of its constituents, based primarily on their volatility. Diesel fuels are on the heavy end of a barrel of crude oil. This gives diesel fuel its high BTU content and power, but also causes problems with diesel vehicle operation in cold weather when this conventional diesel fuel can gel. This is not an issue for gasoline vehicles. A tremendous amount of effort has been spent over the years to understand how to deal with the cold flow properties—or the low temperature operability--of existing petroleum based diesel fuel. The low temperature operability of diesel fuel is commonly characterized by the cloud point, and the cold filter plugging point (CFPP) or the low temperature filterability test (LTFT). In general, Number 2 diesel fuel will develop low temperature problems sooner than will Number 1 diesel fuel. Number 1 diesel fuel is sometimes referred as kerosene. The gelling of diesel fuel in cold climates is a commonly known phenomenon and diesel fuel suppliers, as well as customers and diesel engine designers, have learned over time to manage the cold flow problems associated with Number 2 diesel fuel in the winter time. The leading options to handle cold weather with diesel fuel are: -Blending with kerosene;-Utilization of an additive that enhances cold flow properties;-Utilization of fuel tank, fuel filter or fuel line heaters;-Storage of the vehicles in or near a building when not in use. In most diesel engine systems today, excess diesel fuel is brought to the engine and warm fuel that has come close to the engine is recycled back to the fuel tank. This assists in keeping the fuel from gelling in cold weather. This is, in part, why diesel engines are kept running overnight at truck stops in cold climates (Bickell, 2008; Krishna & Butcher, 2008; Joshi & Pegg, 2007).

3.4.3 The impact of minor components of the vegetable oils in the cold flow properties of biodiesel

The presence of sterol glucosides (SGs), wax, monoglycerides, saturated fatty acids and polymers in both B100 and blends with petrodiesel can limit the application of these fuels due to problems with precipitation. In the last years researches have shown that biodiesel precipitations can arise even if specifications of this biofuel are met. Special attention has been dedicated to the SGs, components commonly found in vegetables and in oils derived from soybean, rapeseed and palm that will be processed to produce biodiesel. Usually the concentration of SGs in vegetable oils is not significant since they are mostly found as sterol glucosides acylated (ASGs). The ASGs have average solubility in vegetable oils, but after transesterification they are broken down chemically by removing the side chain containing the fatty acid and they are converted partially to SGs. This class of compounds is not soluble in biodiesel and its crystallization is extremely slow and depends on temperature, other impurities (as crystallization nuclei) and surface effects. However, even a brand new biodiesel, meeting all the specifications, presents precipitation of SGs after a few days of storage/transport.

The spontaneous clogging of the filters in the production unit or in the supply chain has been observed in Minnesota, United States with a B2 blend of soybean methyl esters and diesel. Several other places in the world have also observed this occurrence with B5 blend. The precipitates do not contain only SGs but also ASGs and other substances. In some cases sources containing higher concentrations of ASGs could be responsible for the deposits. The concentration of SGs and ASGs in vegetable oils depends on the feedstock and on the process used by industry to obtain them. The literature suggests that the highest concentrations of SGs and ASGs will be found in soybean and palm oil. Rapeseed oil usually has low concentrations of these compounds. In order to minimize the effect of SGs and ASGs on the FAME-Biodiesel (Fatty Acid Methyl Esters –biodiesel) and their blends with petrodiesel, special treatments of the esters in an oil refinery, or the use of adsorbents could be possible solutions for a post-processing of FAME (Haupt et al., 2009).

3.5 The effect of the impurities in the quality of biodiesel

Biodiesel is composed by alkyl esters (generally methyl esters) that can be analyzed as tool for controlling the transesterification yield. Low concentration of triacylglycerols is an indicative that transesterification is almost complete. Seen that this kind of reaction is reversible, excess of alcohol must be added to ensure that transesterification will prone to esters production. If great amount of alcohol remains in the biodiesel its flash point decreases and problems with storage and transport can occur. Glycerine is a by-product of transesterification that must be recovered in order to avoid solid in diesel engines. Standard methods that must be employed to determine triacylglycerols, alcohol (methanol), total ester and glycerine are presented in Figure 2.

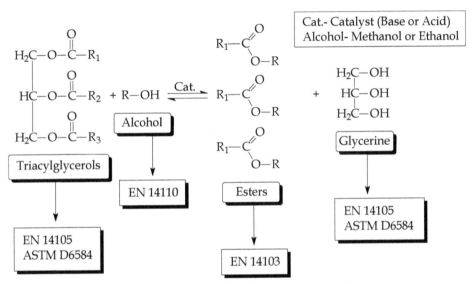

Fig. 2. Methods to quantify some organic impurities present in biodiesel.

Transesterification reaction proceeds in three main steps, shown in Figure 3. Firstly the triacylglycerols are transformed into diacylglycerols and then, these ones are converted into monoacylglycerols, which in turn reacts with alcohol yielding glycerine and an ester. Glycerine can be present in biodiesel in a free form or combined with glycerides. The total glycerine is the sum of these 2 glycerine forms. Maximum limits of methanol, glycerides, free and total glycerine contents in biodiesel, as standard methods for determination of these parameters are shown in Table 3.

All the methods described in Table 3, for determination of the concentration of the organic compounds in biodiesel, employ gas chromatography. So, a typical soybean biodiesel chromatogram, acquired in accordance with EN 14105 standard, is presented in Figure 4. This chromatogram shows the peak of free glycerin (1) and of the internal standards (butanetriol (2) and tricaprine (5)) utilized to quantify free glycerine and mono, di and triglycerides, respectively. It is also observed the regions where the methyl esters (3), monoglycerides (4), diglycerides (6) and triacylglycerols (7) are eluted. This standard method was developed for rapeseed methyl esters determination, but they have been applied successfully for the same determination in soybean and sunflower derivate. In Brazil, Resolution ANP nº7 demands the method validation when EN 14105 is employed to analyze biodiesel samples derived from feedstocks other than rapessed, or when biodiesel was produced from by ethylic route.

$$\text{Triacylglycerol} \quad + \quad R^* OH \quad \xrightleftharpoons{\text{Cat.}} \quad \text{Diacylglycerol} \quad + \quad RCOOR^*$$

$$\text{Diacylglycerol} \quad + \quad R^* OH \quad \xrightleftharpoons{\text{Cat.}} \quad \text{Monoacylglycerol} \quad + \quad RCOOR^*$$

$$\text{Monoacylglycerol} \quad + \quad R^* OH \quad \xrightleftharpoons{\text{Cat.}} \quad \text{Glycerine} \quad + \quad RCOOR^*$$

Fig. 3. Steps of the transesterification reaction.

| Specification | Standards | | | | | |
| | ASTM D6751-08a | | EN 14214:2008 | | ANP nº7 | |
	Test method	Limit (g/100g)	Test method	Limit (g/100g)	Test method	Limit (g/100g)
Methanol	EN 14110	0,20 max.	EN 14110	0,20 max.	EN 14110	0,20 max.
Free glycerine	ASTM D6584	0,02 max.	EN 14105	0,02 max.	ASTM D6584 EN 14105	0,02 max.
Monoglycerides	-	-	EN 14105	0,80 max.	ASTM D6584 EN 14105	0,80 max.
Diglycerides	-	-	EN 14105	0,20 max.	ASTM D6584 EN 14105	0,20 max.
Triglycerides	-	-	EN 14105	0,20 max.	ASTM D6584 EN 14105	0,20 max.
Total glycerine	ASTM D6584	0,24	EN 14105	0,25 max.	ASTM D6584 EN 14105	0,25 max.

ASTM - American Society for Testing and Materials; ISO - International Standards Organization; ANP - National Agency of Oil, Gas and Biofuels

Table 3. Methods and limits of the impurities present in biodiesel.

3.6 Oxidative stability

The oxidation of fatty acid chain is a complex process proceeded by a variety of mechanisms. Oxidation of biodiesel is due to the unsaturation in fatty acid chain and presence of double bonds in the molecule which offers high level of reactivity with O_2, especially, when it is placed in contact with air/water. The primary oxidation products of double bonds are unstable allylic hydroperoxides which are unstable and easily form a variety of secondary oxidation products. This includes the rearrangement of product of similar molecular weights to give short chain aldehydes, acids compounds and high molecular weight materials.

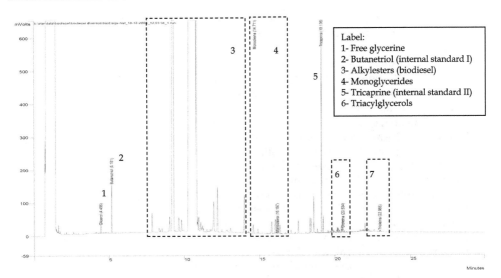

Fig. 4. Soybean biodiesel chromatogram obtained in accordance with EN 14105 standard (Source: Organic Analysis Laboratory- INMETRO - 2008).

Peroxidation occurs by a set of reactions categorized as initiation, propagation, and termination, as shows Figure 5. The reaction mechanism involved in the first step is the removal of hydrogen from a carbon atom to produce a carbon free radical. If diatomic oxygen is present, the subsequent reaction to form a peroxyl radical becomes extremely fast, not allowing significant alternatives for the carbon-based free radical. The peroxyl free radical is not reactive compared to carbon free radical, but is sufficiently reactive to quickly abstract hydrogen from a carbon to form another carbon radical and a hydroperoxide (ROOH). The new carbon free radical can then react with diatomic oxygen to continue the propagation cycle. This chain reaction terminates when two free radicals react with each other to yield stable products.

Fatty oils that contain more poly-unsaturation are more prone to oxidation. Literature reveals the relative rate of oxidation for the methyl esters of oleic (18:1), linoleic (18:2), and linolenic (18:3) acids to be 1:12:25 (Siddharth & Sharma, 2010).

Initiation:	$RH + I \rightarrow R\cdot + IH$
Propagation:	$R\cdot + O_2 \rightarrow ROO\cdot$
	$ROO\cdot + RH \rightarrow ROOH + R\cdot$
Termination:	$R\cdot + R\cdot \rightarrow R\text{--}R$
	$ROO\cdot + ROO\cdot \rightarrow$ Stable Products

Fig. 5. Mechanism of peroxidation of fatty acids

The biodiesel oxidative stability study is a very important parameter to measure the product quality, mainly about its feedstock. This parameter is a measure of time required to reach the point where the oxidation increases sharply. This methodology is useful to determinate the final biodiesel stability under several oxidative conditions. Useful appropriate oxidative automatic techniques are Petrooxy, differential scanning calorimetry (DSC), Pressure

Differential Scanning Calorimetry (PDSC) (Dufaure et al, 1999) and mainly Rancimat technique. At the Rancimat technique, oxidative stability is based on the electrolytic conductivity increase (Hadorn & Zurcher, 1974.). The biodiesel is prematurely aged by the thermal decomposition. The formed products by the decomposition are blown by an air flow (10L/ 110 °C) into a measuring cell that contains bi-distilled, ionized water. The induction time is determined by the conductivity measure and this is totally automatic. Rancimat is the most used technique to determine finalized biodiesel stability, under oxidative accelerated conditions, according to standard EN14112.

At the PetroOxy Technique, the sample is inducted to oxidation through an intense oxygen flow, manipulating by this way the stability conditions through a specific apparatus. The analysis time is recorded as the required time to the sample absorbs 10% of oxygen pressure. The differential scanning calorimetry (DSC) monitors the difference in energy provided/released between the sample (reagent system) and the reference system (inert) as a function of temperature when both the system are subjected to a controlled temperature program. Changes in temperature sample are caused by rearrangements of induced phase changes, dehydration reaction, dissociation or decomposition reactions, oxidation or reduction reaction, gelatinization and other chemical reactions.

The Pressure Differential Scanning Calorimetry (PDSC) is a thermo analytical technique that measures the oxidative stability using a differential heat flow between sample and reference thermocouple under variations of temperatures and pressure. This technique differs from the Rancimat for being a fast method and presents a more variable - the pressure, allowing to work at low temperatures and using a small amount of sample (Candeia, 2009).

3.6.1 Antioxidants used in biodiesel

Most of biodiesel has a lower value of oxidative stability than recommended by current legislation (Ji-Yeon, 2008 & Ferrari, 2009) (Table 4), the soybean derivative has also the same inconvenient. This characteristic is due to the rich composition in mono and polyunsaturated fatty acids from the soybean oil.

Source of Biodiesel	Oxidative Stability (h)
Sunflower	1,17
Jatropha	3,23
Soybean	3,87
Palm	11,00

Table 4. Oxidative stability of biodiesel samples produced from different sources.

Compounds containing allylic and bis-allylic have greater reaction fragility with oxygen due to the formation of stable resonance structure, as shows Figure 6.

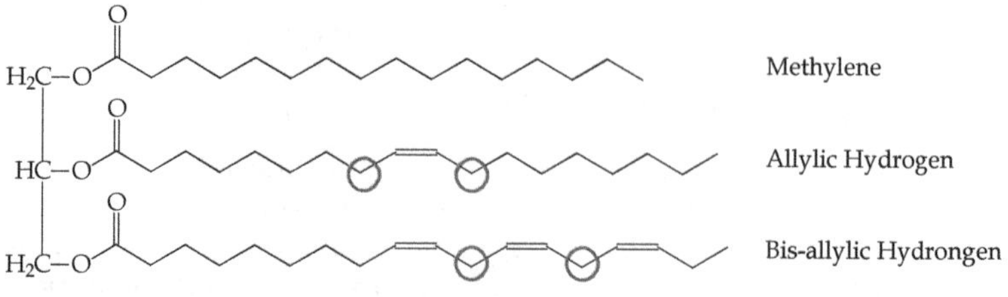

Fig. 6. Methylenic ,Allylic and bis allylic hidrogens at triacylglycerol (Asadukas et al, 2007).

The allylic hydrogen reactivity is 40 times greater than the methylene hydrogen and the bis-allylic is 100 times more reactive than the methylene hydrogen (Knothe, 2007).

The main characteristic for a substance be considered a good antioxidant is its capacity to react with oxygen faster than the biodiesel components, mainly the unsaturated compounds. Moreover, the radicals generated in this reaction have to be stable enough and less reactive with the initial biodiesel components or even with the generated products from the biodiesel reaction.

Denisov & Khudyakov (1987) divide the antioxidants class in four groups:

Group 1 - Inhibitors that terminate chains through reactions with peroxyl radicals, including phenols, aromatic amines, diamines, and aminophenols;

Group 2 - Inhibitors that terminate chains through reactions with alkyl radicals, including stable radicals, quinones, quinone imines, methylenequinones, nitro compounds, and condensed aromatic hydrocarbons (these inhibitors are effective when dissolved oxygen concentration is low);

Group 3 - Agents that decompose peroxides without generating free radicals, including sulfides, disulfides, phosphites, metal thiophosphates, and carbamates;

Group 4 - Complexing agents that deactivate heavy metals are capable of catalyzing hydroperoxide decomposition to free radicals, thereby promoting oxidation, including diamines, amino acids, hydroxy acids and other bifunctional compounds.

Compounds from the group 3 contain sulfur that turn difficult its use as biodiesel for environmental reasons. Compounds from the group two are effective only for oxygen low concentrations.

Actually there are two substances classes are very useful for this purpose: phenols and aromatics amines. These compounds are cheap and very useful at oil and polymers industry and are the most useful at biodiesel industry.

Figure 7 shows the oxidative stability of biodiesel containing different types of phenol antioxidants. Its stability can be up to 5 times higher when *tert*-Butylhydroquinone (TBHQ) is added, Karavalis, 2011.

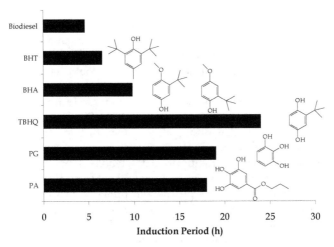

Fig. 7. PA= Propylgallate (3,4,5-trihydroxybenzoate) ; PG= Pyrogallol (benzene-1,2,3-triol) ; BHA= mixture of the isomers 2 and 3-*tert*-butyl-4-hydroxyanisol; BHT= di-*tert*-butyl-metil-phenol (Butylated hydroxytoluene) ; TBHQ= *tert*-Butylhydroquinone.

4. The importance of metrology for biodiesel quality

4.1 Efforts for harmonization of biodiesel standards

In 2006, the Government of Brazil, the European Commission (representing the European Union) and the Government of the United States of America, during trilateral discussions, affirmed their belief that the current market for biofuels is viable. The market will continue to grow within these regions and the international trade in biofuels would increase significantly by the end of this decade (Tripartite Task Force, 2007). However, a potential barrier to global trade in biofuels concerns the differences among the standards describing and ruling their composition and properties.

To overcome these potential barriers, a conference was organized by the European Commission and the European Committee for Standardization (CEN), with the active participation of the U.S. National Institute of Standards and Technology (NIST) and the Brazil's National Institute of Metrology, Standardization, and Industrial Quality (INMETRO). This meeting, held in Brussels in February, 2007, convened a broad range of private-sector biofuels experts and government representatives from the EU, US and Brazil. The participants confirmed that differing standards for biofuels were a potential handicap to the free circulation of biofuels among the three regions.

To support the global trade of biofuels, representatives of Brazil, the EU and the U.S. agreed to promote, whenever possible, the compatibility of biofuels-related standards in their respective regions. Such compatibility would not only facilitate the increasing use of biofuels in each of the regional markets, but also would support both exporters and importers of biofuels by helping to avoid adverse trade implications in a global market. Subsequently, the International Biofuels Forum (IBF) – a governmental initiative among Brazil, China, the European Commission, India, South Africa, and the United States – was launched in March, 2007 to promote the sustained use and production of biofuels around the globe. The IBF also concluded that trade will play an increasing role in providing adequate supplies of biofuels to the markets where the energy demand for transport fuel is rising at an accelerated rate.

In June, 2007, a NIST and INMETRO-sponsored Biofuels Symposium in Washington, DC, convened representatives from Brazil, the EU and the U.S. to build on the work begun in Brussels. These representatives agreed to review existing documentary standards for biofuels and identify areas where greater compatibility could be achieved in the short, medium and long term. According to the tripartite agreement, the standards to be considered were those produced by Brazilian Association for Technical Standards (ABNT), Brazilian Petroleum, Gas and Biofuels Agency (ANP), European Committee for Standardization (CEN) and American Society for Testing and Materials (ASTM International) and in effect before the end of 2007. It was further agreed that only standards pertaining to the biofuels being currently traded – biodiesel and bioethanol – would be addressed; this was further limited to pure biofuels and not to ready-made blends.

Comprised of representatives from the private and public sectors, the Biodiesel Tripartite Task Force and the Bioethanol Tripartite Task Force each started their technical work in July. The immediate task was to classify the various specifications into three categories:

- Category A: specifications that are already similar;
- Category B: specifications with significant differences between parameters and methods, but which might be aligned by work on documentary standards and measurement standards; and

• Category C: specifications with fundamental differences, perhaps due to emissions or environmental regulations within one or more regions, which are not deemed bridgeable in the foreseeable future.

There were commonalities with the approach and methodology used by both of the Task Forces. Each of the two groups assembled and translated existing standards from ABNT, ASTM International and CEN, and the units for specifications were converted to a common basis. Each Task Force first compared the standards as they presently exist. Since it was noted that many parameters were different, the Task force members entered into discussions and negotiations and were able to make specific recommendations to address these differences. They further agreed that these recommendations should be forwarded to standards bodies for consideration and possible implementation. Here, we will only present all biodiesel discussions to compatibility biodiesel standards. Summary results from each group are listed below in Table 5.

4.2 General considerations for biodiesel standards

The current standards established to govern the quality of biodiesel on the market are based on a variety of factors which vary from region to region, including characteristics of the existing diesel fuel standards, the predominance of the types of diesel engines most common in the region, and the emissions regulations governing those engines. Europe, for example, has a much larger diesel passenger car fleet, while United States and Brazilian markets are mainly comprised of heavier duty diesel engines. It is therefore not surprising that there are some significant differences among the three sets of standards.

Category A Similar	Category B Significant Differences	Category C Fundamental Differences
Sulfated Ash	Total glycerol content	Sulfur content
Alkali and alkaline earth metal content	Phosphorus content	Cold climate operability
Free glycerol content	Carbon residue	Cetane number
Copper strip corrosion	Ester content	Oxidation stability
Methanol and ethanol content	Distillation temperature	Mono, di, and tri-acylglycerides
Acid number	Flash point	Density
	Total Contamination	Kinematic viscosity
	Water content and sediment	Iodine number
		Linolenic acid content
		Polyunsaturated methyl ester

Table 5. Classification of the Various Biodiesel Specifications.

Other sources of regional differences in biodiesel standards arise from the following factors. The biodiesel standards in Brazil and the U.S. are applicable for both fatty acid methyl esters (FAME) and fatty acid ethyl esters (FAEE), whereas the current European biodiesel standard is only applicable for fatty acid methyl esters (FAME). Also, the standards for biodiesel in Brazil and the U.S. are used to describe a product that represents a blending component in conventional hydrocarbon based diesel fuel, while the European biodiesel standard describes a product that can be used either as a stand-alone diesel fuel or as a blending component in conventional hydrocarbon based diesel fuel.

It should also be noted that some specifications for biodiesel are feedstock neutral and some have been formulated around the locally available feedstocks. The diversity in these technical specifications is primarily related to the origin of the feedstock and the characteristics of the local markets. Though this currently translates into some significant divergence in specifications and properties of the derived fuels – which could be perceived as an impediment to trade – in most cases it is possible to meet the various regional specifications by blending the various types of biodiesel to the desired quality and specifications.

The Task Force members classify the various specifications according with the limits for each parameter. They have collaboratively assembled a definitive and widely vetted list of Brazilian, EU and US standard specifications that are similar. In addition, they have identified a list of specifications that have significant, but alignable differences. Perhaps even more importantly, some indirect benefits have been derived. There is widespread agreement amongst the participating experts that the discussions and commitment to cross-border cooperation have been a major accomplishment that will support the increase in global trade of biofuels. The experts now have a better understanding of reasons why regional differences exist, and a new atmosphere of collegiality has been created – not only between countries but also between the private and public sector representatives. These positive outcomes foster a working environment that will support ongoing movement towards enhanced compatibility among the biofuels standards.

After the discussions, it was concluded that:

- standardization bodies of the tripartite agreement (i.e., ABNT, ANP, CEN and ASTM International) as a basis for ongoing discussions and cooperation that will promote alignment and mitigate divergence among evolving standards and specifications.
- other members of the International Biofuels Forum as a basis for ongoing discussions on more closely aligning their respective specifications and prioritizing future efforts for maximum impact.
- request the standardization bodies of the Tripartite Agreement to consider adapting existing national standards wherever appropriate. Furthermore the standardization bodies should attempt where possible, when developing and updating their standards on biodiesel from now on to consider the opportunity to align with the other standards in question;
- support efforts to initiate an analysis of the categorized specifications to study trade implications and appropriate next steps for harmonization;
- support the development of internationally-accepted reference methods and certified reference materials for improving the accuracy of measurement results that underpin assessment of product quality, and help facilitate trade.

4.3 Development of Internationally-accepted reference methods and certified reference materials

Beyond the difference among the standards, an unacceptable barrier to trade, are the measurement disagreements between countries. To overcome such problems it is necessary to have an international infrastructure within which it is possible to make comparable measurements (Wielgoz & Kaarls, 2009).

This is true for all areas of measurements including chemical ones. Such a system requires measurement standards that have long-term stability and are internationally recognized. The International System of Units (SI) represents such a system, and by the use of traceable measurements provides an international infrastructure for comparable measurements. This system is demonstrated in Fig. 8.

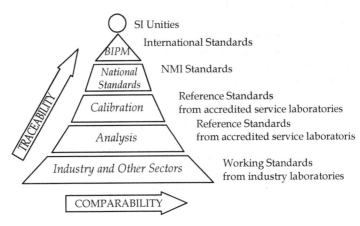

Fig. 8. Traceability scheme.

The International System of Units (SI) is at the top of the system (Dube, 2001). Its units are realized by standards. A measurement is a process, in the course of which the measurand is compared to a standard. For practical measurements, usually a working standard not a primary standard is used. To state the uncertainty of the measurement result, the uncertainty of the value assigned to the working standard must be known. It results from the uncertainty of the comparison measurement of the working standard with the reference standard. The uncertainty of the value assigned to the reference standard results from the uncertainty of the comparison measurement of the reference standard with the primary standard. This chain of comparison measurements is exactly what the definition of the term "traceability" means. If the traceability of a measurement result is guaranteed, its uncertainty can be stated. From this considerations it follows that metrology can provide the tools, necessary to get reliable measurement results.

Metrology is the science of measurement, embracing both experimental and theoretical determinations at any level of uncertainty in any field of science and technology. Within a robust metrological system the values of measurement standards and measurement results are linked *via* comparisons or calibrations which take into account the measurement uncertainty of the linking processes. Measurement uncertainty is the parameter associated with the results of a measurement, that characterizes the dispersion of the values that could reasonably be attributed to the measurand.

The ISO Guide to the Expression of Uncertainty in Measurement (GUM) (BIPM et al., 2008) and the Eurachem (CITAC, 2000) guide on measurement uncertainty provide guidance on the evaluation of measurement uncertainty. The property of the result of a measurement or the value of a measurement standard whereby it can be related to stated references, usually national or international measurement standards, through an unbroken chain of comparisons all having stated uncertainties, is termed (metrological) traceability. Where these stated references are realizations of the SI units the term SI-traceable is used. Traceability is the basis of the comparability of a measurement: whether the result of a measurement can be compared to the previous one, a measurement result a year ago, or to the result of a measurement performed anywhere else in the world. Traceability is most often obtained by calibration, establishing the relation between the indication of a measuring instrument and the value of a measurement standard.

In the field of analytical chemistry the term Certified Reference Material (CRM) is more often used than measurement standard. A CRM is a reference material, accompanied by a certificate, one or more of whose property values are certified by a procedure which establishes traceability to an accurate realization of the unit in which the property values are expressed, and for which each certified value is accompanied by an uncertainty at a stated level of confidence.

Metrological traceability (Eurachem/CITAC, 2003) may also be established to a reference method, defining the measurand and fixing a number of influence parameters, the results of which are expressed in SI units, and an approach that has been documented for the field of laboratory medicine. In the field of metrology in chemistry the role of a National Metrology Institute (NMI) involves: realization, maintenance and dissemination of the units; development and application of primary measurement methods; establishment of traceability structure, guaranteeing the equivalence of measurement standards using programs which facilitate traceable measurements to be achieved, including the provision of certified reference materials, both as pure materials and calibration solutions as well as matrix reference materials for method validation or calibration.

For producers of CRMs, there are three ISO Guides that assist the set-up of a facility to produce and certify RMs and to ensure that the quality of thus-produced CRMs meet the requirements of the end-users (ISO, 2006). ISO Guide 34 (ISO, 2009) outlines the requirements to be met by a CRM producer to demonstrate competence, whereas the Guide 35 provides assistance on how to meet these requirements. At a fairly generic level, this Guide provides models for homogeneity testing, stability testing, and the characterization of the candidate CRM.

ISO Guide 31(ISO, 2000) describes the format and contents of certificates for CRMs. In some ways, this Guide can be seen as an application of the *Guide to the Expression of Uncertainty in Measurement* (GUM) with respect to the peculiarities of the production of CRMs. Where possible, the Guide 35 makes reference to the GUM, as the latter describes in detail how to evaluate measurement uncertainty of a value obtained from measurement. This Guide complements the GUM in a sense that it provides additional guidance with respect to the inclusion of the uncertainties due to the (remaining) batch inhomogeneity and instability of the CRM in the uncertainty of the property values, and the determination of these uncertainty contributions.

Thorough knowledge of the material and its properties, and of the measurement methods used during homogeneity testing, stability testing and characterization of the material, along with a thorough knowledge of the statistical methods (Eurachem/CITAC, 2000), are needed for correct processing and interpretation of experimental data in a typical certification project. It is the combination of these required skills that makes the production and certification of RMs so complex. The greatest challenge in these projects is to combine these skills to allow a smooth implementation of RM certification.

Three categories of values can be assigned for certified reference materials producers: Certified values fulfill the highest standards for reliability. They are traceable to stated references and are accompanied by a GUM (BIPM, JCGM 100:2008) compatible expanded uncertainty statement valid for the entire shelf life of the CRM. Indicative values are not certified due to either a larger uncertainty than required for the intended use or insufficient variety of methods used in the characterization. The information is therefore unsuitable for certification at the accuracy required for certified values. Additional material information are values created during the certification exercise, which are usually the result of one method only and indicate the order of magnitude rather than an accurate value.

In summary, certified values are those values the certifying body is confident in assigning with the highest accuracy, while indicative values display higher uncertainties and/or lack a full traceability statement. This hierarchy in reliability is shown by the fact that only certified values are on the first page of the certificate. It follows that certified values are more assured than indicative values which in turn are more assured than additional material information.

The measurement method used for the homogeneity study should have very good repeatability and selectivity. The main purpose of the homogeneity assessment is, however, to detect unexpected problems, for example due to contamination during packaging (Linsinger et al., 2000, Van der Veen et al., 2000, 2001a, 2001b). To establish the homogeneity, a statistically defined number of bottles is randomly selected and analyzed for all relevant property values. For evaluation of homogeneity results, unifactorial analysis of variance ("one-way ANOVA") (Van der Veen & Pauwels, 2000) is applied. The stability of the reference materials have to be assessed for all parameters, by measuring the property values periodically during the course of the project. In this case, where samples are measured on different days, the selectivity and the reproducibility of the measurement method are of great importance. Therefore, methods for homogeneity and stability studies are not necessarily the same. This is not a problem so long as traceability of the results of the homogeneity and stability studies and characterization to a common reference are established. Such a reference may be a material that is suitable for assessing the various calibrations or results from different measurement methods. Ensuring the traceability of all measurements in a certification project is an important requirement.

For the characterization of the candidate reference material, the producer shall use and document technically valid procedures to characterize its reference materials. It shall comply with the requirements of ISO Guide 35 and ISO/IEC 17025 for testing, calibration and related activities. There are several technically valid approaches for characterizing a reference material. These include carrying out measurements using: a) a single (primary) method in a single laboratory; b) two or more independent reference methods in one or several laboratories; c) one or more methods of demonstrable accuracy, performed by a network of competent laboratories; d) an approach providing method-specific, operationally defined property values, using a network of competent laboratories.

Depending on the type of reference material, its intended use, the competence of the laboratories involved and the quality of methods employed, one approach may be chosen as appropriate. Results obtained from proficiency testing can be used only if the competence of the laboratories involved has been checked and it has been ensured that the measurements done comply with ISO/IEC 17025. The single (primary) method approach shall be carried out only when the procedure and expertise enable it to ensure metrological traceability. More usually, a property value can be reliably assessed when its value is confirmed by several laboratories working independently and using more than one method, for each of which the accuracy has been well established.

Primary methods play an essential role in the practical realization of the base units of the SI and hence in establishing traceability to the SI throughout metrology (Milton, 2001). There are seven SI units (meter, kilogram, second, ampere, Kelvin, mole, candela). They are the essential first link in such a chain of traceability because they do not require prior knowledge of any measurement of the same quantity. Explained briefly, a primary method of measurement allows a quantity to be measured in terms of a particular SI unit without reference to a standard or measurement already expressed in that unit. It is thus, in principle, completely independent of measurements of the same quantity, but calls upon measurements expressed in other units of the SI. By their nature, primary methods are unbiased (their results are accurate) but they may not necessarily be precise. Those primary methods that are at the same time precise are the ones that are useful in practice. Put in a different way, a primary method provides the means to transform the abstract definition of an SI unit into practical measurements made in terms of that unit. This is sometimes referred to as a "realization" of that unit, but this statement can be confusing because it gives the impression that the result is in some way a concrete materialization of the unit. In some cases such a concrete materialization can be made (although it is not possible for the mol unit) but, more importantly, a primary method allows measurements to be made in terms of that unit and this is the main characteristic of a primary method. The potential primary methods available for the chemical area are very few, they are: Gravimetry, Titrimetry, Coulometry, Calorimetry (Differential Scanning Calorimetry), Isotope Dilution Mass Spectrometry (IDMS), Instrumental Neutral Activation Analysis (INAA).

The biggest challenge involving the characterization of a biodiesel CRM is the lack of primary methods for all parameters that involves the biodiesel quality assessment. Therefore, some efforts are being made in that sense. Inmetro and NIST have teamed up to develop a CRM for several parameters for biodiesel derived from soybean and animal fat. The composition of the materials had to be close to specification levels, for those parameters where international agreement on these levels exists. Characterization is carried out for those parameters where SI traceability of the measurement results is possible. NIST and Inmetro reported the results accompanied by a complete uncertainty statement calculated according to the "Guide to the expression of uncertainty in measurement"(GUM) (BIPM, JCGM 100:2008). These CRMs are available at NIST homepage.

Additional effort concerns BIOREMA project (Inmetro et al., 2008) which involved not only Inmetro and NIST, but also several European National Metrology Institutes as Laboratory of Government for Chemistry - LGC and National Physical Laboratory- NPL (England), Van Swinden Laboratory - VSL (Netherlands), Institute for Reference Material and Measurements - IRMM (Belgium). Initially, this group intended to develop a (Certified) Reference Material for biodiesel from rapeseed/canola and conducted homogeneity, stability and characterization studies in test samples to obtain certified values.

However NMI's results for several parameters were not harmonized as expected, so these test samples were not possible to be used as CRM. On the other hand, these results were not dismissed, and they can be used as reference value for intercomparisons. Thus, the

BIOREMA group decided to promote a proficiency testing within Brazilian, American and European testing laboratories. In this proficiency testing, soybean and animal fat biodiesel CRM's, developed by Inmetro and NIST, were included. The agreement of results provided by participating laboratories, and the consensus related to reference values, were perceived as satisfactory. A clear demand was expressed for RMs of biodiesel derived not only from one feedstock but from the various ones. This fact happens due to the considerable difference in chemical and physical properties among biodiesels deriving form the diverse sources, which make harder the accommodation of all aspects relevant in quality control and the validation of analytical methods.

This project concluded that this was a positive example of collaboration among metrology institutes and useful exercise for establishing a common approach to the production of biofuels reference materials. However, for several specifications, the assignment of an SI traceable value still need further research due to the complexity of the material.

5. Conclusion

The use of fuels derived form renewable sources, like crops, microorganisms or animal derivates, bring several advantages related to the environment, economy and the fewer dependence on the main fossil energy source, petroleum. Concerning environmental aspects the main fact that has been constantly emphasized, and actually is one the greatest advantage of using biofuels, is related to the reduction of greenhouse gases exhausts. Since biofuels are derived form renewable sources, the burning of biofuels in internal combustion engines linked to the renovation of those sources which biofuels are derived from, allows the establishment of CO_2 recycling in environment. Moreover, biodiesel production, especially in Brazil, has presented a great social advantage, seen that small farmers are encouraged to produce and furnish the raw material, which in turn stimulate the local economy of small cities and create new job positions.

From all the possible and already studied biodiesel sources, soybean presents great prominence. Soybean is a widespread culture, adapted to cultivation in several climates and countries, there is cultivation technology available and the logistics chain is well established. Otherwise, some aspects related to biodiesel production, economy, supplying and technical-economic availability need more profound studies. Literature points that a change in fatty acids profile of some feedstocks, using biologically modified species, in order to attend both human consumption and its utilization as fuel, would be one the most appropriate alternative for biodiesel replaces efficiently petrodiesel. Other aspect to be considered regards the byproduct glycerine recycling, as transforming it into products with commercial interest. Other point to be addressed is the gradual substitution of feedstocks that today are commodities (like soybean) for alternative raw materials that have less commercial importance, as macauba (*Acrocomia aculeata*) and jatropha (*Jatropha curcas*). This substitution would undervaluate biodiesel production costs. Biodiesel quality control has also crucial importance. Concerning this subject there is a great need for the development of robust and specific analytical methods, since the existing ones are adapted from the petrodiesel quality control. Moreover, in order to ensure biodiesel quality, although some efforts have been done, only two certified reference materials (soybean and animal fat + soybean based ones) are available. Another challenge concerns the development or enhancement for getting more efficient biodiesel production processes, seen that the actual ones, like those employing homogeneous or heterogeneous catalysis, still present several drawbacks.

6. References

Albuquerque, G. A. (2006). *Obtainment and Physical-Chemical Characterization of Canola (Brassica napus)*. M.S. thesis, Centro C. Exatas e da Natureza, Univ. Fed. Paraíba.

Allen, C.A.W., Watts, K.C., Ackman, R.G. & Pegg, M.J. (1999). Predicting the viscosity of biodiesel fuels from their fatty acid ester composition, *Fuel*, Vol. 78, pp. 1319-1326

Antczak, M. S.; Kubiak, A.; Antczak, T. & Bielecki, S. (2009). Enzymatic biodiesel synthesis – Key factors affecting efficiency of the process, *Renewable Energy*, Vol.34, pp.1185-1194

Asadauskas, S.J.; Griguceviciene, A. & Stoncius, A. (2007). Review of late stages of oxidation in vegetable oil lubricant basestocks. *Proceedings of the international conference.*

ASTM D6751-08a. (2008). Standard specification for biodiesel (B100) fuel blend Stock (B100) for middle distillate fuels. In: *ASTM (American Society for Testing and Materials) annual book of standards*. West Conshohocken: ASTM International.

Atadashi, I.M.; Aroua, M.K.; Aziz, A.A. (2011). Biodiesel separation and purification: A Review. *Renewable Energy*, 36, 437-443.

Ataya, F; Dubé, M.A.; Ternan, M. (2007). Acid-Catalyzed Transesterification of Canola Oil to Biodiesel under Single- and Two-Phase Reaction Conditions. *Energy and Fuels*, 21, 2450-2459.

Balat, M & Balat, H. (2008). A critical review of bio-diesel as a vehicular fuel, *Energy conv. Mgmt.*, Vol.49, No.10, pp.2727–2741.

Barnard, M. T.; Leadbeater, N. E.; Boucher, M. T.; Stencel, L. M. & Wilhite, B. A. (2007). Continous flow preparation of biodiesel using microwave heating, *Energy & Fuels*, Vol.21, No.3, pp.1777-1781

Bickell, K. (2008). Cold Flow Properties of Biodiesel and Biodiesel Blends – A Review of Data – University of Minnesota Center for Diesel Research, Available form http://www.biodiesel.org/pdf_files/fuelfactsheets/Cold%20flow.PDF

Biodiesel.ogr. (2009). Biodiesel Basics. *Biodiesel.ogr*, http://www.biodiesel.org/resources/biodiesel_basics/.

Biodieselbr.com. (2011). O que é o Biodiesel. *Biodieselbr.com*, http://www.biodieselbr.com/biodiesel/definicao/o-que-e-biodiesel.htm.

Biopowerlondon.co.uk. (2011). What is Biodiesel. *Biopowerlondon.co.uk. Energy, Power and Fuels*, http://www.biopowerlondon.co.uk/biodiesel.htm.

BIPM, IEC, IFCC, ISO, IUPAC, IUPAP, OIML. (2008). *Guide to the expression of uncertainty in measurement*, first edition, ISO Geneva.

Candeia, R. A.; Silva, M. C. D.; Carvalho F. J. R.; Brasilino, M. G. A.; Bicudo, T. C., Santos, I. M. G. & Souza, A. G. (2009). Influence of soybean biodiesel content on basic properties of biodiesel diesel blends. *Fuel (Guildford)*. Vol. 88, pp. 738-743

Chand, P.; Chintareddy V. R.; Verkade J. G. & Grewell D. (2010). Enhacing biodiesel production from soybean oil using ultrasonics, *Energy & Fuels*, Vol.24, pp.2010-2015

Du, W.; Xu, Y.; Liu, D. & Li, Z. (2005). Study on acyl migration in immobilized lipozyme TL-catalyzed transesterification of soybean oil for biodiesel production, *J Mol Catal B: Enzym.*, Vol.37, pp.68–71

Dube, G. (2001). Metrology in Chemistry – a public task, *Accreditation and Quality Assurance*, Vol. 6, pp. 26-30

Dufare, C.; Thamrin, U. & Mouloungui, Z. (1999). *Thermochim. Acta.* Vol. 388.

Dunn, R.O. & Bagby, M. O. (1995). Low-Temperature Properties of Triglyceride_Based Diesel Fuels: Transesterified Methyl Esters and Petroleum Middle Distillate/Ester Blends, *J. Am. Oil Chem. Soc.,* Vol. 72, pp. 895-904.

EN 14214:2008, (2009). Automotive fuels - fatty acid methyl esters (FAME) for diesel engines - requirement methods. Brussels, Belgium: *European Committee for Standardization (CEN).*

ERM (2008). Application Note 6, *European Reference Material (ERM).* http://www.erm.org.

Eurachem/CITAC. (2000). *Quantifying uncertainty in analytical measurement,* 2nd edition, http://www.eurachem.ul.pt/

Eurachem/CITAC. (2003). Traceability in chemical measurement – a guide to achieving comparable results in chemical measurement, http://www.eurachem.ul.pt/

Evgeniy T. & Denisov, I. V. (1987). Mechanisms of Action and Reactivities of the Free Radicals of Inhibitors. *Chem. Rev.* Vol. 87; pp. 1313-1357

Fangrui, M.A.; Milford, A.H. (1999). Biodiesel production- A Review. *Bioresource Technology,* Vol. 70, pp. 1-15.

Ferella, F.; Mazziotti, G.; De Michelis, I.; Stanisci, V.; Veglio, F. (2010). Optimization of the transesterification reaction in biodiesel production. *Fuel,* 89, 36-42.

Ferrari, R. A. & Souza, W. L. (2009) Avaliação da estabilidade oxidativa de biodiesel de óleo de girassol com antioxidantes. *Quim. Nova.* Vol. 32, No. 1, pp. 106-111 *for midle distillate fuels,* ASTM, West Conshohocken, PA

Fueleconomy.gov. (2011). Biodiesel. *Fueleconomy.gov, US Departament of Energy, US Environmental Protection Agency,* http://www.fueleconomy.gov/feg/biodiesel.shtml.

Fukuda, H.; Konda, A. & Noda, H. (2001). Bioidesel fuel production by transesterification of oils, *J. Biosc. Bioeng.,* Vol.92, pp.405-416

Gerpen, J.V. (2005). Biodiesel processing and production;*Fuel Process Technol,* Vol.86, pp.1097-107

Grunberg, L. & Nissan, A.H. (1949), Mixture law for viscosity, *Nature,* Vol. 164, pp. 799-800

Gryglewicz, S. (1999). Rapeseed oil methyl esters preparation using heterogeneous catalysts, *Bioresour Technol,* Vol. 70, pp.249-253

Hadorn, H. & Zurcher, K. (1974). Zurbestimmung der oxydationsstabilitat von olen und fetten. *Deustsche Ledensmittel Rundschau.* Vol. 70, No. 2, pp. 57-65

Harding, K. G.; Dennis, J. S.; von Blottnitz, H. & Harrison, S. T. L. (2008). A life-cycle comparision between inorganic and biological catalyse for the production of biodiesel, *J. Clean. Produc.,* Vol.16, No.13, pp.1368-1378

Haupt, J., Brankatschk, G. & Wilharm, T. (2009). Sterol Glucoside Content in Vegetable Oils as a Risk for the Production of Biodiesel – Study of The technological Chain Impact – Arbeitsgemeinschaft Qualitätsmanagement Biodiesel e.V. (AGQM) Final Report to American Soybean Association (ASA).

Heck, D.A., Thaeler, J., Howell, S. & Hayes, J.A. (2009). Quantification of the Cold Flow Properties of Biodiesels Blended with ULSD, http://www.biodiesel.org/pdf _files/Cold_Flow_Database_Report.pdf.

Helwani, Z.; Othman, M. R.; Aziz, N.; Fernando, W. J. N. & Kim, J. (2009). Technologies for production of biodiesel focusing on green catalytic techniques: A review, *Fuel Processing Technology*, Vol.90, pp.1502, 1514,

Honda, H.; Kuribayashi, H.; Toda, T.; Fukumura, T.; Yonemoto, T. & Kitakawa, N. S. (2007). Biodiesel production using anionic ion-exchange resin as heterogeneous catalyst. *Bioresource Technology*, Vol. 98, pp. 416–421

ISO 12937:2000. (2000). Determination of Content. Coulometric Karl Fischer Titration Method. *Methods of Test for Petroleum and Its Products: BS 2000-438.*

ISO Guide 31. (2000). Reference Materials – Contents of certificates and labels, 3rd edition, ISO Geneva

ISO Guide 34. (2009). General requirements for the competence of reference material producers, 3rd edition, ISO Geneva

ISO Guide 35. (2006). Reference Materials – General and statistical principles for certification, 3rd edition, ISO Geneva

Jaimasith, M. & Phiyanalinmat, S. (2007). Biodiesel synthesis from transesterification by Clay basic catalyst"; *Chiang Mai J Sci*, Vol.34,No. 2, pp. 201-207

Ji-Yeon P.; Deog-Keun K.; Joon-Pyo L.; Soon-Chul P.; Young-Joo K. & Jin-Suk L. (2008). Blending effects of biodiesels on oxidation stability and low temperature flow properties. *Bioresource Technology*. Vol. 99, pp. 1196–1203

Joshi, R.M. & Pegg, M.J. (2007). Flow Properties of Biodiesel Fuel Blends at Low Temperatures, *Fuels*, Vol. 86, pp. 143-151

Kaieda, M.; Samukawa, T.; Kondo, A. & Fukuda, H. (2001). Effect of methanol and water contents on production of biodiesel fuel from plant oil catalyzed by various lipases in a solvent free system, *J. Biosc. Bioeng.*, Vol.91, No.1, pp. 12-15

Kaieda, M.; Samukawa, T.; Matsumoto, T.; Ban, K.; Kondo, A.; Shimada, Y.; Noda, H.; Nomoto, F.; Obtsuka, K.; Izumoto, E. & Fukuda, H. (1999). Biodiesel fuel production from plant oil catalyzed by rhizopus orizae lipase in a water containing system without an organic solvent, *J. Biosc. Bioeng.*, Vol.88, No.6, pp.627-631

Karavalakis G.; Despina H.; Lida G.; Dimitrios K. & Stamos S. (2011). Storage stability and ageing effect of biodiesel blends treated with different antioxidants. *Energy.* Vol. 36, pp. 369-374.

Kern, D.Q.& Van Nostrand, W. (1948). Heat Transfer Characteristics of Fatty Acids, *Ind, Eng. Chem*, Vol. 41, pp. 2209-2212

Knothe G, Van Gerpen J, Krahl J, editors (2005). *The biodiesel handbook*. Urbana: AOCS Press.

Knothe, G. & Steidley, K. R. (2005). Kinematic Viscosity of Biodiesel Fuel Components and Related Compounds: Influence of Compound Structure and Comparison to Petrodiesel Fuel Components, *Fuel*, 84, 1059–1065.

Knothe, G. (2007). Some aspects of biodiesel oxidative stability. *Fuel Processing Technology.* Vol. 88, pp. 669-667

Knothe, G., (2009). Improving Biodiesel Fuel Properties by Modifying Fatty Ester Composition, *Energy Environ. Sci.*, Vol. 2, pp. 759-766

Knothe, G.; Matheaus, A.C.; Ryan, T.W. (2003), Cetane Numbers of Branched and Straight-Chain Fatty Esters Determined in an Ignition Quality Tester, *Fuel*, Vol. 82, pp. 971-975.

Krawczyk, T., (1996). Biodiesel – alternative fuel makes inroads but hurdles remain. *Inform* 7, 801–829.

Krishna, C.R. & Butcher, T. (2008). Improving Cold Flow Properties of Biodiesel, In: Advanced Energy Conference, 19-20.11.2008, Available from http:www.aertc.org/conference/AEC_Sessions%5CCopy%20of%20Session%202%5CTrack%20B-%20Renewable%5C1

Landommatos, N.; Parsi, M.; Knowles, A. (1996). The effect of fuel cetane improver on diesel pollutant emissions, *Fuel*, Vol. 75, pp. 8-14.

Leadbaeater, N. E. & Stencel, L. M. (2006). Fast, easy, preparation of biodiesel using microwave heating, *Energy Fuels*, Vol.20, No.5, pp.2281-2283

Leclercq, E.; Finiels, A. & Moreau, C. (2001). Transesterification of rapessed oil in the presence of basic zeolites and related solid catalysts, *J. Am. Oil Chem. Soc.*, Vol. 78, pp. 1161-1165

Lee, I., Johnson, L.A. & Hammond, E.G. (1995). Use of branched-chain esters to reduce the crystallization temperature of biodiesel, *J. Am. Oil Chem. Soc.*, Vol. 72, pp. 1155-1160

Li, Q.; Du, W.; Liu, D., (2008). Perspectives of microbial oils for biodiesel production. *Appl. Microbiol. Biotechnol.* 80, 749–756.

Linsinger T.P.J., Pauwels J., Van der Veen A.M.H., Schimmel H., Lamberty A. (2001). Homogeneity and Stability of Reference Materials, *Accreditation and Quality Assurance*, Vol. 6, pp. 20-25

Ma, F. & Hanna, M.A., (1999). Biodiesel production: a review. *Bioresour. Technol.* 70, 1–15.

Mahamuni, N. N. & Adewuyi, Y. G. (2009). Optimization of the Synthesis of biodiesel via Ultrasound-Enhanced Base-Catalyzed Transesterification of Soybean Oil Using a Multifrequency Ultrasonic Reactor, *Energy & Fuels*, Vol.23, No.2757-2766

McCormick, R.L., Graboski, M.S., Alleman T.L. and Herring, A.M. (2001). Impact of biodiesel source material and chemical structure on emissions of criteria pollutants from a heavy-duty engine, *Environ. Sci. Technol.*, Vol. 35, pp. 1742-1747.

Melero, J.A.; Iglesias, J.; Morales, G. (2009). Heterogeneous acid catalysts for biodiesel production: current status and future challenges. *Green Chem.*, Vol. 11, pp. 1285-1308.

Mello, V. M.; Pousa, G. P. A. G.; Pereira, M. S. C.; Dias, I. m. & Suarez, P. A. Z. (2011). Metal oxides as heterogeneous catalysts for esterification of fatty acids obtained from soybean oil. *Fuel Proces. Tech.*, Vol.92, pp.53-57

Mendes, A.A.; Giordano, R.C.; Giordano, R.L.C. & Castro, H.F. (2011). Immobilization and stabilization of microbial lipases by multipoint covalent attachment on aldehyde-resin affinity: Application of the biocatalyst in biodiesel synthesis; *J. Mol. Catlal. B: Enzym*, Vol. 68, pp.109-115

Milton, M.J.T. (2001). Primary Methods for the measurement of Amount of the Substance, *Metrologia* Vol. 38, pp. 289-296.

Mittelbach, M. & Remschmidt, C. (2004), Biodiesel – *The Comprehensive Handbook* M. Mittelbach, Graz, Austria.

Modi, M. K.; Reddy, J. R. C.; Rao, B. V. S. K. & Prasad, R. B. N. (2007). Lipase mediated conversion of vegetables oils in to biodiesel using ethyl acetate as acyl acceptor, *Bioresource. Technology.*, Vol.98, No.6, pp. 1260-1264

Mohamad, I.A.W. & Ali, O.A. (2002). Evaluation of the transesterification of waste palm oil into biodiesel, *Bioresour Technol*, Vol.85, pp.225-256

Monyem, A.; Van Gerpen, J.H.; Canacki, M. (2001). The Effect o Timing and Oxidation on Emissions from Biodiesel. *Fueled Engines, Trans.* , Vol. 44, pp. 35-42

Moser, B.R. (2009). Biodiesel production, properties, and feedstocks. *In Vitro Cell.Dev.Biol. Plant*, Vol. 45, pp. 229-266.

Moser, B.R., (2009). Biodiesel production, properties, and feedstocks. *In Vitro Cell. Dev. Biol.- Plant*. 45, 229-266.

National Biodiesel Board Winter 2007/2008, *Biodiesel Cold Flow Basics* - Information for Petroleum Distributors, Blenders, and End-Users on Issues Affecting Biodiesel in the Winter Months

NIST home page (2011). http://ww.nist.gov/srm

Noureddini, H.; Gao, X. & Philkana, R.S. (2005). Immobilized Pseudomonas cepacia lipase for biodiesel fuel production fromsoybean oil, *Bioresour Technol, Vol.96*, pp.769-77

Raganathan, S. V.; Narasiham, S. L. & Muthukumar, K. (2008). An overview enzymatic production of biodiesel, *Bioresour. Technol.*, Vol.99, No.10, pp.3975-3981

Rakesh S.; Meeta Sharma, S. & Sinharay, R.K. (2007). Jatropha–Palm biodiesel blends: An optimum mix for Asia. *Fuel.* Vol. 86, pp. 1365-1371

Resolution ANP N°7, (2008). ANP, Brazilian National Agency of Petroleum, Natural Gas and Biofuels, "Resolution ANP No 7", from 2008/03/19, http://nxt.anp.gov.br/nxt/gateway.dll/leg/resolucoes_anp/2008/mar%C3%A7o / ranp%207%20-%202008.xml.

Rinaldi R.; Garcia, C.; Marciniuk, L. L.; Rossi, A. V. & Schuchardt U. (2007). Sintese de Biodiesel: Uma proposta contextualizada em experimento para laboratório de química geral. *Quim. Nova*, Vol.30, No.5, pp.1374-1380.

Salis, A.; Pinna, M.; Manduzzi, M. & Solinas, V. (2008). Comparision among immobilised lípases on macroporous polypropilene toward biodiesel synthesis, *J. Molec. Catal. B., Enzim.*, Vol.54, No.1, pp.19-26

Santos, F. F. P.; Rodrigues, S. & Fernandes, F. A. N. (2009). Optimization of production of biodiesel from soybean oil by ultrasound assisted methanolysis, *Fuel Processing Technology*, Vol.90, pp.312-316

Schlink, R. & Faas, B., (2009). Water Content Determination in Biodiesel According to EM ISO 12937. *Metrohm.com*, http://www.metrohm.com/applications/titration/lit/karl-fischer.html.

Sharp, C.A., (1994) Transient Emissions Testing of Biodiesel and Other Additives in a DDC Series 60 Engine. *Southwest Research Report Institute to the National Biodiesel Board.*

Shimada, Y.; Watanabe, Y.; Sugihara, A. & Tominaga, Y. (2002). Enzymatic alcoholysis for biodiesel fuel production and application of the reaction to oil processing, *J Mol Catal B: Enzym* , Vol.17, pp.133–42

Siddharth, J. & Sharma, M.P. (2010). Stability of biodiesel and its blends- A Review. *Renewable and Sustainable Energy Reviews*, Vol. 14, pp. 667-678.

Sirman, M.B.; Owens, E.C.; Whitney, K.A. (2000). Emissions Comparison of Alternative Fuels in an Advanced Automotive Diesel Engine, *SAE Techn. Pap. Ser.* 200-01-2048.

Snåre, M.; Kubic, I.;ková; Arvela, P. M.; Eränen, K.; Warna, J. & Murzin, D. (2007). Production of diesel fuel from renewable feeds: kinetics of ethyl stearate decarboxylation, *Chem. Eng. J.*, Vol.134, No.1, pp.29–34

Suppes, G. J.; Bockwinkel, K.; Lucas, S.; Botts, J. B.; Mason, M. H. & Heppert, J. A. (2001). Calcium carbonate catalyzed alcoholysis of fats and oils, *J. AM. Oil Chem. Soc.*, Vol. 78, No. 2, pp. 139-146

Suppes, G. J.; Dasari, M. A.; Doskocil, E. J.; Mankidy, P. J. & Goff, M. J. (2004). Transesterification of soybean oil with zeolites and metal catalysts; *J. Appl. Catal,.* A., Vol.257, pp. 213-223

Sybist, J.P. & Boehman, A. L., (2003). Behavior of a Diesel Injection System With Biodiesel Fuel, SAE Techn. Pap. Ser. 2003-01-1039

Tat, M.E. & Van Gerpen, J.H. (2003). Measurement of Biodiesel Speed of Sound and Its Impacto on Injection Timing, *National Renewable Energy Laboratory, NREL/SR-510-31462.*

Tomasevic, A. V. & Marinkovic, S. S. (2003). Methanolysis of used frying oils, *Fuel Process Technol.*, Vol.81, No.1–6

Tripartite Task Force. (2007). White Paper on Internationally Compatible Biofuel Standards: Brazil, European Union, and United States of America.

United States Department of Agriculture, (2010). Foreign Agricultural Service, Office of Global Analysis. Oilseeds: World Markets and Trade.

Van der Veen A.M.H., Linsinger T.P.J, Lamberty A., Pauwels J. (2001b). Uncertainty calculations in the certification of reference materials. 3. Stability study, *Accreditation and Quality Assurance*, Vol. 6, pp. 257-263

Van der Veen A.M.H., Linsinger T.P.J, Pauwels J. (2001a). Uncertainty calculations in the certification of reference materials. 2. Homogeneity study, *Accreditation and Quality Assurance*, Vol. 6, pp. 3-7.

Van der Veen A.M.H., Pauwels J. (2000a), Uncertainty calculations in the certification of reference materials. 1. Principles of analysis of variance, *Accreditation and Quality Assurance*, Vol. 5, pp. 464-469
vegetable oils, *J Am Oil Chem Soc*, Vol.80, No.4, pp.367–371

Wang, L. & Yang, J. (2007). Transesterification of soybean oil with nano-MgO or not in supercritical and subcritical methanol, *Fuel*, Vol. 86, No.3, pp. 328-333

Watanabe, Y.; Shimada, Y.; Sugihara, A. & Tominaga Y. (2002). Conversion of degummed soybean oil to biodiesel fuel with immobilized Candida antarctica lipase; *J Mol Catal B: Enzym*, Vol.17, pp.151–155

Wielgoz, R., Kaarls, R. (2009). International Activities in Metrology in Chemistry, *Chimia*,Vol. 63, pp 606-612.

Zhang, Y. & Van Gerpen, J.H. (1996). Performance of Alternative Fuels for SI and CI Engines. *SAE Spec. Publ. SP-* 1160, 1-15.

Zhang, Y.; Dube, M. A.; McLean, D. D. & Kates, M. (2003). Biodiesel production from waste
      cooking oil: 2. Economic assessment and sensitivity analysis, Bioresour Technol.,
      Vol.90, pp.229–240

Zhou, W.; Konar, S. K. & Boocock, D.G.V. (2003). Ethyl esters from the single-phase base-
      catalyzed ethanolysis of vegetable oils, *J Am Oil Chem Soc*, Vol.80, No.4, pp.367–
      371

4

Rationality in the Use of Non Renewable Natural Resources in Agriculture

Rogério de Paula Lana
Universidade Federal de Viçosa
Brazil

1. Introduction

The increase in human population and the demand for life quality have induced the growing production of food and alternative vegetal energy sources in replacement to petrol. Soybean responds to more than 80% of biodiesel production, and will reach 5% inclusion in the fossil diesel in the next years in Brazil. This trend will increase pressure to new areas for soybean production on actually human food production areas, as well on pasture and untouched forests areas.

The progress of agriculture has been based on increase in animals and plants productivity per unit of area, which only has application when land availability is the sole limiting factor. However, the efficiency of use of limiting resources (including water, fertilizers and petrol) has to be considered. This mistaken vision is leading to excessive use of non renewable natural resources and environmental pollution. The reserves of phosphate in the world that can be explored at low cost are enough for 40 to 100 years and the world reserves of potassium are enough for 50 to 200 years. The situation is worse for micronutrients, in which the reserves of copper and zinc are enough for 60 years, manganese for 35 years and selenium for 55 years (Herring & Fantel, 1993; Roberts & Stewart, 2002; Aaron, 2005).

In addition to the depletion of natural reserves, the excessive use of fertilizers can contribute to soil and water courses contamination with nitrate (Angus, 1995; Bumb, 1995), soil acidification (Helyar & Poter, 1989), and emissions of carbon dioxide (CO_2), nitrous oxide (N_2O) and ammonia to the atmosphere. The pollution with nitrate has being an actual preoccupation in Europe and North America. The fertilization with phosphorus and nitrogen cause decrease in water oxygenation by excessive increase in the population of toxic algae in the oceans (Kebreab et al., 2002).

The agriculture participates in 20% of annual increase in the anthropogenic emission of greenhouse gases, mainly CH_4 and N_2O. Approximately 70% of all anthropogenic emission of N_2O is attributed to agriculture. The current methodology used in Canada to estimate the flow of N_2O is based in the direct relation between the emission of N_2O and the application of nitrogen fertilizers (Lemke et al., 1998).

The possible deleterious effects of emissions of N_2O are global warming and catalytic destruction of the ozone chain in the stratosphere, in which the N_2O retains 13 times more heat than methane (CH_4) and 270 times more than CO_2 (Granli & Bockman, 1994). The atmospheric level of N_2O has increased in growing fashion since 1960, associated with increase in utilization of nitrogen fertilizers (Bumb, 1995; Strong, 1995).

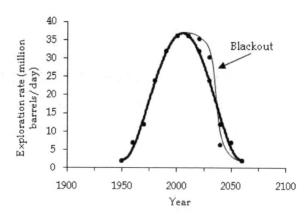

Fig. 1. Hubbert curve of exploration of petrol (non renewable natural resource), and altered curve by artificial maintenance of peak of production.

An worried phenomena about the use of non renewable natural resources can be visualized in the Hubbert curve (by Dr. Marion King Hubbert), which was a Shell Geologist, who predicted in 1956 that the global production of petrol would present a peak in the beginning of the XXI century (Hubbert Peak theory, accessed in March 02, 2009), and the curve of exploration follows the bell shape (Figure 1).

The phenomena observed by Hubbert related to petrol exploration is applicable to any other limiting natural resource, such as fertilizers, soil, and water and, consequently, food production. As more persistent is the maintenance of the maximum exploration of a resource, more drastic is the fall in the exploration of the final reserves in a short space of time, occurring the called blackout or sharply decay in the rate of production (Figure 1). Therefore, after the peak of exploration, if there is no new reserves to be discovered, no alternatives to produce more food without dependence on the available resources, or control of excessive exploration based on efficiency of use of these resources, catastrophic consequences can occur with mankind in some time of this century, as predicted by the Club of Rome in 1972, in the known publication "The limits of growth" (Meadows et al., 1972). The alert of the Club of Rome was based in the model associating accelerated industrialization, rapid population growth, depletion of non renewable natural resources, widespread malnutrition, and environmental pollution.

The objective of this work is to demonstrate the application of saturation kinetic models to improve efficiency of use of non renewable natural resources in agriculture, avoid the complete depletion as predicted by the Hubbert curve, and minimize the problems related to environmental pollution.

2. Population growth curve

The growth curve of populations of life beings in the absence of factors that affects the physical integrity, such as sickness and predation, has sigmoid curve (Gompertz or hyperbolic curve), including latency, exponential growth, plateau and senescence or death. The plateau occurs due to the saturation phenomena associated with depletion of nutrients or in some cases by environmental pollution, which acts in feedback against the uncontrolled growth (Figure 2).

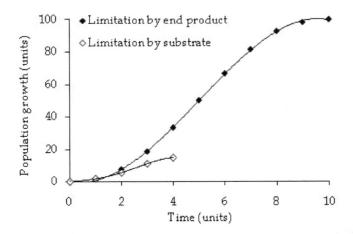

Fig. 2. Theoretical growth curve of life beings as a function of saturation by limitation of nutrients (under nutrition) or by products of metabolism (environmental pollution).

As an example, a bacterium with volume of 1 μm³ and duplication time of 20 minutes has potential to reach a biomass much greater than the earth in only 48 hours or 144 generations (Russell, 2002; p.57-58). Therefore, the archaea and bacteria are the first life beings of the planet, are hungry most of time, and can be the survivors of a biological collapse, such as those that occurred 65,000,000 years ago with the dinosaurs and 250,000,000 years ago, when more than 90% of life beings were extinct, leading to the formation of petrol reservoirs that are being explored actually.

Another example of cessation of growth and death of population by environmental pollution is in the silage production, where the bacteria die and the nutrients are conserved to be used by ruminants, as a consequence of the acidity caused by accumulation of fermentation end products – the volatile fatty acids.

Speaking of food production crisis lead we back to the Malthus theory, which although there are some conceptual errors, it will threaten the humanity and all life beings forever. According to Malthus, the population growth curve follow geometric progression and the food production arithmetic progression, leading to the crisis of food supply in some situations or in some periods of our existence (Thomas Malthus, accessed in March 02, 2009).

However, both population growth and food production follow a sigmoid curve up to the plateau or in form of a bell or double sigmoid over time (the second goes down hill, similar to the first in a mirror). The population growth curve is cumulative, as a consequence of the sum of the annual growth rates (Figure 3), which depend on the annual rate of food production (productivity), that by its time is consequence of the annual rate of soil utilization and exploration of non renewable natural resources (fertilizers and petrol), that follow the Hubbert curve. Changes in these curves can be caused by men or naturally, with discoveries of new food production technologies, population death (caused by diseases, wars, predations, among others), proliferation of plagues and diseases in plants, climatic changes, among others.

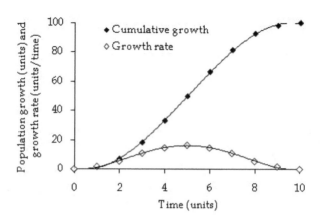

Fig. 3. Theoretical population growth (cumulative and growth rate) as function of time.

Studies of these phenomena lead us to understand the need for rational use of non renewable natural resources. The kinetic saturation models are important tools generated by science in order to evaluate the efficiency and allow the rational use of non renewable natural resources (Lana, 2005; Lana et al., 2005; Lana, 2007a,b; Lana et al., 2007a,b; Lana, 2008). As a result, they can avoid the complete depletion of the resources and the collapse in food and energy supply, with dramatic consequences for our civilization, as predicted by Malthus, Hubbert and Club of Roma.

3. Models of biological responses to nutrients

The first studies on the limiting factors in the plants growth were developed by Carl Sprengel in 1826 and 1828, and by Liebig in 1840, leading to the rejection of humus theory and formulation of Law of minimum (van der Ploeg et al., 1999). The Law of minimum or Law of response is associated with the absence of nutrient replacement, linear response in production by increase in the quantity of the limiting factor and a maximum plateau of response, in which the plants do not respond satisfactorily anymore to the limiting nutrient.

A posterior mark was the Law of diminishing return of Mitscherlich (1909). The convex exponential equation of Mitscherlich, with a model that includes the maximum asymptotic yield, allows calculating the optimum economic level of fertilization, based in the benefit-cost ratio.

The Michaelis-Menten model (Michaelis & Menten, 1913) was developed to describe the enzymatic kinetic in the beginning of the 20 century. The Lineweaver-Burk model (Lineweaver & Burk, 1934), an equation of the linear regression of the reciprocal of Y (enzymatic activity) as a function of the reciprocal of X (concentration of substrate), was used to obtain the kinetic constants of the Michaelis-Menten model: k_s (the amount of substrate needed to reach half of maximum enzymatic activity) and k_{max} (maximum enzymatic activity).

Later, researchers verified that the microbial growth rate was dependent of substrate concentration and both were related to the saturation kinetic typical of enzymatic systems (Monod, 1949; Russell, 1984).

Although the use of saturation kinetic model to explain the nutrients responses by the superior forms of life is not being adopted (Morgan et al., 1975), the Michaelis-Menten model allows to explain the curvilinear relationship of plants and animals to the nutrients and the model of Lineweaver-Burk allows to obtain the kinetic constants, k_s (the amount of substrate needed to reach half theoretical maximum response in rate of growth or production of milk, wool, eggs, among others) and k_{max} (theoretical maximum response in rate of growth or production), according to Lana et al. (2005).

The responses of plants and animals to nutrients as saturation phenomena have important implications in addition to calculation of the rate of decreasing economical return and estimates of nutrients recommendations, such as the consciousness about the excessive use of non renewable natural resources; soil, water, and air pollution; and global warming.

The knowledge about the efficiency of utilization of fertilizers in agriculture will play an important role in the political decisions about the rational use of non renewable natural resources in the future. The natural fertilizer sources have to be used with maximum efficiency and with minimum negative effects in the environment.

4. Marginal response or Law of diminishing return in plants

Recommendations of fertilization are mostly based in the method of calculation of nutrients requirements of a culture and the mineral contribution of the soil. The fertilizers are then calculated to supply the deficiencies. This method allows recommendation of the lower level that maximize the production. However, the method does not indicate changes in the recommendation based on changes in the costs of nutrients and grains. Also, it does not give direct information of the effect of application of other level than the recommended one (Makowski et al., 1999).

It has being utilized a variety of empirical models to predict the responses to nutrients and to calculate the optimum levels of nutrients. Among then, it is included the model of Mitscherlich, square root (Mombiela et al., 1981; Sain & Jauregui, 1993), exponential, linear-plus-plateau, linear-plus-hyperbola, quadratic and quadratic-plus-plateau (Cerrato & Blackmer, 1990; Bullock & Bullock, 1994; Makowski et al., 1999, 2001).

The use of saturation kinetics to explain the nutritional responses to nutrients by superior life beings are rarely employed (Morgan et al., 1975). The model of Michaelis-Menten has not being evaluated to make recommendations of fertilization. This model has a great potential in recommendation of use of nutrients in agriculture, by considering the efficiency of use of nutrients and the Law of diminishing return, as observed by Mitscherlich (1909). This model can aggregate important concepts such as responses to different levels of nutrients, benefit-cost ratio, efficiency of use of nutrients, rationality of use of non renewable natural resources and consciousness about environmental pollution.

Linear regressions of reciprocal of plants responses as a function of reciprocal of nutrients supply, methodology known as data transformation of Lineweaver-Burk (Lineweaver & Burk, 1934; Champe & Harvey, 1994), were proposed by Lana et al. (2005) as follow:

$$1/Y = a + b * (1/X)$$

where:
Y = responses of plants (grain yield, x 1,000 kg/ha),
a = intercept,

b = coefficient of linear regression,

X = amount of nutrient (kg/ha/year).

The theoretical maximum grain production (k_{max}) is obtained by the reciprocal of intercept (1/a). The amount of nutrient (X) needed to reach half of theoretical maximum response (k_s) is obtained by the model presented above, replacing Y by 1/a x 50(%) x 0.01, or dividing the coefficient of the linear regression by the intercept (b/a).

The efficiency of use of fertilizers is calculated dividing the accretion in grain production ($Y_2 - Y_1$) by the accretion in fertilization ($X_2 - X_1$), from a specific level of fertilizer in relation to the previous level.

Simulations of biological responses to nutrients in the absence or presence of a second limiting nutrient are presented in Table 1 and Figure 4, in which are expected changes in the maximum yield (k_{max}) and k_s of the first limiting nutrient (increase, no effect or decrease). The Figure 4A illustrates four kind of responses in production and models of double-reciprocal are presented in Table 1 and Figure 4B, demonstrating the combination of two values of k_{max} by two of k_s.

The best effects that a second limiting nutrient can cause are by increasing k_{max}, decreasing k_s, or both changes that is even better. However, the most common kind of response is by increasing both k_{max} and k_s. Increase in k_{max} by increase in productivity with a second nutrient lead to increase the efficiency of use of the first limiting nutrient (Figure 4C), but this benefit decreases sharply by increase in the amount of the first limiting nutrient, especially when k_s is low.

Equation	Symbol	Intercept (a)	Coefficient (b)	r^2	k_s	k_{max}
1	O	0.8163	79.789	1.00	98	1.2
2	△	0.9195	39.591	1.00	43	1.1
3	□	0.4082	39.894	1.00	98	2.4
4	◇	0.4768	19.483	1.00	41	2.1

Table 1. Constants of linear regression of reciprocal of grain production (x1,000 kg/ha) as a function of reciprocal of amount of fertilizers (kg/ha/year) in hypothetic situations of high or low values of the saturation constants k_s (kg of fertilizer/ha) and k_{max} (x1,000 kg/ha) – see Figure 4B

The plants responses to fertilization depend on soil fertilization, in which high responses occur when soil fertility is low (Figure 5A) and in low level of fertilization, that is the main factor that affects the efficiency of use of fertilizers (Figure 5B).

Equations of data transformation of Lineweaver-Burk were used to explain the effect of fertilization and the effect of a second factor in the yield, k_s, k_{max} and efficiency of use of fertilizers in soybean, bean, wheat and cotton production (Tables 2, 3 and 4).

When limestone was the second factor, there was change in k_s and k_{max} in 34 and 85%; -75 and -10%; and 33 and 22% for soybean fertilized with P_2O_5 (Table 2). Limestone as a second factor changed k_s and k_{max}, respectively, in -55 and -12% for wheat fertilized with P_2O_5, and in 9 to 87% in cotton fertilized with K_2O.

As seen above, increase or decrease in k_{max} is associated with the same effect in k_s, but increase in k_{max} associated with exaggerated increase in k_s is not desirable because it requires more fertilizer to reach the plateau. In other words, the greater values of k_s present greater response to the use of fertilizers in high level of fertilization, but it cannot be advantageous due to the increase in the cost of fertilization.

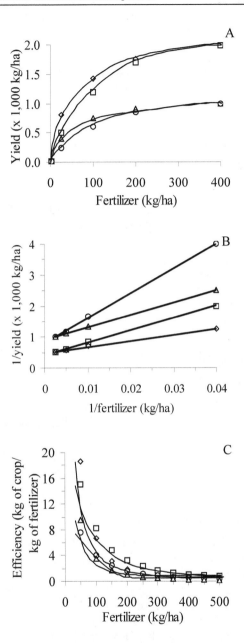

Fig. 4. Biological responses to nutrients as a function of a second limiting nutrient (A) - control (O), decrease in k_s ($\triangle$), increase in k_{max} ($\square$) and decrease in k_s and increase in k_{max} ($\diamond$); reciprocal of production as a function of reciprocal of fertilizer level – plot of Lineweaver-Burk (B); and effect of a second limiting nutrient in the efficiency of use of the first one (C)

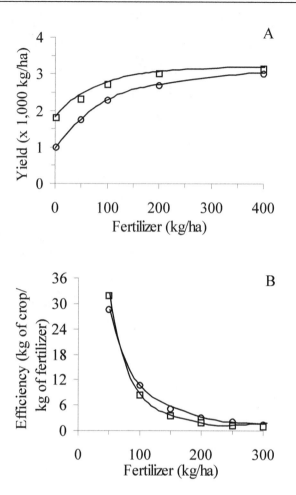

Fig. 5. Plants responses to fertilizers in low (O) and high (□) soil fertility (A); and the effect of soil fertility in the efficiency of use of fertilizer (B)

In the case of soybean (Table 2), considering US$1.208/kg of P_2O_5 and US$0.178/kg of soybean, it is necessary 6.8 kg of soybean to pay 1 kg of fertilizer. Therefore, it is viable to use 50 to 100 kg of P_2O_5 in the absence of limestone and 100 kg of P_2O_5 in the presence of limestone. Above 150 kg of P_2O_5, although in some cases there was still response to fertilizer, especially in high values of k_s, the response is not viable economically.

In bean production (Table 3), the second factor (P_2O_5) increased the k_{max} of nitrogen from 0.1-0.5 to 1.0-1.6 x 1,000 kg/ha of bean, but also increased the k_s (1 to 13 and 17 to 29 kg/ha of nitrogen). When the second factor was nitrogen, this increased the k_{max} of P_2O_5 from 0.7-0.8 to 1.5 x 1,000 kg/ha of bean, but also increased the k_s (5 to 15 and 136 to 199 kg/ha of P_2O_5). In the second case, the high values of k_s for P_2O_5 caused low improvement in the efficiency of use of this fertilizer in low level of fertilization (Table 4). The extra production in this case cannot be enough to pay the extra amount of fertilizers.

Product	Fertilizer (kg/ha/year)	Second factor[1]	Intercept (a)	Coefficient (b)	r^2	k_s[2]	k_{max}[3]	Source of data[4]
Soybean	P_2O_5	-	0.7536	57.766	1.00	77	1.3	1
		+	0.4096	42.198	1.00	103	2.4	1
Soybean	P_2O_5	-	0.3502	30.524	0.98	87	2.9	2
		+	0.3801	8.2987	0.99	22	2.6	2
Soybean	P_2O_5	-	0.3103	3.6726	0.68	12	3.2	2
		+	0.2535	3.9962	0.53	16	3.9	2
Wheat	P_2O_5	-	0.4169	174.48	1.00	419	2.4	1
		+	0.4781	91.00	1.00	190	2.1	1
Cotton	K_2O	-	0.622	4.6865	0.91	7.5	1.6	1
		+	0.3284	2.7052	0.97	8.2	3.0	1

	Fertilizer (kg/ha)	Efficiency of use of fertilizers (kg of grains/kg of fertilizer)[5]					
		50[2]	100	150	200	250	300
Soybean	P_2O_5 - Lim[1]	10.0	4.5	2.5	1.6	1.1	0.8
	+ Lim	16.0	8.1	4.9	3.3	2.4	1.8
Soybean	P_2O_5 - Lim	15.5	9.7	5.6	3.7	2.6	1.9
	+ Lim	15.4	6.6	2.7	1.5	1.0	0.7
Soybean	P_2O_5 - Lim	28.0	5.5	2.1	1.1	0.7	0.5
	+ Lim	37.0	8.2	3.2	1.7	1.1	0.7
Wheat	P_2O_5 - Lim	5.0	4.1	3.4	2.9	2.4	2.1
	+ Lim	8.3	5.7	4.0	3.0	2.3	1.8
Cotton	K_2O - Lim	10.5	2.0	0.7	0.4	0.2	0.2
	+ Lim	20.8	4.0	1.5	0.8	0.5	0.3

[1] Limestone: without (-) or with (+) 4,000 to 7,000 kg/ha; [2] Kg of fertilizer/ha - P_2O_5 or K_2O; [3] x1,000 kg/ha of grain; [4] 1 = Malavolta (1989), p.61, 275 and 283; 2 = Oliveira et al. (1982), p.36; [5] Considering US$1.208/kg of P_2O_5 and US$0.178/kg of soybean, is necessary 6.8 kg soybean to pay one kg of fertilizer. Efficiency lower than 6.8 kg of soybean/kg of P_2O_5 is not viable. These calculations can be used to choose the level of fertilization.

Table 2. Changes in the constants of linear regression of the reciprocal of grain production (x1,000 kg/ha) as a function of the reciprocal of amount of fertilizer (kg/ha/year), by the second factor, and the respective efficiency of use of fertilizers (kg of grains/kg of fertilizer)

5. Marginal response in bovines

The weight gain in growing bovines in pasture in the dry season is curvilinear as a function of supplement supply, based on corn and soybean meal, in which the supplement conversion (kg of supplement/kg of accretion in weight gain) becomes worse with increase in the supplementation (Lana et al., 2005; Keane et al., 2006; Lana, 2007b) (Figure 6).

The milk production by supplemented cows in pasture or in feedlot is also curvilinear as a function of increase in the concentrate supply, based on corn and soybean meal (Figure 7A), in which the marginal increase in milk production per kg of concentrate decreases with increase in the amount of concentrate (Bargo et al., 2003; Pimentel et al., 2006a; Sairanen et al., 2006; Lana et al., 2007a,b), as shown in Figure 7B, and in some studies the milk response to concentrate was satisfactory only up to 2-4 kg of concentrate/animal/day (Fulkerson et al., 2006).

The curvilinear response can also be verified with specific nutrients, such as the observed positive curvilinear response in milk production and negative curvilinear response in the efficiency of use of nitrogen by increasing the dietary crude protein content from 11 to 19% in cows with mean production of 38 kg of milk/day (Baik et al., 2006). In the third experiment of Figure 7, in addition to decreasing response in milk production, there was decreasing response in body weight variation with increase in the concentrate level (0.20, 0.12, and 0.095 kg of body weight gain per additional kilogram of concentrate intake; Teixeira et al., 2006).

Fertilizer (kg/ha/year)	Second factor	Intercept (a)	Coefficient (b)	r^2	k_s [1]	k_{max} [2]	Sorce of data [3]
	P_2O_5(kg/ha)						
	0	2.044	2.794	0.82	1	0.5	
N	40	0.782	8.516	1.00	11	1.3	1
	80	0.710	6.718	0.99	9	1.4	
	100	0.630	8.205	1.00	13	1.6	
	P_2O_5(kg/ha)						
	0	10.764	183.36	0.23	17	0.1	
N	50	2.5229	46.235	0.78	18	0.4	2
	150	1.3364	15.623	0.99	12	0.7	
	250	0.9539	28.056	0.98	29	1.0	
	N (kg/ha)						
P_2O_5	0	1.3812	6.7411	0.98	5	0.7	1
	30	0.8181	8.8241	1.00	11	1.2	
	60	0.6842	10.186	1.00	15	1.5	
	N (kg/ha)						
P_2O_5	0	1.3257	180.38	0.95	136	0.8	2
	50	1.0402	90.586	1.00	87	1.0	
	120	0.6684	132.74	1.00	199	1.5	

[1] Kg of fertilizer/ha; [2] Ton of grain/ha; [3] 1 = Bolsanello et al. (1975) and Oliveira et al. (1982), p.155; 2 = Malavolta (1989), p.273.

Table 3. Changes in constants of linear regression of reciprocal of bean production (x1,000 kg/ha) as a function of reciprocal of amount of fertilizer (kg/ha/year), by a second factor

According to the Biotechnology and Biological Sciences Research Council (1998), formerly known as AFRC (Agricultural and Food Research Council), all currently feed systems calculate the dietary requirements of energy and protein to meet the animals needs for maintenance and production. However, in practice, the situation is different, because there is no need for the farmer to meet the cow's nutritional requirements if it is against the economical interest. So, it is evident that studies in animal response to increasing levels of concentrate or specific nutrients are needed, as suggested by Lana (2003; p.87).

Although the animal's responses to nutrients are curvilinear, the daily weight gains estimated by the level 1 of NRC (1996) of beef cattle are linear as a function of intakes of metabolizable energy and protein (Figure 8A). In the same way, the milk production estimated by the model CNCPS 5.0 as a function of intakes of metabolizable energy and protein, and model NRC (2001) of dairy cattle as a function of intakes of net energy for lactation and metabolizable protein, were linear by using increasing levels of concentrate

Fertilizer (kg/ha)	Second factor	Efficiency of use of fertilizers (kg of grains/kg of fertilizer)[1]					
		50[2]	100	150	200	250	300
N	P$_2$O$_5$ (kg/ha)						
	0	1.1	0.1	0.0	0.0	0.0	0.0
	40	10.9	2.1	0.8	0.4	0.3	0.2
	80	11.5	2.0	0.8	0.4	0.2	0.2
	100	14.2	2.9	1.1	0.6	0.4	0.3
N	P$_2$O$_5$ (kg/ha)						
	0	0.6	0.2	0.1	0.0	0.0	0.0
	50	2.8	0.9	0.4	0.2	0.1	0.1
	150	4.5	1.3	0.5	0.3	0.2	0.1
	250	7.9	3.0	1.3	0.7	0.5	0.3
P$_2$O$_5$	N (kg/ha)						
	0	5.4	0.6	0.2	0.1	0.1	0.0
	30	12.2	2.0	0.7	0.4	0.2	0.2
	60	15.3	2.9	1.1	0.6	0.4	0.3
P$_2$O$_5$	N (kg/ha)						
	0	4.0	2.3	1.5	1.1	0.8	0.6
	50	6.9	3.3	1.9	1.2	0.9	0.6
	120	6.0	4.0	2.9	2.1	1.7	1.3

[1] Considering US$0.966/kg of N, US$1.208/kg of P$_2$O$_5$ and US$0.36/kg of bean, it is necessary 2.68 and 3.36 kg of bean to pay 1 kg of N or P$_2$O$_5$. Efficiency worse than 2.68 or 3.36:1 for N or P$_2$O$_5$ is not economically desirable. These calculations can be used to choose the level of fertilization. [2] Level of fertilizer (kg/ha) - N in the first two cases or P$_2$O$_5$ in the last two cases.

Table 4. Efficiency of use of fertilizers (kg of bean/kg of fertilizer) calculated with base in the equations of Table 3

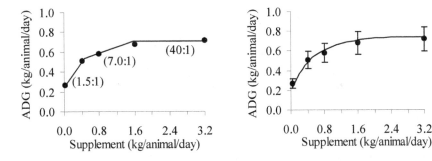

Fig. 6. Body weight gain (BWG) of growing bovines in pasture during the dry season, as a function of daily intake of supplement with 24% CP, in which the values among parenthesis represent the differential in kilograms of supplement given daily divided by the differential in weight gain, in relation to the previous treatment (Lana, 2005; Lana et al., 2005)

(Figure 8B), as suggested by Lana (2005; p.290-291) and Lana (2007b; p.39 a 43). Therefore, in order to these systems be compatible with the tropical conditions, in which it is more evident the curvilinear responses to nutrients, it is necessary modifications in future versions, by adopting models of saturation kinetics.

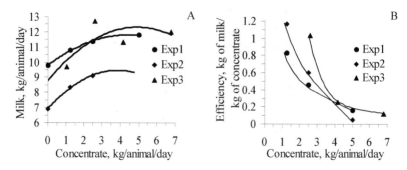

Fig. 7. Production of milk (A) and efficiency of use of concentrate (B) as a function of intake of increasing level of concentrate in three experiments (Pimentel et al., 2006b, 2006c; Teixeira et al., 2006)

6. Production versus productivity

Two studies were conducted to evaluate the factors that affect milk production in Brazil, on farmer level or by state of federation (Guimarães et al., 2008; Lana et al., 2009). In the first case, data were collected from fifty producers that sell milk for a dairy plant in the south region of Rio de Janeiro state, including data of daily milk production by producer, with the respective data of production per cow and per hectare, farmer size and size area designated to the herd, total of milking cows and herd size, and breed (Table 5). In the second case, data were collected from EMBRAPA and IBGE in the years of 2004-2006, in which the emphasis was in milk production per state instead of production per producer.

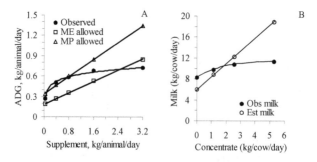

Fig. 8. Mean daily weight gain of steers in pastures, observed and estimated by level 1 of NRC (1996) as a function of intake of metabolizable energy and protein in the supplement (A); and observed milk production (mean of data of Figure 7A) and estimated by CNCPS 5.0 and NRC (2001) as a function of intakes of metabolizable energy or net energy of lactation, respectively, and metabolizable protein (B)

Production (Kg of milk/ producer/day)	Number of producers	Production per producer (Kg of milk/day)	Area for the herd (ha)	Herd (number of animals)	Milking cows (number)
Up to 150	6	117	37	56	14
151-300	12	238	70	73	28
301-600	14	451	106	94	43
601-1200	14	821	111	159	66
1201-2400	3	1667	316	633	132
2401-4800	1	4000	300	1800	300
Kg of milk/ producer/day	Milking cows/ha	Milk (Kg/ha/day)	Milk (Kg/milking cow/day)	Milk (Kg/total herd/day)	Milking cows/total herd
Up to 150	0.38	3.17	8.33	2.08	0.25
151-300	0.40	3.40	8.57	3.27	0.38
301-600	0.41	4.27	10.53	4.77	0.45
601-1200	0.59	7.41	12.51	5.15	0.41
1201-2400	0.42	5.27	12.66	2.63	0.21
2401-4800	1.00	13.33	13.33	2.22	0.17

Table 5. Number of producers, daily mean milk production by producer, area for the herd, number of animals and milking cows in the herd, and productivity indexes per area and per cow, as a function of production levels

The milk production in farmer level (first case) ranged from 60 to 4000 kg/producer/day. The increase in milk production was highly correlated with the number of milking cows (r = 0.94), followed by moderate correlation with the size of pasture (r = 0.67) and, surprisingly, the productivity per cow and per unit of area did not correlate with the milk production per producer (r = 0.11 and 0.06, respectively; Table 6). In the nation level (second case), the result repeated, in which there was high correlation of milk production/state/year with the total of milking cows in relation to productivity of milk/km²/year and milk/cow/year (r = 0.95, 0.55 and 0.51, respectively; Table 7).

Parameter	Correlation (r)	Parameter	Correlation (r)
Total animals in the herd	0.94	Milk (kg/ha)	0.13
Total milking cows	0.93	Milk (kg/cow/day)	0.11
Area for the herd (ha)	0.67	Milking cows/ha	0.06
Total area in the farm	0.20	Milk (kg/total of animals/day)	-0.11

Table 6. Linear correlation of daily milk production by producer with: total of animals in the herd, total of milking cows, area for the herd, farmer size and some productivity indexes (daily milk production per hectare and per cow, milking cows per hectare and daily milk production per total animals in the herd)

Therefore, the milk production by Brazilian farmers is more dependent on the farmer size and pasture extension, than on productivity indexes, and similar effect occur with crop production (Lana, 2009). Agricultural showed the same results verified with dairy cattle, in which cultivated area generally presents more than 90% correlation with crop production (Table 8 and Figure 9). Then, the concepts about agricultural production need to be revised, facing the actual problems related with the inadequate use and depletion of the non renewable natural resources, and environmental pollution.

Parameters	Liters of milk/state/year	Total of milked cows	Liters of milk/km²/year	Liters of milk/cow/year
Total of milking cows	0.95			
Liters of milk/km²/year	0.55	0.39		
Liters of milk/cow/year	0.51	0.31	0.88	
Surface of state (in km²)	0.11	0.21	-0.37	-0.26

Table 7. Linear correlation (r) of annual milk production by Brazilian states with total of milked cows, liters of milk/km²/year, liters of milk/cow/year and surface of the state (in km²).

Item	n	Production (ton/year) X planted area (ha)	Production (ton/year) X productivity (ton/ha)	productivity (ton/ha) X planted area (ha)
Corn	110	0.95	0.35	0.14
Bean[1a]	110	0.87	0.18	0.02
Bean[1b]	110	0.96	0.42	0.31
Sugarcane	101	0.99	0.34	0.30
Coffee	95	0.98	0.07	0.00
Banana[2]	90	0.96	-0.02	-0.14
Cassava	86	0.98	0.21	0.09
Orange[3]	84	0.91	0.32	0.03
Rice[4a]	76	0.88	0.22	-0.05
Rice[4b]	71	0.91	0.53	0.38
Tomato	51	0.98	0.46	0.37
Coconut	18	0.95	0.31	0.12
Potato[5a]	12	0.99	0.47	0.42
Potato[5b]	11	0.88	0.22	-0.19
Potato[5c]	10	0.99	-0.20	-0.29
Mean		0.94	0.26	0.10

Source: Lana & Guimarães (2010); n = number of municipalities; 1a = first harvest; 1b = mean of second and third harvest; 2 = racemes instead of ton (1,000 kg); 3 = number (x1000) of oranges instead of ton (1,000 kg); 4a = rice planted in wet land; 4b = rice planted in dry land, without or with irrigation; 5a,b,c = harvest 1, 2 and 3, respectively; Source of data: www.cidadesnet.com.br (year of 2003).

Table 8. Linear correlation (r) among some variables related to agricultural production, in municipalities of Zona da Mata and Central of Minas Gerais state, Brazil.

7. Conclusions

The agriculture progress is based in improvements of animals and plants productivity per unit of area, which is only applicable when land is the limiting factor, but other factors are emerging as limiting, such as water, fertilizer and petrol.

Models of saturation kinetics are important tools to improve the efficiency and decrease costs of utilization of non renewable natural resources in agriculture, allowing the conservation of these resources for the future generations, and decreasing the negative impacts in the environment.

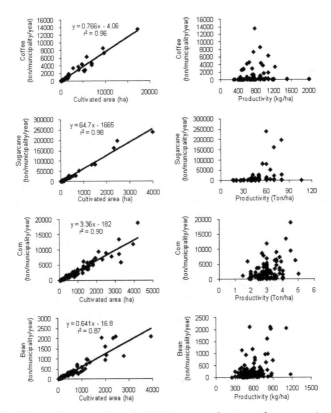

Fig. 9. Effect of cultivated area and productivity on production of some main cultures (coffee, sugarcane, corn and bean), in municipalities of Zona da Mata and Central of Minas Gerais state, Brazil.

8. References

Aaron, S. (2005). *Some Statistics on Limited Natural Resources*, 31.07.2006, available from Http://scotaaron.com/resources2.html

Angus, J.F. (1995). Modeling N Fertilization Requirements for Crops and Pasture, In: *Nitrogen Fertilization in the Environment*, P.E. Bacon, (Ed.), 109-127, Marcel Dekker inc., ISBN 0-8247-8994-6, New York, NY, United States

Baik, M.; Aschenbach, J.R.; Vandehaar, M.J. & Liesman, J.S. (2006). Effect of Dietary Protein Levels on Milk Production and Nitrogen Efficiency in Dairy Cattle. *Journal of Dairy Science*, Vol.89, Suppl. 1, pp.81, ISSN 0022-0302

Bargo, F.; Muller, L.D.; Kolver, E.S. & Delahoy, J.E. (2003) Invited Review: Production and Digestion of Supplemented Dairy Cows on Pasture. *Journal of Dairy Science*, Vol.86, No.1, pp.1-42, ISSN 0022-0302

Biotechnology and Biological Sciences Research Council (1998). *Responses in the Yield of Milk Constituents to the Intake of Nutrients by Dairy Cows*, CAB International, ISBN 0-8519-9284-6, Wallingford, United Kingdon

Bolsanello, J.; Vieira, C.; Sediyama, C.S. & Vieira, H.A. (1975). Ensaios de Adubação Nitrogenada e Fosfatada no Feijão (*Phaseolus vulgaris* L.) na Zona Metalúrgica de Minas Gerais. *Revista Ceres*, Vol.22, pp.423-430, ISSN 0034-737X

Bullock, D.G. & Bullock, D.S. (1994). Quadratique and Quadratic-Plus-Plateau Models for Predicting Optimal N Rate for Corn. A comparison. *Agronomy Journal*, Vol.86, pp.191-195, ISSN 1435-0645

Bumb, B.L. (1995). World Nitrogen Supply and Demand: an Overview, In: *Nitrogen Fertilization in the Environment*, P.E. Bacon, (Ed.), 1-40, Marcel Dekker inc., ISBN 0-8247-8994-6, New York, NY, United States

Cerrato, M.E. & Blackmer, A.M. (1990). Comparison of Models for Describing Corn Yield Response to Nitrogen Fertilizer. *Agronomy Journal*, Vol.82, pp.138-143, ISSN 1435-0645

Champe, P.C. & Harvey, R.A. (1994). *Biochemistry*, J.B. Lippincott Company, ISBN 0-3975-1091-8, Philadelphia, PA, United States

Fulkerson, W.J.; Nandra, K.S.; Clark, C.F. & Barchia, I. (2006). Effect of Cereal-Based Concentrates on Productivity of Holstein-Friesian Cows Grazing Short-Rotation Ryegrass (*Lolium multiflorum*) or Kikuyu (*Pennesitum clandestinum*) Pastures. *Livestock Science*, Vol.103, pp.85-94, ISSN 1871-1413

Granli, T. & Bockman, O.C. (1994). Nitrogen Oxide from Agriculture. *Norwegian Journal of Agricultural Sciences*, Vol.12, pp.7-127, ISSN 0801-5341

Guimarães, G.; Lana, R.P.; Guimarães, A.V.; Santos, M.A.; Andrade, F.L.; Fialho, C. & Castro, T.R. (2008). Sustentabilidade da Agricultura Familiar na Produção de Leite, *Proceedings of 10° Minas Leite*, EMBRAPA, CDD 637.1, Juiz de Fora, MG, Brazil

Helyar, K.R.; Poter, W.M. (1989). Soil Acidification, its Measurement and the Processes Involved, In: *Soil acidity and plant growth*, A.D. Robson, (Ed.), 61-100, Academic Press, ISBN 0-1259-0655-2, Sydney, Australia

Herring, J.R. & Fantel, R.J. (1993). Phosphate Rock Demand into the Next Century: Impact on World Food Supply. *Nonrenewable Resources*, Vol.2, No.3, pp.226-246, ISSN 0961-1944

Keane, M.G.; Drennan, M.J. & Moloney, A.P. Comparison of Supplementary Concentrate Levels with Grass Silage, Separate or Total Mixed Ration Feeding, and Duration of Finishing in Beef Steers. *Livestock Science*, Vol.103, pp.169-180, ISSN 1871-1413

Kebreab, E.; France, J.; Mills, J.A.N.; Allison, R. & Dijkstra, J. (2002). A Dynamic Model of N Metabolism in the Lactating Cow and an Assessment of Impact on N Excretion on the Environment. *Journal of Animal Science*, Vol.80, pp.248-259, ISSN 0021-8812

Lana, R.P. (2003). *Sistema Viçosa de Formulação de Rações*, Editora UFV, ISBN 978-85-7269-314-1, Viçosa, MG, Brazil

Lana, R.P. (2005). *Nutrição e Alimentação Animal (Mitos e Realidades)*, Suprema Gráfica, ISBN 978-85-9050-672-0, Viçosa, MG, Brazil

Lana, R.P. (2007a). Plants Responses to Nutrients Follow a Michaelis-Menten Relationship, *Proceedings of ASA, CSSA, SSSA International Annual Meetings*, ASA-CSSA-SSSA, ISSN 1529-9163, New Orleans, United States

Lana, R.P. (2007b). *Respostas Biológicas aos Nutrientes*, Editora CPD, ISBN 978-85-9050-673-7, Viçosa, MG, Brazil

Lana, R.P. (2008). Plants Responses to Nutrients Follow the Saturation Kinetic Typical of Enzyme Systems: Biological, Economical and Environmental Implications. *Online Journal of Biological Sciences*, Vol.8, No.1, pp.19-24, ISSN 1608-4217

Lana, R.P. (2009). Uso Racional de Recursos Naturais não Renováveis: Aspectos Biológicos, Econômicos e Ambientais. *Revista Brasileira de Zootecnia*, Vol.38, pp.330-340, Supplement Especial, ISSN 1806-9290

Lana, R.P.; Abreu, D.C.; Castro, P.F.C. & Zamperline, B. (2007a). Milk Production as a Function of Energy and Protein Sources Supplementation Follows the Saturation Kinetics Typical of Enzyme Systems, *Proceedings of 2nd International Symposium on Energy and Protein Metabolism and Nutrition*, European Association for Animal Production, ISBN 978-90-8686-041-8, Vichy, France

Lana, R.P.; Abreu, D.C.; Castro, P.F.C.; Zamperlini, B. & Souza, B.S.B.C. (2007b). Kinetics of Milk Production as a Function of Energy and Protein Supplementation. *Journal of Animal Science*, Vol.85, Suppl. 1, pp.566, ISSN 0021-8812

Lana, R.P.; Goes, R.H.T.B.; Moreira, L.M.; Mancio, A.B. & Fonseca, D.M. (2005). Application of Lineweaver-Burk Data Transformation to Explain Animal and Plant Performance as a Function of Nutrient Supply. *Livestock Production Science*, Vol.98, pp.219-224, ISSN 0301-6226

Lana, R.P.; Guimarães, G.; Guimarães, A.V. & Santos, M.A. (2009). Factors Affecting Milk Production in Brazil. *Journal of Dairy Science*, Vol.92, Suppl. 1, pp.427, ISSN 0022-0302

Lana, R.P. & Guimarães, G. (2010). Produção Média e Correlações entre Algumas Variáveis que Afetam a Produção de Diversas Culturas em Municípios da Zona da Mata e Central de Minas Gerais, In: *2° Simpósio Brasileiro de Agropecuária Sustentável – Anais de Resumos Expandidos*, R.P. Lana & G. Guimarães, (Eds.), ISSN 2176-0772, Viçosa, MG, Brazil

Lemke, R.L.; Izaurralde, R.C.; Malhi, S.S.; Arshad, M.A. & Nyborg, M. (1998). Nitrous Oxide Emissions from Agricultural Soils of the Boreal and Parkland Regions of Alberta. *Soil Science Society of America Journal*, Vol.62, pp.1096-1102, ISSN 0361-5995

Lineweaver, H. & Burk, D. (1934). The Determination of Enzyme Dissociation Constants. *Journal of the American Chemical Society*, Vol.56, pp.658-666, ISSN 0002-7863

Makowski, D.; Wallach, D. & Meynard, J.-M. (1999). Model of Yield, Grain Protein, and Residual Mineral Nitrogen Responses to Applied Nitrogen for Winter Wheat. *Agronomy Journal*, Vol.91, pp.377-385, ISSN 1435-0645

Makowski, D.; Wallach, D. & Meynard, J.-M. (2001). Statistical Methods for Predicting Responses to Applied Nitrogen and Calculating Optimal Nitrogen Rates. *Agronomy Journal*, Vol.93, pp.531-539, ISSN 1435-0645

Meadows, D.H.; Meadows, D.L.; Randers, J. & Behrens III, W.W. (1972). *The Limits of Growth. A Report for the Club of Rome's Project on the Predicament of Mankind*, Universe Books, ISBN 0-87663-165-0, New York, NY, United States

Michaelis, L. & Menten, M.L. (1913). Kinetics of Invertase Action. *Biochemistry Zournal*, Vol.49, pp.333-369, ISSN 1470-8728

Mitscherlich, E.A. (1909). Das Gesetz des Minimuns und das Gesetz des Abnehmenden Bodenertrages. *Landwirtschaftliches Jahrbuch*, Vol.38, pp.537-552, ISSN 0005-7150

Malavolta, E. (1989). *ABC da Adubação*, Editora Ceres, ISBN 85-318-0002-1, São Paulo, SP, Brazil

Mombiela, F.; Nicholaides III, J.J. & Nelson L.A. (1981). A Method to Determine the Appropriate Mathematical form for Incorporating Soil Test Levels in Fertilizer Response Models for Recommendation Purposes. *Agronomy Journal*, Vol.73, pp.937-941, ISSN 1435-0645

Monod, J. (1949). The Growth of Bacterial Cultures. *Annual Review of Microbiology*, Vol.3, pp.371-394, ISSN 0066-4227

Morgan, H.P.; Mercer, L.P. & Flodin, N.W. (1975). General Model for Nutritional Responses of Higher Organisms. *Proceedings of the National Academy of Sciences*, Vol.72, No.11, pp.4327-4331, ISSN 0027-8424

NRC. *Nutrient Requirements of Beef Cattle*. (1996). National Academy Press, ISBN 0-309-06934-3, Washington, DC, United States

NRC. *Nutrient Requirements of Dairy Cattle*. (2001). National Academy Press, ISBN 0-309-06997-1, Washington, DC, United States

Oliveira, A.J.; Lourenço, S. & Goedert, W.J. (1982). *Adubação Fosfatada no Brasil*, EMBRAPA-DID, CDD 631.850981, Brasília, DF, Brazil

Pimentel, J.J.O.; Lana, R.P.; Zamperlini, B.; Paulino, M.F.; Valadares Filho, S.C.; Teixeira, R.M.A. & Abreu, D.C. (2006a). Milk Production as a Function of Nutrient Supply Follows a Michaelis-Menten Relationship. *Journal of Dairy Science*, Vol.89, Suppl. 1, pp.74-75, ISSN 0022-0302

Pimentel, J.J.O.; Lana, R.P.; Zamperlini, B.; Valadares Filho, S.C.; Abreu, D.C.; Silva, J.C.P.M. & Souza, B.S.B.C. (2006b). Efeito do Teor de Proteína e Níveis de Suplementação com Concentrado na Produção e Composição do Leite em Vacas Leiteiras Confinadas, *Proceedings of 43ª Reunião Anual da Sociedade Brasileira de Zootecnia*, Sociedade Brasileira de Zootecnia, ISSN 1983-4357, João Pessoa, PB, Brazil

Pimentel, J.J.O.; Lana, R.P.; Teixeira, R.M.A.; Zamperlini, B.; Sobreira, H.F.; Paulino, M.F.; Leão, M.I. & Oliveira, A.S. (2006c). Produção de Leite em Função de Níveis de Suplementação com Concentrado para Vacas Leiteiras sob Pastejo, *Proceedings of 43ª Reunião Anual da Sociedade Brasileira de Zootecnia*, Sociedade Brasileira de Zootecnia, ISSN 1983-4357, João Pessoa, PB, Brazil

Roberts, T.L. & Stewart, W.M. (2002). Inorganic Phosphorus and Potassium Production and Reserves. *Better Crops*, Vol.86, No.2, pp.6-7, ISSN 0006-0089

Russell, J.B. (1984). Factors Influencing Competition and Composition of the Ruminal Bacterial Flora, In: *The Herbivore Nutrition in the Subtropics and Tropics*, F.M.C. Gilchrist & R.I. Mackie, (Eds.), 313-345, Science Press, ISBN 10 0907997031, Craighall, South Africa

Russell, J.B. (2002). *Rumen Microbiology and its Role in Ruminant Nutrition*, James B. Russell, Ithaca, NY, United States

Sain, G.E. & Jauregui, M.A. (1993). Deriving Fertilizer Recommendations with a Flexible Functional Form. *Agronomy Journal*, Vol.85, pp.934-937, ISSN 1435-0645

Sairanen, A.; Khalili, H. & Virkajarvi, P. (2006). Concentrate Supplementation Responses Of The Pasture-Fed Dairy Cow. *Livestock Science*, Vol.104, No.3, pp.292-302, ISSN 1871-1413

Strong, W.M. (1995). Nitrogen fertilization of upland crops, In: *Nitrogen Fertilization in the Environment*, P.E. Bacon, (Ed.), 129-169, Marcel Dekker inc., ISBN 0-8247-8994-6, New York, NY, United States

Teixeira, R.M.A; Lana, R.P.; Fernandes, L.O.; Veloso, R.G.; Ferreira, M.B.D. & Paiva, V.R. (2006). Efeito da Adição de Concentrado em Dietas de Vacas Gir Leiteiro Confinadas sob a Produção de Leite, *Proceedings of 43ª Reunião Anual da Sociedade Brasileira de Zootecnia*, Sociedade Brasileira de Zootecnia, ISSN 1983-4357, João Pessoa, PB, Brazil.

Van Der Ploeg, R.R.; Böhm, W. & Kirkham, M.B. (1999). On the Origin of the Theory of Mineral Nutrition of Plants and the Law of the Minimum. *Soil Science Society of America Journal*, Vol.63, pp.1055-1062, ISSN 0361-5995

5

Chemical Conversion of Glycerol from Biodiesel into Products for Environmental and Technological Applications

Miguel Araujo Medeiros[1], Carla M. Macedo Leite[2]
and Rochel Montero Lago[2]
[1]*Universidade Federal do Tocantins,*
[2]*Universidade Federal de Minas Gerais,*
Brasil

1. Introduction

Currently, fossil fuels represent over 80% of energy consumption in the world. However, due to environmental and geopolitical issues the development of new energy sources is mandatory. For example, only the Middle East holds 63% of global reserves, which directly influences in the final price of fuel.

In developed nations there is a growing trend towards employing modern technologies and efficient bioenergy conversion using a range of biofuels, which are becoming cost competitive with fossil fuels (Puhan et al., 2005). In Brazil, this work is focused on the production of bioethanol and biodiesel.

There are discussions around the world on the feasibility of using renewable fuels, which may cause a much smaller impact to global warming, because the balance of CO_2 emissions decreases when using these fuels. (Demirbas, 2008)

In 1997 at a meeting in Kyoto, Japan, many of the developed nations agreed to limit their greenhouse gas emissions, relative to the levels emitted in 1990. In this occasion Brazil established social and environmental policies to collaborate with those global goals (Puhan et al., 2005). An example is the biodiesel program which in 2008 implemented the use of B2 (2% biodiesel into conventional diesel). In other countries, like Germany, it is possible to supply only with B100 biodiesel (100% biodiesel).

1.1 Biodiesel

Biodiesel, a renewable biofuel produced from biomass, is biodegradable and does not cause significant contamination with emissions containing sulfur or aromatics. Biodiesel, is an viable alternative for compression-ignition engines (Puhan et al., 2005), in total or partial substitution of fossil diesel (Chiang, 2007).

The use of biodiesel as fuel should occupy a prominent place in the world, with a market that is booming because of its enormous contribution to the environment, such as qualitative and quantitative reduction of environmental pollution (Ferrari et al., 2005). Furthermore, this fuel is a strategic source of renewable energy to replace petroleum products.

Biodiesel is fuel produced mainly by transesterification of vegetable oils, but can also be obtained by the reaction of animal fat (Pinto et al. 2005; Puhan et al., 2005; Chiang, 2007) soybean (Costa Neto & Rossi, 2000), Cotton (Pinto et al., 2005; Puhan et al., 2005), castor bean (Pinto et al., 2005), canola (Pinto et al., 2005; Catharino et al., 2007; Kocak et al., 2007; Puhan et al., 2005), palm (Pinto et al., 2005; Catharino et al., 2007; Puhan et al., 2005), sunflower (Pinto et al., 2005; Catharino et al., 2007; Puhan et al., 2005; Costa Neto & Rossi, 2000), peanut and babassu.

Synthesis of biodiesel can be accomplished by using acid, basic (Costa Neto & Rossi, 2000; Puhan et al., 2005; Chiang, 2007; Pinto et al., 2005) or enzymes (Talukder et al., 2007; Schuchardt, 1990) catalysts or even in supercritical methanol (Puhan et al., 2005).

1.2 Biodiesel production

Transesterification (Figure 1) is the reaction of triglycerides with an alcohol to form esters and glycerol (Chiang et al., 2007; Georgogianni et al., 2007; Krishna et al., 2007; Wu et al., 2007; Talukder et al., 2007; Aparício et al., 2007; Zuhair, 2005; Vicente et al., 2005; Medeiros et al., 2008; Stern & Hillion, 1990; Freedman et al., 1984; Encinar et al, 2002; Vicente et al., 2006; Bunyakiat et al., 2006; Karinen & Krause, 2006). This process decreases the viscosity of the oil and transforms the large, branched molecular structure of bio-oils into smaller molecules, of type required in regular diesel engines.

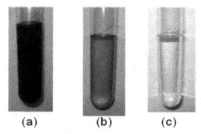

Fig. 1. Synthesis of biodiesel by transesterification of triglyceride.

In the transesterification for biodiesel production, a large amount of glycerol as a byproduct (about 10% compared to the mass of ester produced) (Puhan et al., 2005; Medeiros et al., 2010) is produced. The separation step of glycerol can be accomplished by decanting, in which the lower phase has the glycerol, the catalyst of the process (usually homogeneous and high polar character), alcohol and oil residue without reacting (crude glycerol, Figure 2, a). The biodiesel separates from the upper stage, almost pure.

The transesterification using methanol is the most used process around the world (Chiang, 2007) offering several advantages, such as: (i) small volume of alcohol recovery, (ii) lower cost of alcohol compared to ethanol (not in Brazil) and (iii) shorter reaction times (Pinto et al., 2005). The use of ethanol proves more advantageous, when considering its lower toxicity.

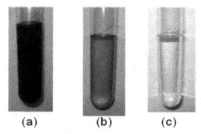

(a)   (b)   (c)

Fig. 2. (a) crude glycerol, (b) pre-purified glycerol, (c) glycerol purified.

1.3 Use of the glycerol

The investigation of new uses for glycerol is critical for the success of the biodiesel program, especially in relation to the crude glycerol, which has few direct uses and market value marginalized. Currently, the demand for purified glycerol PA for the pharmaceuticals, food additives, personal care (Puhan et al., 2005), industry is supplied by the petrochemical industry.

The biodiesel production will produce a large increase in, the amount of glycerol in the market, causing a decrease in the prices significantly, in the world. In the European Union, for example, the price of glycerol, in 1995 was € 1500 t⁻¹ and reduced to 330 € t⁻¹ in 2006 (Puhan et al., 2005). In Brazil, in 2005 the price of glycerol reached € 1270 t⁻¹, but already in 2007 the price dropped to 720 € t⁻¹. And in regions close to the price of biodiesel plants did not exceed € 300 t⁻¹, in 2010.

Different routes have been investigated to transform this glycerol to new products and new applications. Some of these processes are listed in Table 1.

Process	Conditions	Products	Ref.
Polymerization	$T = 210\text{-}230°C$; reduced pressure (~0,3atm); 0,5-1,5% NaOH.	Cyclic polymers.	(Blytas & Frank, 1993)
Pyrolise	$T = 650°C$.	CO; acetaldehyde; acrolein.	(Chiang, 2007)
steam reforming	$T = 200\text{-}250°C$; 1% cat. Níquel-Raney (Ni-Sn).	50-70% H_2; 30-40% CO; 2-11% of alkanes.	(Stein et al., 1983)
Esterification	$T = 200\text{-}240°C$; 0,1-0,3% NaOH; t =0,5 h; 100% methanol.	Carbohydrates and esters.	(Noureddini & Medikonduru, 1997)
Oxidation	$T = 50°C$; Pd/C (5-8% of Pd), t = 8h; pH = 5-11.	Dihydroxyacetone.	(Garcia et al., 1995)
Etherization	$T = 90°C$; 1-7,5% amberlist 15; t = 2-3h.	70% of 3-tert-butoxi-1,2-propanodiol (*mono ether*). 87% of mono ether.	(Klepácová et al., 2003, 2006)
Oligomerization	$T = 260°C$; 2% $Mg_{25}Al_{20}$ (cat.); t = 8h.	65% of diglycerol; 20% of triglycerol and 15% of tetraglycerol.	(Barrault et al., 2004)

Table 1. Conversion of glycerol to different products.

Table 1 can be summarized in Scheme 1, which shows some reactions that originate from glycerol.

Oxidation products of glycerol, for example, can be used in cosmetics and pharmaceuticals intermediates (Davis et al., 2000; Pachauri & He, 2006; Krishna et al., 2007) and even suntan lotion (Kimura, 1993).

The products of oligomerization of glycerol can be used as additives for cosmetics and foods, the raw material for resins and foams (Shenoy , 2006; Lemke, 2003; Werpy, 2004; Pagliaro & Rossi, 2008), lubricants (Pagliaro & Rossi, 2008), cement additives (retains moisture) and are synthetic intermediates and possible substitutes of polyols, e.g. polyvinyl alcohol, in some applications (Werpy, 2004; Pagliaro & Rossi, 2008; Medeiros et al., 2008).

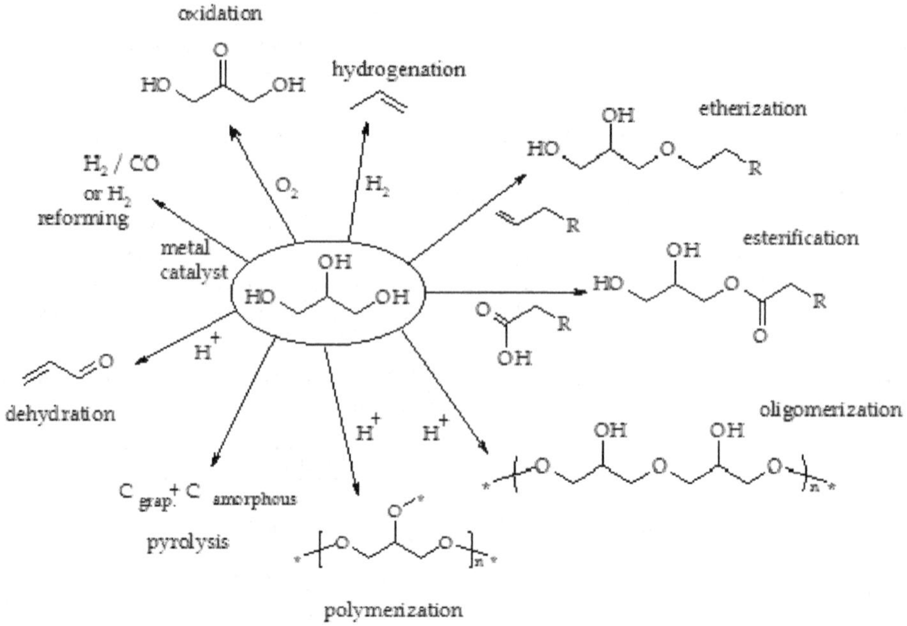

Scheme 1.

It is noteworthy that many of the applications mentioned for the glycerol require high degree of purity, which for glycerol derived from biodiesel requires several stages of treatment, increasing its cost. The main impurities in the glycerol from biodiesel is methanol or ethanol, water, inorganic salts and catalyst residues, free fatty acids, unreacted mono, di and triglycerides and various other matter organic non-glycerol (MONG) (Pagliaro & Rossi, 2008). Thus, it is necessary to develop new routes for the consumption of glycerol from biodiesel.

In this chapter, will be treated the transformation of glycerol based on production of ethers from condensation of two (or more) glycerol's molecules. One of the mechanisms Favorable for the formation of ethers is by alcohol protonation (ROH_2^+), followed by condensation of other alcohol and water loss. The condensation reaction of glycerol (Scheme 2), is usually catalyzed by acids or bases producing small polymers called oligomers and water. Along the text will be described the oligomers, polymers and carbons obtained from polyglycerol and its applications.

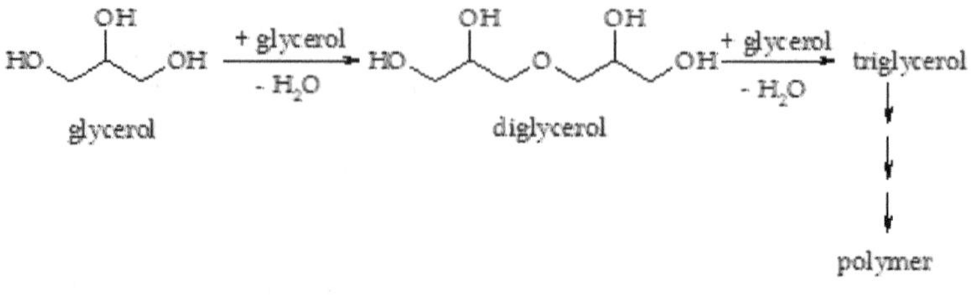

Scheme 2.

2. Oligomers

Oligomerization of glycerol (Scheme 2) is an alternative to the use of byproduct of biodiesel, because their products have wide application. For a better understanding of oligomerization (and polymerization), was accompanied through ESI-MS (Electrospray Ionization Mass Spectrometry in the positive ion mode) a typical reaction of oligomerization - glycerol PA catalyzed by 1% H_2SO_4 at 280°C/2h, in reflux.

Analysis of the sample (2h) is shown in Figure 3. The presence of an intense ion of m/z 93 (protonated glycerol = [glycerol + H]$^+$) is clearly noticeable indicating the subsistence of glycerol in the reaction medium even after 2 h reaction.

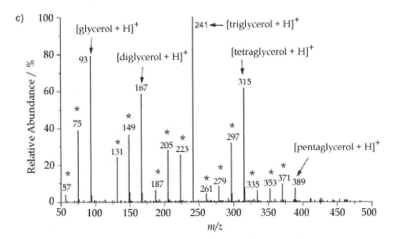

Fig. 3. ESI(+)-MS of the acid-catalyzed oligomerization of glycerol conducted in aqueous medium at 280 °C, 2 h. The ions marked with an asterisk (*) refer to dehydration products.

A remarkable presence of an ion of m/z 167 is also noticed in Figure 3. This corresponds to the protonated form of diglycerol, i.e. [(glycerol)$_2$ – H_2O], formed under these reaction conditions via the condensation of two molecules of glycerol and loss of water. This condensation can occur via the primary or secondary hydroxyl groups at the glycerol molecule to yield linear (α, α -diglycerol) and branched (α, β -diglycerol; β, β -diglycerol) isomers, as displayed in Scheme 3.

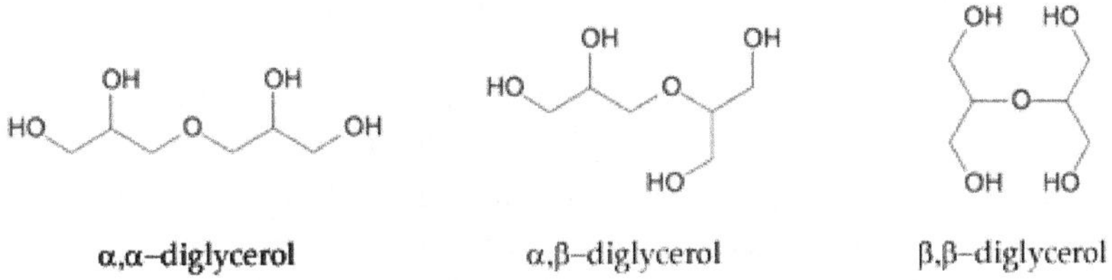

α,α–diglycerol   α,β–diglycerol   β,β–diglycerol

Scheme 3.

Across of the fragmentation of the ion of m/z 167 are yield mainly product ions from losses of one or two molecules of water (m/z 149 and 131, respectively) besides to other product ions, such as [glycerol + H]+ (m/z 93), [glycerol – H_2O + H]+ (m/z 75), and [glycerol – 2 H_2O + H]+ (m/z 57). To illustrate the formation of such fragments, the dissociation pathways for protonated α, α-diglycerol are shown in Scheme 4.

In the Table 2, are showed ions ascribed to be the protonated forms of products formed by successive dehydrations of di, tri, tetra and pentaglycerol.

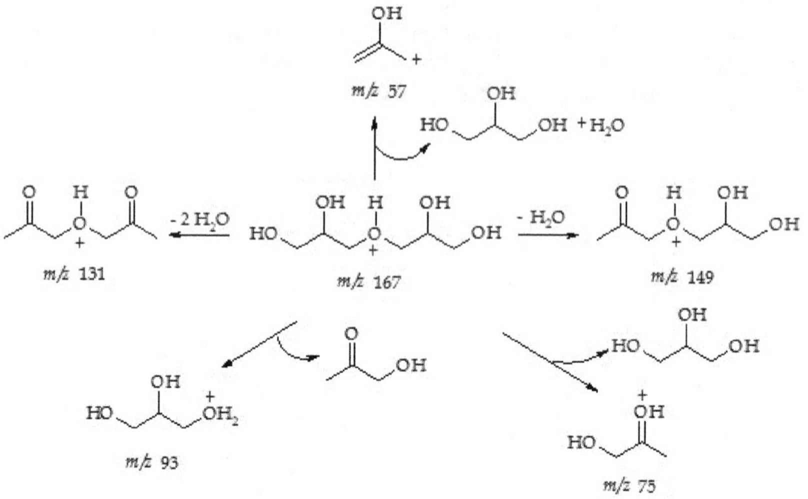

Scheme 4.

Primary Oligomers (m/z of the protonated forms)	Dehydration Products (m/z of the protonated forms)
diglycerol (167)	[diglycerol – H_2O] (149) [diglycerol – 2 H_2O] (131)
triglycerol (241)	[triglycerol – H_2O] (223) [triglycerol – 2 H_2O] (205) [triglycerol – 3 H_2O] (187)
tetraglycerol (315)	[tetraglycerol – H_2O] (297) [tetraglycerol – 2 H_2O] (279) [tetraglycerol – 3 H_2O] (261)
pentaglycerol (389)	[pentaglycerol –H_2O] (371) [pentaglycerol – 2 H_2O] (353) [pentaglycerol – 3 H_2O] (335)

Table 2. Primary products (diglycerol, triglycerol, tetraglycerol and pentaglycerol) and their dehydration products formed upon acid-catalyzed oligomerization of glycerol at 280°C. All these products were observed as their protonated forms in the ESI(+)-MS (Fig. 3).

These findings thus indicate that under acidic medium and heating, olygomers can easily lose one or two molecules of water to form a myriad of isomeric products. Scheme 5 shows, for instance, products possibly formed as a result of the mono-dehydration of diglycerol, such as the cyclic species **1a-c** (their formation have been reported by Barrault and coworkers (Barrault et al., 2004, 2005) that submitted glycerol to similar reaction conditions than those employed herein) besides the acyclic carbonyl compounds **2a-b** and the alkene **2c**.

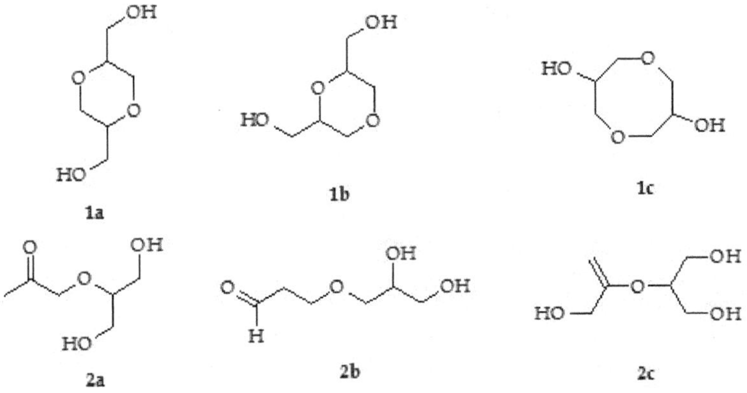

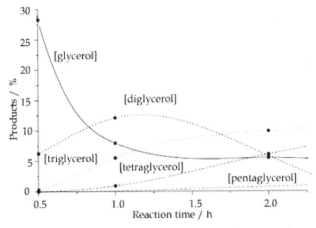

Scheme 5.

All the products resulting from the mono-dehydration of diglycerol, including the ones shown in Scheme 3 (**1a-c** and **2a-c**), possess the same chemical formula ($C_6H_{12}O_4$) and bear similar functional groups (especially hydroxyl substituents). Hence, these protonated molecules lose mainly water and other small molecules, being unfeasible the unambiguous characterization of a particular product based exclusively on your fragmentation profile.

Fig. 4. Fractions of the ions of m/z 167, 241, 315, and 389 as a function of reaction time. Each fraction was calculated as the quotient ratio between the absolute intensity of one of such ions and the sum of the absolute intensities of the whole set of ions.

In Figure 4, the fraction of the ions of m/z 93, 167, 241, 315 and 389 (protonated glycerol, di, tri, tetra and pentaglycerol, respsctively), given as a quotient ratio between the absolute intensity of one of such ions and the sum of the absolute intensities of the whole set of ions, are plotted against the reaction time. These results show that after 2 h reaction more than 90% of glycerol is consumed. Furthermore, during the first 30 min a relatively high concentration of diglycerol is formed. At longer reaction times, however, its concentration decreases whereas the amount of the heavier oligomers (tri, tetra and pentaglycerol) concomitantly increases. The result shows that glycerol is continuously converted into the heavier oligomeric compounds.

3. Polymerization of glycerol

In open system, the polycondensation (condensation of many molecules to create larger molecules - polymers) that the glycerol suffers in the presence of H_2SO_4 at 150 ° C, is a type of polymerization in which mingle the three stages: initiation, propagation and termination, which are characteristic of polymerization reactions (Mano & Mendes, 1999). The condensation polymerization, when employ monomers (molecules susceptible to undergo polymerization) with more than two functional groups (glycerol has three OH groups), tends to form crosslinked or branched polymers (structures with crosslinks between chains). In this case, the polymerization is complex because it is formed gel (polymer molecular weight too large), in the same setting of the sol (the fraction that remains soluble and can be extracted from the middle). As the sol will turn into gel, the mixture becomes increasingly viscous until elastic consistency, and finally rigid. In this transformation of glycerol in hard polymer, the catalyst concentration has an important role. An example of the participation of the catalyst in the polymerization of glycerol is shown in Figure 5, in which the viscosities of solution reaction is monitored by 60 minutes, with different concentrations of catalyst (0.5, 1, 3 and 5 mol%) .

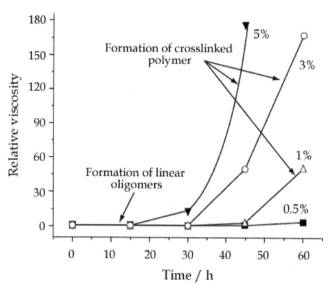

Fig. 5. Variation of relative viscosity of the solution for the polymerization of glycerol, with 0.5,1, 3 and 5 mol% H_2SO_4 (viscosity values are relative to the glycerol).

The curves shown in Figure 4 indicate a significant increase in viscosity of the solution, by varying the mole percentage of catalyst of 0.5-5%. However, this increase is gradual, as it rises the concentration of H_2SO_4. It is interesting to note that the system promoted by 5 mol% of catalyst is very active, because it took only 45 minutes to produce a solid polymer (unable to measure the viscosity, since the material solidified, Figure 5), whereas in other systems it took at least 120 minutes.

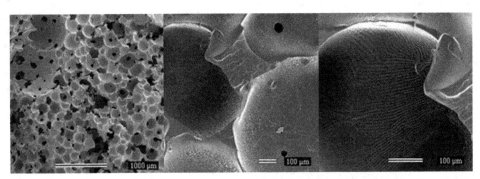

Fig. 6. Images of scanning electron microscopy (SEM) of polyglicerol with 5 mol% of H_2SO_4, after 45 min of reaction.

The system promoted by 0.5, 1 and 3 mol% H_2SO_4 show similar viscosity curves (Figure 4) but with varying slopes (the higher the catalyst concentration, greater the slope of the viscosity). With the increase of H_2SO_4 concentration (0.5-3%) is expected larger number of simultaneous condensation, therefore, the selectivity of the catalyst to reactive hydroxyl groups of glycerol decreases, leading to complex structures, which offer increased viscosity of the solution (in 60 minutes: system promoted by 0.5 mol% → relative viscosity of 4,1%; 1 mol% → relative viscosity of 51 e 3 mol% → relative viscosity of 169 times that of glycerol.

In the first 15 minutes of reaction, the viscosity of the medium practically does not change. It is believed that during this period, is occurring the formation of linear oligomers and products of dehydration. However, as the polymerization reaction progresses, the ethers formed formed become larger and more complex, mainly due to the formation of branches and some bonds between parallel chains of oligomers and/or polymers. And this is the increase size and complexity of structures of the ethers formed which increased the solution viscosity, reaching 169 times the viscosity of glycerol in just 60 minutes (system promoted by 3 mol% H_2SO_4), because the move of the structures is becoming increasingly difficult.

To confirm the nature (thermoplastic or thermosetting) polymeric material formed by polymerization of glycerol, are carried out two separate tests: heating in the direct flame of a Bunsen burner, to ensure that it is malleable (suffers fusion) or undergo thermal decomposition, and washing the polymer in solvents with different polarities (hexane, THF and ethanol). The results of these tests showed that all the polymers (0.5, 1, 3 and 5 mol% H_2SO_4) are thermosets, because not suffers fusion, but rather, thermal decomposition and did not dissolve in any solvent tested.

The polyglycerol may be used as a substitute for thermosetting phenolic resins, used in home utensils as well as controlled release fertilizers.

4. High surface area carbons

The thermosetting polymers have the property that thermally decomposes, producing carbon in quantities that can vary with the degree of crosslinking of the polymer. As previously discussed, the degree of crosslinking (polymerization in all directions, linking parallel chains of the polymer) is influenced by the concentration of the catalyst. Hence, a polymer obtained with 5 mol% H_2SO_4 is more reticulated and produces more carbon than a polymer with only 1 mol% catalyst.

The thermal decomposition of polyglycerol (obtained with 5 mol% H_2SO_4) yields 16% of carbon (relative to initial mass of polymer), which has extremely low surface area (2 m^2g^{-1}). For environmental applications, are typically used carbonaceous materials with high surface area. Thus, it was necessary to increase the surface area of carbons derived from polyglycerol, performing physical activation (850°C) with a flow of CO_2 by 3, 5, 10, 15 and 18 h (Figure 7).

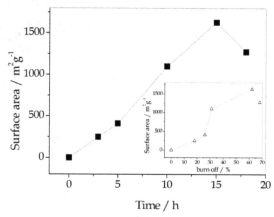

Fig. 7. Surface area of carbons derived from polyglycerol. Detail: surface area as a function of burn off.

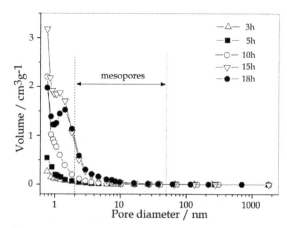

Fig. 8. Distribution of pores in activated carbonaceous material for 3, 5, 10, 15 and 18 h.

The Figure 7 shows a gradual increase in surface area of carbonaceous material up to 1830 m2 g-1, 15 h of activation. After that time, the surface area begins to decrease, reaching a value of 1275 m2 g-1 at 18 h. A similar behavior, but not linear is observed for the surface area as a function of burn off (mass loss of carbon during activation) (detail of Figure 7). The analysis of distribution of pores indicates that the materials are essentially microporous (internal diameter of less than 2 nm) (Figure 8).

As the surface area of carbonaceous material derived from polyglycerol increased with activation time by 15 hours, tests were made to adsorb organic contaminants (methylene blue) during the activation process (0, 3, 5, 10 and 15 h). The results of these tests are shown in Figure 9.

It is evident the relationship between the activation time of the carbonaceous material (surface area) and the adsorption of organic contaminant. During the activation process of the carbonaceous material, there was a gradual increase of its surface area, which is intimately related to the growth of its adsorption capacity. The sample of the material that was activated for 15 hours showed better results in the removal of organic contaminant (in 20 minutes, the removal of contaminant was 90%, while others samples took at least 60 minutes to obtain the same results).

5. Vermiculite composites/activated carbon

In order to facilitate the application of carbonaceous material in environmental problems, was produced a composite based in vermiculite clay and activated carbon, derived from the polyglycerol. This composite was designed, considering (i) some properties of the expanded vermiculite clay (Figure 10), which has low cost and ability to float in water, (ii) large adsorption capacity of Activated carbon derived from polyglycerol and (iii) facility removal of the composite in case of water application, requiring only one net.

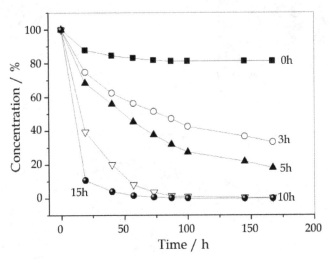

Fig. 9. Adsorption of methylene blue by carbonaceous material derived from the polyglycerol at different activation times (0, 3, 5, 10 and 15 h).

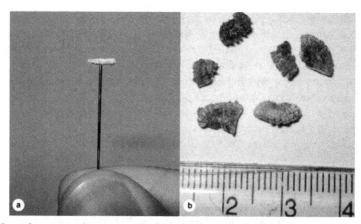

Fig. 10. (a) Sample vermiculite *in natura*; (b) Sample expanded vermiculite.

The composite vermiculite/carbon is prepared the same way that the pure carbon, except that the clay expanded vermiculite (EV) is added before the initial stage of polymerization of glycerol, which will occur on the surface of clay.

The best condition for prepare of the composite (GVE4), which has carbon content of 25% (compared to the mass of the composite) is 3 mol% H_2SO_4 and 580°C/3h and ratio (by mass) glycerol/VE = 4. This condition was obtained after tests with different reaction conditions. The images of scanning electron microscopy (SEM) for pure EV and composite (GVE4) showed significant differences in their surfaces (Figure 11).

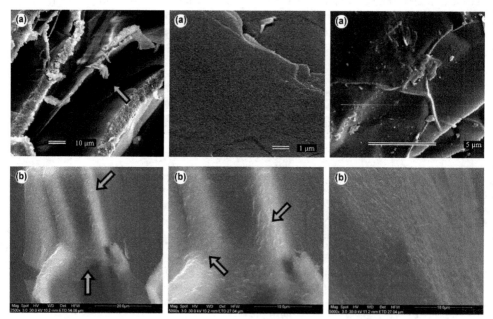

Fig. 11. SEM images of: (a) EV; (b) GVE4.

The SEM images for pure EV (Figure 11 (a)) show regular flat surfaces with an interlamellar space between 10-100 μm with some fragments attached to the edges (arrow). Figure 11 (b) shows large amounts of regular deposits on the EV layers with materials connecting some layers.

To obtain a good adsorbent material, the composite GVE4 was submitted to physical activation with CO_2 for periods of 0.5 (GVE4CA0.5); 1 (GVE4CA1), 2 (GVE4CA2) and 4 hours (GVE4CA4). Table 3 shows the surface area and burn off the composite GVE4, activated at different times.

To observe the data presented in Table 3 and Figure 12, perceives a linear increase in surface area, depending on the activation time, until the limit value of 835 m^2g^{-1} (2h of activation), when the value of surface area begins to decrease to 143 m^2g^{-1}, 4h of activation. A similar performance is observed for the surface area as a function of burn off (detail of Figure 12).

Sample	Burn off/ %	Surface Area / m^2g^{-1}
GVE4CA0	0	9
GVE4CA0.5	23,4	387
GVE4CA1	44,2	648
GVE4CA2	58,0	835
GVE4CA4	73,0	146

Table 3. Data for surface area GVE4CA0, GVE4CA0.5, GVE4CA1, GVE4CA2 and GVE4CA4, obtained by BET method.

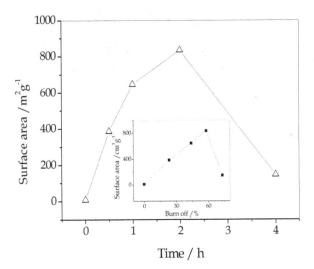

Fig. 12. Surface area of the activated composites. Detail: surface area as a function of burn off.

SEM images presented in Figure 13 show how the carbon deposits on the surface of the composites was changed during activation. After the first hour of activation, the amount of material deposits on the surface of the composite is significantly lower when compared to the composite without activation (Figure 11 (b)), because of the oxidizing action of CO2,

850°C/1h. SEM images show that 2 h of activation are sufficient to make large part of the surface of the EV is exposed, reducing the carbonaceous deposits, although the surface area is the largest obtained (835 m²g⁻¹). But it's after 4 h of activation that the composite loses most part of the carbon deposits and therefore reduces the surface area to only 146 m²g⁻¹.

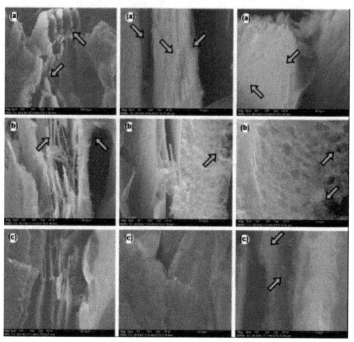

Fig. 13. SEM images of GVE4, actived by: (a) 1h; (b) 2h and (c) 4h.

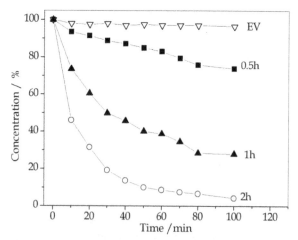

Fig. 14. Adsoption of methilene blue by EV and composites actived by 0.5, 1 and 2h.

After the activation process of the composite GVE4, samples of 0.5, 1 and 2h of activation were tested as adsorbents for organic contaminants (methylene blue) (figure 14). It is possible notice that the EV practically no adsorbs or reacts with the contaminant during all the test period. The composite GVE4CA2 adsorbs 90% of the dye in only 60 minutes, tending to 100% until the end of the test (100 min). It is also notable, the action of the composite GVE4CA1, which absorbs more than 50% of the dye in the first 30 minutes of testing, tending to an equilibrium around 70% of contaminants adsorbed to the end of the test. Already the composite GVE4CA0.5, has unsatisfactory result, with only 20% of adsorbing dye after 100 minutes of testing.

6. Conclusions

Glycerol is a very versatile chemical species which can produce different materials to distinct applications. In this chapter, we discussed some possibilities for the glycerol that boost its use in the production of polymers and adsorbents for organic contaminants.
Study of oligomerization of glycerol, by ESI-MS, is an important step in understanding how the molecules glycerol were initially organized to enable the formation of thermosetting polymers and, later, special carbonaceous materials.
The preparation of carbonaceous materials from glycerol, for environmental applications is a way to consume an important portion of glycerol introduced in the market from the production of biodiesel.

7. Acknowledgements

UFMG, UFT, Capes, CNPq and Fapemig.

8. References

Aparício, C., Guignon, B., Rodriguez-Antón, L.M., Sanz, P.D. *J. Agric. Food Chem.*, 55, 7394-7398, 2007
Barrault, J., Clacens, Y., Pouilloux, Y. *Topics in Catalysis*, 27, 137-142, 2004.
Barrault, J., Jerome, F., Pouilloux, Y. *Lipid Technol.*, 17, 131-135, 2005.
Blytas, G.C., Frank, H. US Pat. US5401860, 1993 – atribuída à Shell Oil Company.
Bunyakiat, K., Makmee, S., Sawangkeaw, R., Ngamprasertsith, S. *Energy Fuels*, 20, 812-817, 2006.
Catharino, R.R., Milagre, H.M.S., Saraiva, S.A., Garcia, C.M., Schuchardt, U., Eberlin, M.N., Augusti, R., Pereira, R.C.L., Guimarães, M.J.R., de Sá, G.F., Caxeiro, J.M.R., Souza, V. *Energy Fuels*, 21, 3698-3701, 2007.
Chiang, W. H. *Biodiesel – Sebrae*, 2007
Costa Neto, P.R., Rossi, L.F.S. *Quím. Nova*, 23 (4), 531-537, 2000.
Davis, W. R., Tomsho, J., Nikam, S., Cook, E. M., Somand, D., Peliska, J. A. *Biochemistry*, 39, 14279–14291, 2000.
Dorado, M.P., Ballesteros, E., Mittelbach, M., López, F.J. *Energy Fuels*, 18, 1457-1462, 2004.
Encinar, J.M., Gozález, J.F., Rodrigues, J.J., Tejedor, A. *Energy Fuels*, 18, 443-450, 2002.
Ferrari, R.A., Oliveira. V.S., Scabio, A. *Quim. Nova*, 28, 19-23, 2005.
Freedman, B., Pryde, E.H., Mounts, T.L. *J. Am. Oil Chem. Soc.*, 61, 1638, 1984.
Garcia, R., Besson, M., Gallezot, P. *Applied Catalysis A: Gen.*, 127, 165-176, 1995.

Georgogianni, K.G., Kontominas, M.G., Tegou, E., Avlonitis, D, Gergis, V. *Energy Fuels*, 21, 3023-3027, 2007.

Huber, G.W., Shabaker, J. W., Dumesic, J. A. *Science* 300(5628), 2075-2077, 2003.

Karinen, R.S., Krause, A.O.I. *Appl. Cat. A: Gen.*, 306, 128-133, 2006.

Kimura, H., *Appl.Catal. A: Gen.*, 105 (2), 147-158, 1993.

Klepácová, K., Miravec, D., Hájeková, E., Bajus, M. *Petroleum and Coal*, 45, 54-57, 2003.

Klepácová, K., Miravec, E., Bajus, M. *Chem. Pap.*, 60(3), 224,230, 2006.

Kocak, M.S., Ileri, E., Utlu, Z. *Energy Fuels*, 21, 3622-3626, 2007.

Krishna, C.R., Thomassen, K., Brown, C., Butcher, T.A., Anjom, M., Mahajant, D. *Ind. Eng. Chem. Res.*, 46, 8846-8851, 2007.

Lemke, D.W. US Patent US 6620904 (2003).

Mano, E. B.; Mendes, L. C.; *Introdução a Polímeros*, Blücher: São Paulo, 1999.

Medeiros, M.A., Oliveira, L.C.A., Gonçalves, M., Araújo, M.H., Lago, R.M. *Estudo por ESI-MS das reações de oligomerização catalítica do glicerol*, abstract XXI SICAT, Málaga, 2008.

Medeiros, M.A.; Oliveira, D.L.; Sansiviero, M.T.C.; Araujo, M.H. ; Lago, R.M. ; *J. Chem. Technol. Biotechnol.* 2010, 85, 447.

Noureddini, H., Medikonduru, V. *J. Am. Oil Chem. Soc.*, 74(4), 419-425, 1997.

Ott, L., Bicker, M., Vogel, H. *Green Chem.*, 8, 214-221, 2006.

Pachauri, N.; He, B. *Encontro Anual da American Society Biological and Agricultural Engineering*, 2006, *Paper Number: 066223*.

Pagliaro, M., Rossi, M. *The Future of Glycerol* – New uses of a versatile raw material, RSC Publishing: Cambridge, 2008.

Pinto, A.C., Guarieiro, L.L.N., Rezende, M.J.C., Ribeiro, N.M., Torres, E.A., Lopes, W.A., Pereira, P.A.P., Andrade, J.B. *J. Braz. Chem. Soc.*, 16, 1313-1330, 2005.

Puhan, S., Vedaraman, N., Rambrahaman, B.V., Nagarajan, G. 2005. Mahua (*Madhuca indica*) seed oil: a source of renewable energy in India. J Sci Ind Res 64:890–896.

Schuchardt, U., Sercheli, R., Vargas, R.M. *J.Braz. Chem. Soc.*, 9, 199-210, 1998.

Shenoy, M. A., Sabnis, A., D'Melo, D. J., *Pigment & Resin Technology*, 35, 326–333, 2006.

Stein, Y.S., Antal, M. J. J., Jones, M. *J. Anal. and Appl. Pyrolysis*, 4(4), 283-296, 1983.

Stern, R., Hillion, G. Eur. Pat. EP356317, 1990.

Talukder, M.M.R., Beatrice, K.L.M., Sond, O.P., Wu, S.P.J.C., Won, C.J., Chow, Y. *Energy Fuels*, 21, 2007.

Vicente, G., Martinez, M., Aracil, J. *Energy Fuels*, 20, 394-398, 2006.

Vicente, G., Martinez, M., Aracil, J., Estaban, A. *Ind. Eng. Chem. Res.*, 44, 5447-5454, 2005.

Werpy, T., Petersen, G. *Top Value Added Chemicals From biomass*, U.S. Department of Energy. Oak Ridge, Richland, 2004.

Wu, H., Fu, Q., Giles, R., Bartle, J. *Energy Fuels*, 21, 2007.

Zuhair, S.A., *Biotechnol. Prog.*, 21, 1442-1448, 2005.

6

Use of Soybean Oil in Energy Generation

Roberto Guimarães Pereira[1], Oscar Edwin Piamba Tulcan[2],
Valdir de Jesus Lameira[3], Dalni Malta do Espirito Santo Filho[4]
and Ednilton Tavares de Andrade[5]
[1]Fluminense Federal University/TEM/PGMEC/MSG, Niterói, RJ
[2]National University of Colombia, Bogota
[3]INESC, Coimbra
[4]LAFLU/DIMEC/INMETRO, RJ
[5]Fluminense Federal University/TER/PGMEC
[1,4,5]Brazil
[2]Colombia
[3]Portugal

1. Introduction

This chapter deals with the possibility of using soybean oil in energy generation. The environmental, energetic and social-economic aspects are discussed. The steps for obtaining biodiesel from soybean oil are presented as well as the characterization of soybean oil and soybean biodiesel. Results for performance and emissions of using soybean oil and soybean biodiesel in a stationary engine are also presented.

Vegetable oils are obtained predominantly from grains of different plant species. The oil extraction can be made by physical process (pressing) or chemical (solvent). The solvent extraction produces better results, but the more traditional way is physical extraction, which uses mechanical and hydraulic presses to crush the grains. A mixed extraction (mechanical/solvent) can also be done. Selecting the type of extraction depends on two factors: the productive capacity and oil content.

Soybean (*Glycine max* (L.) Merrill) is a very versatile grain that gives rise to products widely used by agro-chemical industry and food industry. Besides is a raw material for extraction of oil for biofuel production. Soybean has about 25% of oil content in grain.

In the agribusiness world, soybean production is, among the economic activities in recent decades, the most prominent. This can be attributed to several factors, such as structuring of a large international market related to trade in products of soybean, oilseed consolidation as an important source of vegetable protein and increased development and delivery of technologies that made possible the expansion of soy exploration for various regions of the world. The largest producers of soybeans are: United States, Brazil, Argentina, China and India.

One possible use of vegetable oil is in the power generation engines. The vegetable oil can be used directly in diesel engines, preferably mixed with diesel. It may also undergo a chemical reaction (transesterification), yielding biodiesel and glycerol. In literature, several works are related to the use of vegetable oil and biodiesel for power generation, as evidenced below.

A review of the use of vegetable oils as fuel in compression ignition (CI) engines is presented (Hossain & Davies, 2010). The review shows that a number of plant oils can be used satisfactorily in CI engines, without transesterification, by preheating the oil and/or modifying the engine parameters. As regards life-cycle energy and greenhouse gas emission analyses, these reveal considerable advantages of raw plant oils over fossil diesel and biodiesel.

It is pointed out (Grau et al., 2010) that straight vegetable oil can be used directly in diesel engines with minor modifications. It is proposed a small-scale production system for self-supply in agricultural machinery.

It is emphasized (Misra & Murthy, 2010) that the ever increasing fossil fuel usage and cost, environmental concern has forced the world to look for alternatives. Straight vegetable oils in compression ignition engine are a ready solution available, however, with certain limitations and with some advantages.

It is presented (Sidibé et al., 2010) a literature review on the use of crude filtered vegetable oil as a fuel in diesel engines. It is emphasized the potential and merits of this renewable fuel. Typically, straight vegetable oils produced locally on a small scale, have proven to be easy to produce with very little environmental impact. However, as their physico-chemical characteristics differ from those of diesel oil, their use in diesel engines can lead to a certain number of technical problems over time.

A review on the utilization of used cooking oil biodiesel is presented (Enweremadu & Rutto, 2010). There were no noticeable differences between used cooking oil biodiesel and fresh oil biodiesel as their engine performances, combustion and emissions characteristics bear a close resemblance.

A review on biodiesel production, combustion, emissions and performance is shown (Basha et al., 2009). A vast majority of the scientists reported that short-term engine tests using vegetable oils as fuels were very promising, but the long-term test results showed higher carbon built up, and lubricating oil contamination, resulting in engine failure. It was reported that the combustion characteristics of biodiesel are similar as diesel. The engine power output was found to be equivalent to that of diesel fuel.

An overview of political, economic and environmental impacts of biofuels is presented (Demirbas, 2009a). Biofuels provide the prospect of new economic opportunities for people in rural areas in oil importer and developing countries. Renewable energy sources that use indigenous resources have the potential to provide energy services with zero or almost zero emissions of both air pollutants and greenhouse gases. Biofuels are expected to reduce dependence on imported petroleum with associated political and economic vulnerability, reduce greenhouse gas emissions and other pollutants, and revitalize the economy by increasing demand and prices for agricultural products.

It is emphasized (Sharma & Singh, 2009) that Biodiesel, a renewable source of energy seems to be an ideal solution for global energy demands.

The soy, which cultivation is widespread in the world, can be used to produce vegetable oil in order to be used as fuel, according some following examples.

It is pointed out (Liu et al., 2008) that Soybean (*Glycine max* (L.) Merrill), one of the most important crops in China, has been known to man for over 5000 years. The largest production areas in China are in the Northeast China's three provinces, where soybean is spring seeded and grown as a full-season crop.

A study about large-scale bioenergy production from soybeans in Argentina is presented (van Dam et al., 2009), showing the potential and economic feasibility for national and international markets.

Inedible vegetable oils and their derivatives can, also, be used as alternative diesel fuels in compression ignition engines (No, 2011).

The advantages of using vegetable oil as fuel are evidenced in the literature (Alonso et al. 2008; Misra & Murthy, 2010) as described below.

One of the main advantages of vegetable oils is its life cycle, as it is a closed cycle. Crops take CO_2 via photosynthesis from the atmosphere. Oil is extracted from these crops which can be used directly as fuel or, after the pertinent transformations, a fuel can be obtained which, when burned, generates CO_2 that can be absorbed by the plants.

Environmental advantages of vegetable oils are: minor influence on the greenhouse effect when used instead of fossil fuels; biodegradability; lower sulfur and aromatic content; at lower percentages of vegetable oil blends with diesel have shown better results than the fossil diesel in terms of engine performance and exhaust emissions.

Energetic advantages of vegetable oils are: renewable energy source; reduction of the dependence on fossil fuels; positive energy balance; the fuel production technology is simple and proven; heating values of various vegetable oils are nearly 90% to those of diesel fuel; higher flash point allows it to be stored at high temperatures without any fire hazard; additional oxygen molecule in its chemical structure helps in combustion process.

Social-economic advantages of vegetable oils are: use of marginal land for energy purposes; maintains employment and income levels in rural areas; avoids population migrations; encourages job creation in various agro-industries; contributes to the creation of new jobs; straight vegetable oils are available normally in rural area where its usage is advantageous especially in smaller engines in agricultural sector; improve the living conditions of the rural people and offer greater income opportunities through enhanced rural employment; plant leafs and cake can be used as organic manure which can be source of additional income farmers; selected crops can be grown on arid and semi-arid lands which are presently not cultivable; having carbon credit value (Kyoto protocol).

The main disadvantages of vegetable oils, compared to diesel fuel, are higher viscosity, lower volatility, and the reactivity of unsaturated hydrocarbon chains. The problems meet in long term engine use (Misra & Murthy, 2010). The land required for commercial production is vast.

Many countries have potential to produce biodiesel from vegetable oil. The top 10 countries in terms of absolute potential are: Malaysia, Indonesia, Argentina, USA, Brazil, Netherlands, Germany, Philippines, Belgium and Span (Sharma & Singh, 2009).

Some reasons for sufficiently strengthen the program for the use of biofuels are: the variation in prices for oil, which in recent years has fluctuated between $ 40 and 150 per barrel, the political and social pressure in order to reduce the emission of gases that cause global warming, the development of the market for carbon credits, the need to strengthen the agriculture industry and the possibility of strengthening the energy matrix, reducing dependence on foreign energy sources and giving relief to the trade balance with clear effects on the macroeconomics of the country.

The possibility to produce hydrocarbons from different raw material to replace the diesel makes the definition of biodiesel a complex and legal nature. Many documents of an academic nature define biodiesel as a monoalkyl ester of vegetable oil or animal fat, but official documents and international standards are more specific, defining the process by which one can obtain the ester and the characteristics that it must have to be considered biodiesel.

The regulations for biodiesel have been developed in different countries where its use is permitted. In the U.S. the standard for biodiesel is set by the technical standard ASTM D 6751, the European Union is related with the standard EN 14214 and in Brazil is set in the ANP (National Petroleum Agency) No. 07 from 19.03.2008.

Biodiesel can be produced from different oilseeds, according to the design of the plant, market conditions and availability of raw material in the region. Each oilseed production has a different culture method and a different destination. The disposal of oils for biodiesel production must take into account beyond the capacity of oil production, market competitiveness in relation to the cost of oil and the price of a barrel of fuel. Biodiesel can be obtained, too, from used cooking oil, animal fats and algae.

Just as vegetable oil, the use of biodiesel as fuel in partial or total replacement to diesel has many advantages that have been highlighted in the literature (Demirbas, 2009b; Pereira et al., 2007).

Environmental advantages of biodiesel are: greenhouse gas reductions; biodegradability; higher combustion efficiency; improved land and water use; carbon sequestration; lower sulfur content; lower aromatic content; less toxicity.

Energetic advantages of biodiesel are: supply reliability; higher flash point; reducing use of fossil fuels; ready availability; renewability.

Social-economic advantages of biodiesel are: sustainability; fuel diversity; increased number of rural manufacturing jobs; increased income taxes; increased investments in plant and equipment; agricultural development; international competitiveness; reducing the dependency on imported petroleum.

2. Theoretical foundations

All vegetable oils consist primarily of triglycerides. The triglycerides have a three-carbon backbone with a long hydrocarbon chain attached to each of the carbons. These chains are attached through an oxygen atom and a carbonyl carbon, which is a carbon atom that is double-bonded to second oxygen. The differences between oils from different sources relate to the length of the fatty acid chains attached to the backbone and the number of carbon–carbon double bonds on the chain. Most fatty acid chains from plant based oils are 18 carbons long with between zero and three double bonds. Fatty acid chains without double bonds are said to be saturated and those with double bonds are unsaturated (Misra & Murthy, 2010).

In general, vegetable oils are made especially of fatty acids with chains between 12 and 24 carbons: Lauric (C12:0); Myristic (C14:0); Palmitic (C16:0); Palmitoleic (C16:1); Stearic (C18:0); Oleic (C18:1); Linoleic (C18:2); Linolenic (C18:3); Arachidic (C20:0); Gadoleic (C20:1); Behenic (C22:0); Erucic (C22:1); Lignoceric (C24:0). The proportions of the fatty acid composition can be determined by gas chromatography method.

Triglycerides are hydrocarbons with physical and chemical characteristics that can be classified as liquid fuels in the majority.

Vegetable oils have a high heating value, near the heating value of conventional diesel fuel, which makes them an important energy resource.

To improve the properties of vegetable oils and in order to use them as substitutes for diesel fuel, triglycerides are converted into esters (biodiesel) by transesterification process, modifying the composition of molecules and changing the characteristics of the fluids. The composition of the resulting esters has approximately the same proportion of fatty acids present in oils before the transformation process.

Biodiesel is known as monoalkyl, such as methyl and ethyl, esters of fatty acids. Biodiesel can be produced from a number of sources, including recycled waste vegetable oil, oil crops and algae oil. Biodiesels play an important role in meeting future fuel requirements in view of their nature (less toxic), and have an edge over conventional diesel as they are obtained from renewable sources (Demirbas, 2009b).

The transesterification process is used to transform triglycerides into esters, or biodiesel. In the process of transesterification, the triglycerides found in different kinds of oils and fats react with alcohol, usually methanol or ethanol to produce esters and glycerin. For the reaction to occur it is necessary to use a catalyst. Processes performed the supercritical conditions of methanol transesterification can be conducted without the catalyst. (Demirbas, 2003; Saka & Kusdiana, 2001)

In the transesterification process, a triglyceride molecule reacts with an alcohol molecule causing the separation of one of the fatty acids of the triglyceride, producing a diglyceride and an ester. This diglyceride reacts with a second molecule of alcohol that takes another fatty acid, forming a second ester and a monoglyceride. Finally a third molecule of alcohol reacts with the monoglyceride, forming the third ester and a molecule of glycerin. The reactions occurring are reversible, and the stoichiometric ratio is three moles of alcohol for each mole of oil being processed. The reaction can be carried out with concentrations of alcohol in excess, as this reduces time and increases the conversion efficiency of the process. (Lang et al., 2001)

The transesterification process using methanol and base catalyst is the most commonly used to produce biodiesel. The catalysts commonly used are sodium hydroxide (NaOH) or potassium hydroxide (KOH). The catalyst is diluted in alcohol and then added the oil. The product is the ester (biodiesel) and crude glycerin. The glycerin is separated from the ester by decanting or centrifuging.

Repeated washing processes are performed by adding acidified water in the reaction products. This mixture is stirred lightly and serves to remove residual glycerine soap, catalyst and serves as a neutralizing agent of the fuel. The washing process is repeated until the biodiesel becomes clear. The main drawbacks of the process are the presence of water on some of the reagents and the high level of free fatty acids in the raw material. In both cases, the transesterification reaction is replaced by a saponification reaction.

The transesterification process can also be developed using acid catalysts and no homogeneous catalysts. In the case of acid catalysts, the process times are longer, but do not have drawbacks with the water content and free fatty acids. In the case of no homogeneous catalysts, these bring benefits in: reducing the washing process; product separation and reuse of catalysts.

Methanol and ethanol are produced on an industrial scale and their use in transesterification reactions has been reported.

The biodiesel used in several countries of Europe and in the United States is a mixture of methyl esters. Methanol is usually obtained from non-renewable fossil fuels, but can also be obtained by distillation of wood; this route, however, produces smaller quantities. The technology of biodiesel production using methanol is fully understood, however, this route has the disadvantage that methanol is extremely toxic.

The transesterification using ethanol is more difficult because the use of alcohol, even if anhydrous, involves problems in the separation of glycerin from the reaction medium. However, the use of ethanol is advantageous, since it is produced on a large scale in countries like Brazil and the United States, being from a renewable source of energy, resulting in environmental gains could generate carbon credits. As for the difficulties in the separation of phases in reactions employing ethanol in biodiesel synthesis, they can be bypassed by adjustments in reaction conditions.

3. Soybean oil and soybean biodiesel obtaining and characterization procedures

The fuel blends used were prepared at the Thermo-sciences Laboratory of Mechanical Engineering Department at Fluminense Federal University (UFF).

It was used Oxx as the nomenclature to designate mixtures of soybean oil with diesel, being xx the percentage by volume of vegetable oil added to diesel. The following mixtures were used: O5; O10; O15 and O20, besides pure diesel (O0).

It was used Bxx as the nomenclature to designate mixtures of biodiesel with diesel, being xx the percentage by volume of biodiesel added to diesel. The following mixtures were used: B5; B10; B15; B20; B50 and B75, besides pure diesel (B0) and pure biodiesel (B100).

Diesel fuel used is from the Laboratory of Distributed Power Generation at Fluminense Federal University. This fuel is used as reference for the tests.

The soybean oil used in the tests was obtained in the food market and the biodiesel used was produced at the Thermo-sciences Laboratory of Mechanical Engineering Department at Fluminense Federal University.

Refined oils are free of substances inhibiting the transesterification process. They have low amount of free fatty acids, less than 0.5%. The beta-carotene and phosphatides of the raw material are eliminated in the process of bleaching and degumming the oil. Thus the processing of these oils for conversion to biodiesel is a process that does not present major technical difficulties. The process is performed at atmospheric pressure.

3.1 Basic procedures for biodiesel obtaining

A guide for biodiesel obtaining from refined oils is given as follows.

Raw material:

- 100 mL of refined vegetable oil (commercial oil, whose acidity is less than 0.5%, degummed, microfiltered, deodorized and bleached);
- 25 mL of anhydrous methyl alcohol;
- 1 g of potassium hydroxide;

Steps:

1. Put 100 mL of vegetable oil in a glass container and lead to heating to a temperature of 45°C;
2. Place the container on the balance. Tare, and put 1g of KOH in it and close the container to avoid hydration of the reagent;
3. Measure a volume of 25 mL of anhydrous methyl alcohol using a test tube (methanol is toxic and must be handled with care in appropriate place);
4. Mixing the methanol with KOH to achieve a uniform solution of methoxide;
5. Add the methoxide to the solution of vegetable oil and stir for a period of two hours;
6. Turn off the mixer and check whether it produces a phase separation, it can be seen that a fluid dark (glycerin) deposits at the bottom of the container;

7. After checking the reaction, put the fluid in the decanting funnel, allow the glycerin settle for a period exceeding 30 minutes. Separate the glycerin and ester (biodiesel) produced in clean containers;

8. Wash the decanting funnel to remove the glycerin stuck on the walls and then fill with the ester obtained;

9. Perform the washing process of ester, adding 30 mL of water, preferably hot (50-60°C), inside the funnel, stirring to ensure the contact of two fluids;

10. Decant the water and remove it from the funnel;

11. Perform the washing process three more times to ensure complete removal of the glycerin;

12. Perform the drying process putting the ester in an oven, heating up to 110 ° C for 10 minutes

13. Cool and bottle the product (biodiesel)

3.2 Characterization of soybean oil and soybean biodiesel and stationary engine tests

The properties of soybean oil, soybean biodiesel and diesel were determined at the Thermosciences Laboratory and Rheology Laboratory of Fluminense Federal University. The heating values were determined at the Laboratory of fuels at the National University of Colombia. The characterization tests followed the standards, as detailed in Table 1.

The tests conducted in the stationary engine were made at constant speed of 3600 rpm and variable power in the Fluminense Federal University

For each fuel tested, a test was performed and repeated. Performance data and emissions were measured continuously during the test and the series of data were analyzed to obtain values representative of engine performance

PROPERTY	STANDARD
Viscosity	ASTM D 445 Standard Test Method for Kinematic Viscosity of Transparent and Opaque Liquids
Density	ASTM D 4052 Density and Relative Density of Liquids by Digital Density Meter.
Flash point	ASTM D 93 Standard Test Methods for Flash Point by Pensky-Martens Closed Cup Tester
Cloud point	ASTM D2500 Test Method for Cloud Point of Petroleum Products
Pour point	ASTM D97 Test Method for Pour Point of Petroleum Products
Copper strip corrosion	ASTM D130 Test Method for Copper strip corrosion of Petroleum Products
Heat of combustion	ASTM 240 Test Method for heat of combustion of Petroleum Products.

Table 1. Technical standards associated with the characterization tests

The stationary engine used (Figure 1) is formed by an engine, a generator and a control panel, with the possibility of producing electricity at 115V and 230V. The generator has a control system to regulate the motor rotation. The characteristics of the diesel engine are: 3600rpm; four-stroke; direct injection; one cylinder; air cooling system; 0.211L displacement volume; 2.0kW maximum output; 1.8kW nominal power; 2.5L fuel capacity and 47 kg weight.

The engine was modified in order to have a fuel consumption control by gravity, changing the original fuel tank by a remote tank, being possible to be placed on a balance

The electrical load was simulated on a load bank, where 150W power lamps were activated to modify the load (Figure 1). The measurements of instantaneous power, current frequency, voltage and electrical current were made using a measuring device (CCK 4300) manufactured by CCK Automation Ltda (São Paulo, Brazil).

Emissions were measured using the gas analyzer Greenline 8000 built by Eurotron Instrument S.A. The equipment has measurement system of gas concentration by non-dispersive infrared (NDIR) and electrochemical method, in addition to measuring temperature, pressure and temperature of gases. The equipment has RS232 communication system for data acquisition and algorithms for calculating the efficiency indicators for different fuels. The resolution and the error limits of the equipment for measured gases are: electrochemical CO, 1 ppm and ± 10 ppm; NDIR CO_2, 0.01% and ± 0.3%; electrochemical NO, 1 ppm and ± 5 ppm; electrochemical NO_2, 1 ppm and ± 5 ppm and calculated NOx, 1 ppm. The SO_2 measurements were made by the method of molecular absorption spectroscopy (Tulcan, 2009)

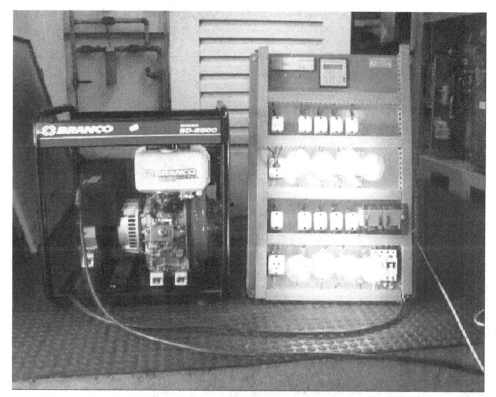

Fig. 1. Stationary engine and control panel (operating).

4. Energy generation using soybean oil and soybean biodiesel

The yield of the process of producing biodiesel from soybean was 0.91L of biodiesel per 1L of used oil.

Table 2 shows the properties of diesel, soybean oil and soybean biodiesel, obtained in accordance with the standards shown in Table 1.

The expanded uncertainty of measurements are: density = ± 0.00008 kg/L; viscosity = ± 0.006 mm²/s; flash point = ± 2.1 °C; cloud point = ± 1.5°C and pour point = ± 1.8°C (Santo Filho, 2010a; Abreu, 2010).

The weight composition of fatty acids found in soybean oil was: C16:0=11.6%, C16:1=0.1%, C18:0=3.2%, C18:1=20.4%; C18:2=59.7% and C18:3=5%.

Based on the physicochemical characterization performed in the soybean oil and soybean biodiesel, some correlations were determined.

PROPERTY	DIESEL	SOYBEAN OIL	SOYBEAN BIODIESEL
Density kg/L (20 °C)	0.85519	0.92037	0.88230
Viscosity mm²/s (40 °C)	4.689	30.787	4.161
Flash point °C	82	332	150
Cloud point °C	2	-2	0
Pour point °C	-12	-14	-6
Copper strip corrosion	1a	1b	1a
Higher heating value kJ/kg	42800	-	41685

Table 2. Properties of diesel, soybean oil and soybean biodiesel.

The equation that best describes the behavior of flash point for blending diesel-soybean biodiesel is (Tulcan, 2009):

$$FP(mixture) = \left[(FP(a) - FP(d)).(\%mixture)^{\left(\frac{FP(a)}{FP(d)}\right)} \right] + FP(d) \qquad (1)$$

where: FP (a) is the flash point of the additive (biodiesel) and FP (d) is the flash point of diesel.

The best fitting function representing the behavior of soybean oil density is, as follows (Santo Filho et al., 2010b):

$$\rho = -0.00069T + 0.93420 \qquad (2)$$

where: ρ is the density of soyben oil (kg/L) and T is the temperature (°C).

The best fitting function representing the behavior of soybean biodiesel density is, as follows (Santo Filho et al., 2010c):

$$\rho = -0.00073T + 0.89691 \qquad (3)$$

where: ρ is the density of soyben biodiesel (kg/L) and T is the temperature (°C).

The behavior of soybean oil viscosity is, as follows (Santo Filho, 2010a):

$$v = (0.1115)e^{(953.1562/(T+129.5732))} \tag{4}$$

where: the viscosity of soyben oil is given in mm²/s and the temperature (T) is in (⁰C).
The behavior of soybean biodiesel viscosity is, as follows (Santo Filho, 2010a):

$$v = (0.1246)e^{(609.1440/(T+133.6409))} \tag{5}$$

where: the viscosity of soyben biodiesel is given in mm²/s and the temperature (T) is in ⁰C.
Figures 2-7 show the values of specific fuel consumption (SFC) and emissions of NO, NO_x, CO, CO_2 and SO_2 for diesel, soybean biodiesel and mixtures of diesel-soybean oil and diesel-soybean biodiesel. The reported values represent the average for four values of load (400W, 700W, 1000W and 1300W).
Figure 2 shows the behavior of SFC for diesel-soybean oil mixtures and diesel-soybean biodiesel blends. The specific fuel consumption is lower for mixtures of 5% soybean oil than that for diesel. For this percentage, the oil has an oxygenating effect which improves engine performance, with an average of 1.9% decrease from the SFC. For larger percentages of mixture, the SFC increases, indicating a drop in engine performance. This is a consequence of lower heating value of soybean oil and of the increasing of the difficulties to burn fuel in the combustion chamber, requiring more fuel. For mixtures of 20% soybean oil, the increase in SFC is 4.5%. In the case of diesel-soybean biodiesel blends, a slight decrease in the SFC can be observed for smaller proportions of the mixture (5% to 10% soybean biodiesel). This decrease is due to the oxygenating capacity of biodiesel. For larger values of the mixture (15% to 100% soybean biodiesel), the SFC increases. This increase in SFC is due to the lower heating value biodiesel, requiring more fuel. As the mixing ratio increases, the specific fuel consumption increases. The SFC hits an increase of 14% for the use of pure biodiesel compared to diesel.
Figures 3 and 4 present the average values of NO and NO_x emissions for different mixing ratios of soybean oil with diesel and soybean biodiesel with diesel. The NO and NO_x emissions increase with the use of blends up to 10% of soybean oil in diesel. From this amount of mixture, the NO and NO_x emissions decrease reflecting a decrease in the temperature of combustion chamber. However, the values are still higher than the emissions of diesel. For diesel-soybean biodiesel blends, the production of NO and NO_x is higher for blends superior to 5% than for diesel. Emissions of NO and NO_x grow rapidly for blends above 15% of soybean biodiesel. For mixtures between 20% and 75% soybean biodiesel, the emission levels of NO are an average of 200 ppm. For pure biodiesel the emission levels fall, this is a consequence of the low temperature in the combustion chamber due to the lower heating value of fuel.
Figure 5 shows the mean values of CO for soybean oil and soybean biodiesel blended in diesel. It may be noted that in proportions of up to 5% of soybean oil, the mixtures have an advantage in relation to diesel. For higher proportions, the CO emission increases to a level of 405 ppm for the mixture with 20% of soybean oil, 18% higher than the levels achieved by diesel emissions. The increase in the amount of CO shows a less efficient combustion. In the case of diesel-soybean biodiesel blends, it can be observed that the emission of CO for mixtures was lower than for diesel. The CO emission decreases by increasing the proportion of soybean biodiesel in the blend. In the case of pure biodiesel, the reduction in CO emission was 21% compared with diesel.

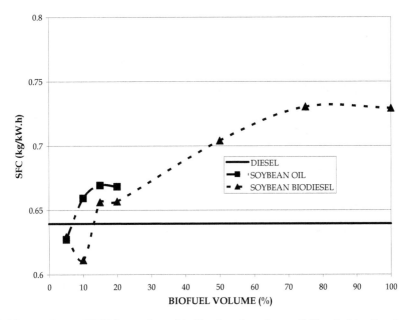

Fig. 2. Mean values of SFC for soybean biodiesel and soybean oil blended in diesel.

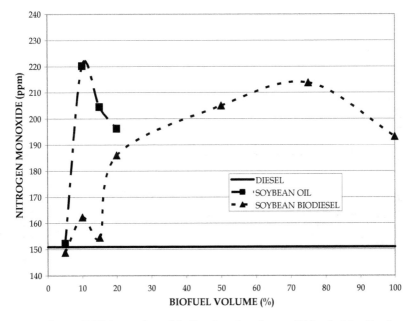

Fig. 3. Mean values of NO for soybean biodiesel and soybean oil blended in diesel.

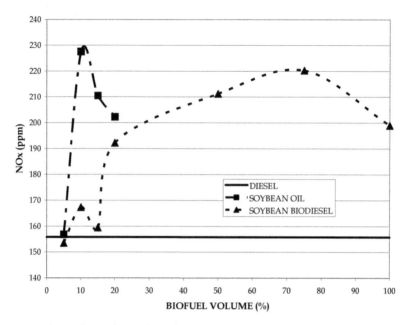

Fig. 4. Mean values of NO_x for soybean biodiesel and soybean oil blended in diesel.

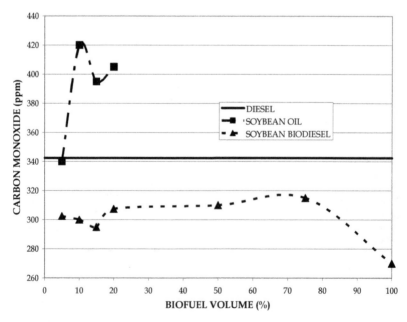

Fig. 5. Mean values of CO for soybean biodiesel and soybean oil blended in diesel.

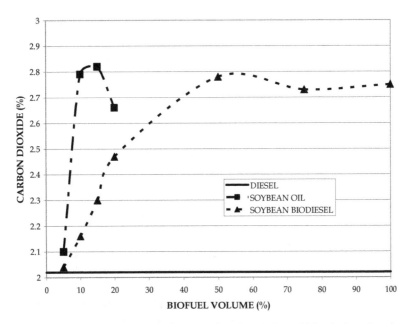

Fig. 6. Mean values of CO_2 for soybean biodiesel and soybean oil blended in diesel.

Figure 6 shows the CO_2 average emissions for soybean oil and soybean biodiesel blended in diesel. It can be observed that the percentages of CO_2 emissions for the mixtures are always higher than for diesel. Compared with diesel, a mixture of 15% of soybean oil increases the production of CO_2 by 40% and the mixture of 20% of soybean oil by 31%. The production of CO_2 increases with the addition of soybean biodiesel reaching a value of 2.77% to 50% soybean biodiesel, representing an increase of 37% compared to diesel. From this value a slight reduction in CO_2 occurs.

Figure 7 shows average results of the SO_2 emission obtained by the method of molecular absorption spectroscopy. In the figure it can be observed that, in some cases, the addition of soybean oil increases the production of SO_2. The production of SO_2 is caused by oxidation of sulfur in the fuel. Although the addition of soybean oil reduces the presence of sulfur in fuel, the cause of production of sulfur oxides may be due to rising temperatures in the combustion chamber and the consequent degradation of lubricating oils in the engine and volatilization of no burning fuel inside the combustion chamber. For fuel mixtures in proportions greater than 10% of soybean oil, the levels of SO_2 emissions begin to decrease. For mixtures of 15% to 20% of soybean oil, the levels of SO_2 emissions are lower than for diesel. In the case of 20% soybean oil, the reduction of SO_2 is 69%. For diesel-soybean biodiesel blends, the emission of sulfur dioxide, in some cases, is higher than for diesel, as also happened in the case of mixtures with soybean oil. Moreover, one can observe that the emission of SO_2 decreases for fuel mixtures in proportions greater than 50% of soybean biodiesel. In the case of 100% soybean biodiesel, the reduction of SO_2 is 71%. The presence of SO_2 in the burning of pure biodiesel can be due to burning of lubricating oil and residual diesel into the combustion chamber.

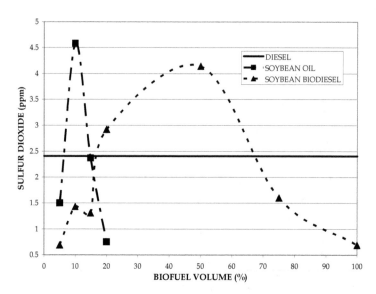

Fig. 7. Mean values of SO$_2$ for soybean biodiesel and soybean oil blended in diesel.

Figures 8 to 13 show the emission behavior as a function of load in the stationary engine for mixtures of 20% soybean oil and 20% soybean biodiesel with diesel.

The behavior of the SFC is similar in the cases of diesel and the mixtures 20% soybean oil-diesel and 20% soybean biodiesel-diesel. As shown in Figure 8 the value of SCF for the fuels studied decreases with increasing load.

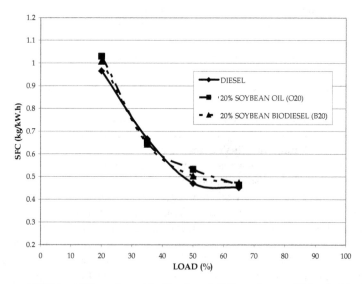

Fig. 8. Values of SFC for 20% soybean biodiesel and 20% soybean oil blended in diesel.

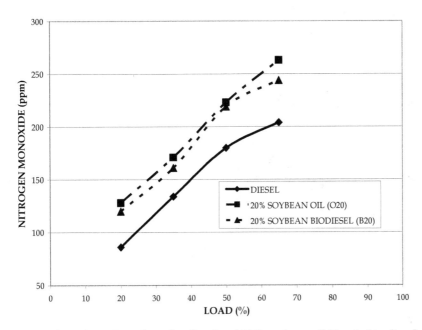

Fig. 9. Values of NO for 20% soybean biodiesel and 20% soybean oil blended in diesel.

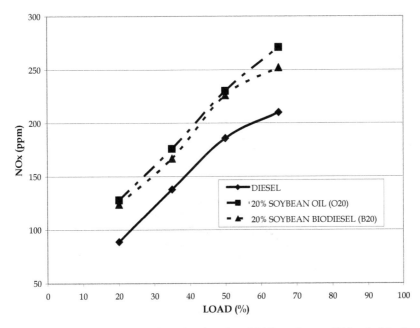

Fig. 10. Values of NO$_x$ for 20% soybean biodiesel and 20% soybean oil blended in diesel.

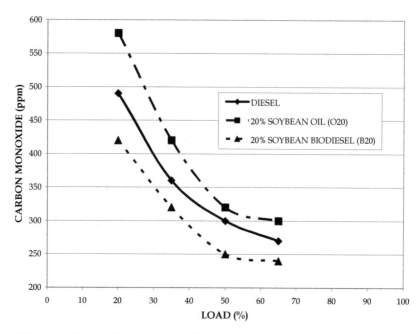

Fig. 11. Values of CO for 20% soybean biodiesel and 20% soybean oil blended in diesel.

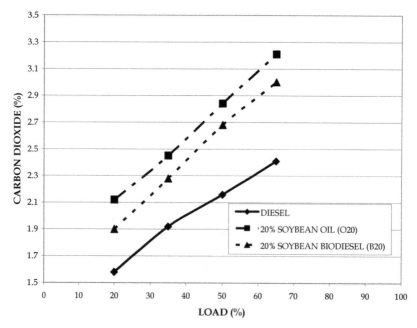

Fig. 12. Values of CO_2 for 20% soybean biodiesel and 20% soybean oil blended in diesel.

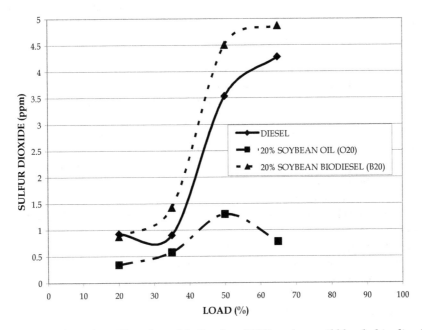

Fig. 13. Values of SO_2 for 20% soybean biodiesel and 20% soybean oil blended in diesel.

The Figures 9 and 10 show that the emission of NO and NO_x, for mixtures of 20% soybean oil-diesel and 20% soybean biodiesel-diesel, are higher than those of diesel at all loads studied, probably due to increased temperature in the combustion chamber.

In Figure 11 it can be observed that the CO emissions decrease with increasing load for all fuels tested, showing a better combustion in these cases. The lower values of CO occur with the 20% blend of soybean biodiesel (B20).

As shown in Figure 12, as the load increases the amount of CO_2 emissions increases for all fuels studied.

The emissions of SO_2 with the load are shown in Figure 13. The lowest emissions occur at lower loads.

5. Conclusion

Soybean oil and soybean biodiesel can be added to diesel fuel to be burned in combustion engines. These compounds have an oxygenate capacity that is useful to improve engine performance, but this ability only gives you an edge when the mix ratio is 5% for vegetable oil and 10% for biodiesel. The gains made in reducing the SFC using the oxygenating additives affect about 2% in the case of 5% soybean oil blended with diesel and about 4.5% for 10% soybean biodiesel blended with diesel. Using a larger proportion of mixture generates increases in SFC by 9% on average when pure biodiesel is used, and 3% when mixture of 20% soybean oil is used.

The emission of NO and NO_x increases with the addition of oxygenated components (vegetable oil and biodiesel). The use of 20% soybean oil blended with diesel (O20) increases the NO emission by 30%. But the use of pure soybean biodiesel (B100) promotes an increase in NO emission by 28%.

The addition of soybean oil in diesel reduces emissions of CO only for mixtures of up to 5% soybean oil. In the case of biodiesel addition in diesel, CO emissions decrease with the mixture reaching a 21% reduction, when pure soybean biodiesel (B100) is used.

The addition of soybean oil and soybean biodiesel in diesel increases the emission of CO_2 which however is compensated by the absorption of CO_2 by the plants (raw material for production of vegetable oil and biodiesel).

The addition of soybean oil and soybean biodiesel in diesel reduces the sulfur content in fuel and consequently reduces the emission of sulfur dioxide.

Soybean oil can be successfully applied in CI engine blending with diesel up to 20% of soybean oil. Soybean oil can also be converted in biodiesel and applied in CI engines neat or blended with diesel in any proportion. Concerning the exhaust emissions it is better use soybean biodiesel or blends of soybean biodesel with diesel instead of blends of soybeans oil with diesel.

The possibility of using soybean oil in power generation, leads to the concept of energy farms soybeans that can be an opportunity for farmers who can not meet the quality standards required for selling the soybeans to food industry.

6. Acknowledgment

The authors are grateful to the National Research Council of Brazil, CNPq, for the financial support received.

7. References

Abreu, F.L.B. (2010). Power generation and analysis of emissions in stationary engine using biodiesel, blends of biodiesel and blends of biodiesel containing additive, produced via ethylic route and its characterization, (in portugues), *Ph.D. Thesis*, Fluminense Federal University, Niterói, RJ, Brazil, December 2010

Alonso, J.; Sastre, J.; Ávila, C. & López, E. (2008). A Note on the Combustion of Blends of Diesel and Soya, Sunflower and Rapeseed Vegetable Oils in a Light Boiler. *Biomass and Bioenergy*, Vol. 32, Issue 9, (September 2008), pp. 880-886,ISSN 0961-9534

Basha, S.; Gopal, K. & Jebaraj, S. (2009). A Review on Biodiesel Production, Combustion, Emissions and Performance. *Renewable and Sustainable Energy Reviews*, Vol. 13, Issues 6-7, (August-September 2009), pp. 1628-1634, ISSN 1364-0321

Demirbas, A. (2003). Biodiesel Fuels from Vegetable Oils via Catalytic and non-Catalytic Supercritical Alcohol Transesterifications and other Methods and Survey. *Energy Conversion and Management*, Vol 44, Issue 13, (August 2003), pp. 2093-2109, ISSN 0196-8904

Demirbas, A. (2009a). Political, Economic and Environmental Impacts of Biofuels: A Review. *Applied Energy*, Vol. 86, Supplement 1, (November 2009), pp. S108-S117, Bio-fuels in Asia, ISSN 0306-2619

Demirbas, A. (2009b). Biofuels Securing the Planet's Future Energy Needs. *Energy Conversion and Management*, Vol. 50, Issue 9, (September 2009), pp. 2239-2249, ISSN 0196-8904

Enweremadu, C. & Rutto, H. (2010). Combustion, Emission and Engine Performance Characteristics of Used Cooking Oil Biodiesel -A Review. *Renewable and Sustainable Energy Reviews*, Vol. 14, Issue 9, (December 2010), pp. 2863-2873, ISSN 1364-0321

Grau, B.; Bernat, E.; Antoni, R.; Jordi-Roger, R. & Rita, P. (2010). Small-Scale Production of Straight Vegetable Oil from Rapeseed and its use as Biofuel in the Spanish Territory. *Energy Policy*, Vol. 38, Issue 1, (January 2010), pp. 189-196, ISSN 0301-4215

Hossain, A. & Davies, P. (2010). Plant Oils as Fuels for Compression Ignition Engines: A Technical Review and Life-Cycle Analysis. *Renewable Energy*, Vol. 35, Issue 1, (January 2010), pp. 1-13, ISSN 0960-1481

Lang, X.; Dalai, A.; Bakhshi, N.; Reaney, M. & Hertz, P. (2001) Preparation and Characterization of Bio-Diesels from Various Bio-Oils. *Bioresource Technology*, Vol. 80, Issue 1, (October 2001), pp. 53-62, ISSN.0960-8524

Liu, X.; Jin, J.; Wang, G. & Herbert, S.(2008). Soybean Yield Physiology and Development of High-Yielding Practices in Northeast China. *Field Crops Research*, Vol. 105, Issue 3, (February 2008), pp. 157-171, ISSN 0378-4290

Misra, R. & Murthy, M. (2010). Straight Vegetable Oils usage in a Compression Ignition Engine - A Review. *Renewable and Sustainable Energy Reviews*, Vol. 14, Issue 9, (December 2010), pp. 3005-3013, ISSN 1364-0321

No, S. (2011), Inedible Vegetable Oils and their Derivatives for Alternative Diesel Fuels in CI Engines: A Review. *Renewable and Sustainable Energy Reviews*, Vol. 15, Issue 1, (January 2011), pp. 131-149, ISSN 1364-0321

Pereira, R.; Oliveira, C.; Oliveira, J; Oliveira, P., Fellows, C. & Piamba, O. (2007). Exhaust Emissions and Electric Energy Generation in a Stationary Engine using Blends of Diesel and Soybean Biodiesel. *Renewable Energy*, Vol. 32, Issue 14, (November 2007), pp. 2453-2460, ISSN 0960-1481

Saka, S. & Kusdiana,D.(2001) Biodiesel Fuel from Rapeseed Oil as prepared in Supercritical Methanol. *Fuel*, Vol. 80, Issue 2, (January 2001), pp. 225-231, ISSN.0016-2361

Santo Filho, D. M. E. (2010a). Metrology applied to analysis of biodiesel (in portugues), *Ph.D. Thesis*, Fluminense Federal University, Niterói, RJ, Brazil, July 2010

Santo Filho, D; Abreu, F; Pereira, R.; Santos Junior, J; Siqueira, J.; Ferreira, P.; Barbosa, T.; Lima, L. & Baldner, F. (2010b) The Influence of the Addition of Oils in the Diesel Fuel Density. *Journal of ASTM International (Online)*, Vol. 7, Issue 8, (September 2010), pp. 1-9, ISSN 1546-962X

Santo Filho, D. ; Abreu, F.; Pereira, R..; Siqueira, J.; Santos Junior, J. & Daroda, R. (2010c). Characterization of Density of Biodiesel from Soybean, Sunflower, Canola and Beef Tallow in Relation to Temperature, Using a Digital Density Meter, with a Metrological Point of View. *Journal of ASTM International (Online)*, Vol. 7, Issue 2, (February 2010), pp. 1-6, ISSN 1542-962X

Sharma, Y. & Singh, B.(2009). Development of Biodiesel: Current Scenario. *Renewable and Sustainable Energy Reviews*, Vol. 13, Issues 6-7, (August-September 2009), pp. 1646-1651, ISSN 1364-0321

Sidibé, S.; Blin, J; Vaitilingom, G. & Azoumah. Y. (2010). Use of Crude Filtered Vegetable Oil as a Fuel in Diesel Engines State of the Art: Literature Review. *Renewable and*

Sustainable Energy Reviews, Vol. 14, Issue 9, (December 2010), pp. 2748-2759, ISSN 1364-0321

Tulcan, O. E. P., Performance of a stationary engine using different biofuels and emissions evaluation (in portugues), *Ph.D. Thesis*, Federal Fluminense University, Niterói, RJ, Brazil, November 2009

van Dam, J.; Faaij, A.; Hilbert, J.; Petruzzi, H. & Turkenburg, W. (2009). Large-Scale Bioenergy Production from Soybeans and Switchgrass in Argentina: Part A: Potential and Economic Feasibility for National and International Markets. *Renewable and Sustainable Energy Reviews*, Vol. 13, Issue 8, (October 2009), pp. 1710-1733, ISSN 1364-0321

7

Machine Vision Identification of Plants

George. E. Meyer

University of Nebraska, Department of Biological Systems Engineering,
USA

1. Introduction

Weedy and invasive plants cost Americans billions of dollars annually in crop damage and lost earnings. Various Western states have reported annual weed control costs in the hundreds of millions of dollars. Herbicides account for more than 72 per cent of all pesticides used on agricultural crops. $4 billion was spent herbicides in the US in 2006 and 2007 (Grube, et al, 2011). The USDA Economic Research Service reported that adoption of herbicide-tolerant soybeans had grown to 70% from 1996 to 2001, yet significant impacts on farm financial net returns attributable to adoption has yet to be documented. Nebraska is part of regional strategic pest plan published in 2002. During 2001, 97% of the soybean acres in Nebraska were treated with herbicides. One means of improving economic benefit is to develop more efficient management inputs, which may be accomplished with better selection of the kind of pesticide and/or site-specific application of pesticides. Moreover, measuring the impact of various management inputs often depends on manual visual assessment and perhaps this could be automated. One method for estimating impact on crop yield loss includes counting weeds per length of row or determining weed populations by species. In order to improve the weed suppression tactics, accurate mapping and assessment of weed populations within agricultural fields is required. See Figure 1. Weed mapping and taxonomy are major activities and species type found in all regions, which cover much broader ecological areas other than farm fields. These are shown by active websites in Nebraska, Iowa, Pennsylvania, Montana, Nevada, Colorado, and California, as examples. Weed and invasive species mapping also has international implications, (Montserrat, et al, 2003). Efforts of this type support integrated pest management (IPM) programs of both Crops and Risk (CAR) and Risk Avoidance and Mitigation (RAMP) which involve profitability and environmental stewardship and risk management, by providing a tool for timely acquisition of weed information. Research in this area promotes an interdisciplinary, IPM systems approach to weed mapping. There is high labor cost associated with the manual scouting of fields to obtain such maps.

2. Spatial variability of weed populations

Weeds are present in every field and lawn every year. The severity of the weed population is determined by local management practices such as the previous crop in the rotation and the herbicide use. According to a 2002 North Central strategic plan, tillage remained a major tool for controlling perennials, although the dilemma is that tillage contributes to soil erosion. Weed spatial distributions are unique, with monocot infestations more patchy than dicots (Mortensen et al., 1992 and Johnson et al., 1993, 1995). Monocots differ architecturally from dicots. Most weeds are serious competitors for moisture and soil nutrients. By first classifying the weed as either a monocot or dicot, a herbicide could be selected that most effectively controls that type of plant, resulting in better application efficiencies. Most post-emergent herbicides are selective in controlling one plant type or the other. Wiles and Schweizer (1999, 2002) researched the spatial distribution of weed seed banks using soil samples to map locations of weed seed banks in a given field. Seed banks have been found distributed in a patchy manner. Using the maps as a guide, farmers could treat just the weed patches with minimal amounts of the appropriate chemical. Site-specific weed management could mean a significant reduction in herbicide use, which saves the farmer money and benefits the environment. However, a large number of soil and plant samples are needed to get an accurate map—and that can be costly.

Stubbendick, et al (2003) provided a comprehensive compendium of weedy plants found across the Great Plains of the United States. Color plates were provided of canopy architecture and sometimes close-ups of individual leaves, flowers, and fruit. A hand drawing of canopy architecture was also given. In order to recognize a particular species, one needs to understand the concept of inflorescence and various plant taxonomy terms. There are many existing plant image databases around the United States. However, their suitability as reference images has yet to be determined for machine vision applications. An important application using machine vision is site-specific or spot herbicide application systems to reduce the total amount of chemical applied (Lindquist et al., 1998, 2001 a,b; Medlin, et al, 2000). Therefore, a major need for improved weed IPM and ecological assessment of invasive plant species is the development of a low-cost, but high resolution, machine vision system to determine plant incidence, even when imbedded with other plants, and to identify the species type. Machine vision systems should assist in the creation of plant field maps, leading to valid action thresholds (National Roadmap for IPM 2004).

3. Machine vision

Field plants, residue, and soil ecosystems are very complex, but, machine vision technology has the potential to systematically unravel and identify plants using optical properties, shape, and texture of leaves (Meyer et al., 1998). Considerable research has been reported using optical or remote sensing sensors to identify crop health by surface reflectance of green plants in agricultural fields (Gausman et al., 1973; Tucker, et al, 1979; Gausman et al., 1981; Thomas, et al, 1988; Storlie et al., 1989, Tarbell and Reid, 1991.; Franz et al., 1991b; and others). Hagger, et al (1983, 1984) reported the first prototype, reflectance-based plant sensor for spraying weeds. Hummel and Stoller (2002) evaluated a later commercial weed sensing system and noted their problems. Tian, et al (1999) developed a simple weed seeker in Illinois. Unfortunately, subsequent optical, non-image, sensor-based weed seekers and spot sprayers have not gained commercial acceptance for various reasons: first, single-element optical sensors can change the size of their field of view based on lens properties and distance to a target. Secondly, sensed reflectance properties may change according to the spatial contents of target components within the field of view Woebbecke, et al (1994); and finally, these sensors therefore may not always distinguish conclusively between crop, weed, or soil residue background. The voltage signal originating from an optical diode or transistor along with the Gaussian lens system used creating the field of view is a weighted average-problem, where the proportions of contributing reflectance and spatial contents are unknown. That problem can be solved only by spatial image analysis.

Image analysis is a mathematical process to extract, characterize, and interpret tonal information from digital or pixel elements of a photographic image. The amount of detail available depends on the resolution and tonal content of the image. The process is iterative, starting with large features followed by more detail, as needed. However, shape or textural feature extraction first requires identification of targets or Regions of Interest (ROI). These regions are then simply classified as green plants or background (soil, rocks, and residue). ROI's can be also identified with supervised control of the camera or field of view (Woebbecke, et al, 1994, Criner, et al, 1999), using a supervised virtual software window, cropping of selected areas, or unsupervised crisp or fuzzy segmentation procedures. ROI's are then binarized to distinguish target and background. Binarized images are then used for shape analysis or boundary templates for textural feature analysis. The binary image is combined with tonal intensity images of the targets (Gerhards and Christensen, 2003, Meyer et al., 1999; Kincaid and Schneider, 1983; Jain, 1989; Gonzalez and Woods, 1992; and others). Machine vision offers the best potential to automatically extract, identify, and count target plants, based on color, shape, and textural features (Tillett et al. 2001). However, directing the image analysis process toward the classical botanical taxonomic, plant identification approach has previously required considerable supervised human intervention. A major problem is the presentation of plant features including individual leaves and canopy architecture to a discrimination or classification system. Camargo Neto, et al (2004 a,b; 2005) presented a combination of traditional image processing techniques, fuzzy clustering, pattern recognition, and a fuzzy inference neural network to identify plants, based on leaves. A particular difficult problem was the development of an algorithm to extract individual leaves from complex canopies and soil/residue color images.

If image vegetative/background classification is to be useful for plant species identification, a separated plant region of interest (ROI) must be found to provide important canopy information needed to discriminate at the very least, broadleaf versus grass species (Woebbecke et al., 1995a; Meyer et al., 1998). Four basic steps for a computerized plant species classification system were presented by Camargo Neto (2004). The first step is creating a binary image which accurately separates plant regions from background. The second step is to use the binary template to isolate individual leaves as sub images from the original set of plant pixels (Camargo Neto, et al, 2006a). A third step was to apply a shape feature analysis to each extracted leaf (Camargo Neto, et al, 2006b). The fourth and final step was to classify the plant species botanically using additional leaf venation, textural features acquired during the previous steps (Camargo Neto and Meyer, 2005). Machine vision plant image analysis has been greatly enhanced through the introduction of the automatic color and focusing digital camera (Meyer, et al, 2004). Digital cameras when run in the automatic mode make decisions on "best picture", and thus are extremely popular as consumer products.

4. Vegetation indices

The use of vegetation indices in remote sensing of crop and weed plants is not new. It represents the first step shown in Figure 2. Studies for crop and weed detection have been performed using different spectral bands and combinations for vegetative indices (Woebbecke et al. 1995b, El-Faki, et al., 2000ab, Marchant et al., 2004; Wang et al., 2001, Lamm et al., 2002; Mao et al., 2003; Yang et al., 2003). Color vegetation indices utilize only the red, green and blue spectral bands. The advantage of using color indices is that they

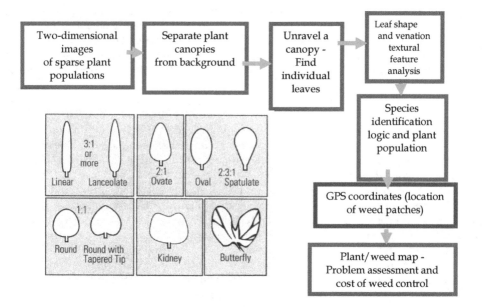

Fig. 1. A strategic approach to weed assessment.

accentuate a particular color such as plant greenness, which should be intuitive by human comparison. Woebbecke et al. (1995a) was one of the first researchers to test vegetation indices that were derived using color chromatic coordinates and modified hue for distinguishing green plant material in images from bare soil, corn residue, and wheat straw residue. Woebbecke's indices (without row and column indices of each pixel) included:

$$\text{Color indices: } (r - g, g - b, \frac{g - b}{r - g}, \text{ and } 2 \cdot g - r - b) \qquad (1)$$

where: r, g, and b are known as the chromatic coordinates (Wyszecki and Stiles, 1982), given as:

$$r = \frac{R^*}{R^* + G^* + B^*}, g = \frac{G^*}{R^* + G^* + B^*}, \text{ and } b = \frac{B^*}{R^* + G^* + B^*} \qquad (2)$$

and: R^*, G^*, and B^* are normalized RGB values (0 to 1), defined as:

$$R^* = \frac{R}{R_m}, G^* = \frac{G}{G_m}, \text{ and } B^* = \frac{B}{B_m}$$

R, G, and B are the actual pixel values obtained from color images, based on each RGB channel or band.

R_m, G_m, and B_m = 255, are the maximum tonal value for each primary color.

Woebbecke discovered that the excess green vegetation index (ExG = $2 \cdot g - r - b$) provided an interesting near-binary, tonal image outlining a plant region of interest. Woebbecke's excess green (ExG) index has been widely cited in the literature and has been tested in recent studies (Giltelson et al., 2002; Lamm et al., 2002; Mao et al., 2003; and others). ExG plant regions of interest could then be completely binarized using a selected contrast threshold value for each image. Thus, an important condition was the selection of the threshold value. Mao et al. (2003) subsequently tested several indices: ExG, normalized difference index (NDI), and the modified hue for separating plant material from different backgrounds (soil and withered plant residue). In his study, the ExG index was found superior to the other indices tested. A critical step was to select a manual threshold value to binarize the tonal image into a black and white image.

Other color vegetation indices have been reported for separating plants from soil and residue background in color images. For example, the normalized difference vegetation index (NDI) by Perez et al. (2000) uses only the green and red channels and is given as:

$$NDI = \frac{G - R}{G + R} \tag{2}$$

Perez's NDI was improved by adding a one, and then multiplying by a factor of 128. Hunt, et. al (2005) developed a vegetation index, known as the Normalized Green-Red Difference Index (NGRDI) for their model airplane photography for assessing crop biomass. Zhang, et al (1995) and Gebhardt, et al (2003) also used various RGB transforms for their plant image segmentation step.

Color indices have been suggested to be less sensitive to in lighting variations, and may have the potential to work well for different residues backgrounds (Campbell, 1996). However, a disproportionate amount of redness from various lighting sources may overcast a digital image, making it more difficult to identify green plants with simple RGB indices (Meyer et al, 2004b). For example, image redness may be related to digital camera operation and background illumination, but may also be related to redness from the soil and residue itself. An alternate vegetative index called excess red $(ExR = 1.4 \cdot r - g)$ was proposed by Meyer et al.(1998a), but was not tested until later studies.

Meyer and Camargo Neto (2008) reported on the development of an improved color vegetation index: Excess Green minus Excess Red (ExG-ExR). This index does not require a threshold and compared favorably to the commonly used Excess Green (ExG), and the normalized difference (NDI) indices. The latter two indices used an Otsu threshold value to convert the index near-binary to a full-binary image. The indices were tested with digital color images of single plants grown and taken in a greenhouse and field images of young soybean plants. Vegetative index accuracies were compared to a hand extracted plant regions of interest using a separation quality factor algorithm. A quality factor of one represented a near perfect binary match of the computer extracted plant target compared to the hand extracted plant region. The ExG-ExR index had the highest quality factor of 0.88 $\pm$ 0.12 for all three weeks, and soil-residue backgrounds for the greenhouse set. The ExG+Otsu and NDI-Otsu indices had similar quality factors of 0.53 $\pm$ 0.39. and 0.54 $\pm$ 0.33 for the same set, respectively. Field images of young soybeans against bare soil gave quality factors for both ExG-ExR and ExG+Otsu around 0.88 $\pm$ 0.07. The quality factor of NDI+Otsu using the same field images was 0.25 $\pm$ 0.08. ExG-ExR has a fixed, built-in plant-background zero threshold, so that it does not need Otsu or any user selected threshold value. The ExG-ExR index worked especially well for fresh wheat straw backgrounds, where it was generally 55 per cent more accurate than the ExG+Otsu and NDI+Otsu indices. Once a binary plant region of interest is identified with a vegetation index, other advanced image processing operations may be applied, such as identification of plant species such as would be needed for strategic weed control.

Near-Infrared (NIR) along with color bands have been used in vegetative indices for satellite remote sensing applications. However, NIR is less human intuitive, since the human eye is not particularly sensitive to the NIR spectrum which begins with red light. The human eye is only able to discern color (retinal sensors called cones). The eye also contains rods which are essentially receptive to small amounts of blue light that may exist after sundown. NIR is also not readily available with an RGB color digital camera. NIR usually requires a special monochromatic camera with a silicon-based sensor that can detect light up to one micron in wavelength with an NIR band pass filter. Hunt, at al (2011) has experimented with extracting near infrared out of RGB digital cameras. They developed a low-cost, color and color-infrared (CIR) digital camera that detects bands in the NIR, green, and blue. The issue still remains as to how does one verify the accuracy of infrared-image-based vegetative index without comparison to vegetation observed in a corresponding color visual image? So, the verification process of existence of plant material either returns to color images or some other non-optical method.

Two additional problems tend to exist with previous research regarding vegetative indices (a) the disclosure of the manual or automatic threshold used during the near-binary to binary conversion step, and (b) generally, the lack of reporting of vegetation index accuracy. Gebhardt, et al (2003) suggested that it was not necessary to classify vegetation on a pixel basis with digital imaging. However, if there are too many plant pixels mixed up with background pixels, accuracy may be reduced. Hague, et al (2006) suggested a manual comparison of vegetative areas from high resolution photographs. To date, very few vegetative index studies have reported validation accuracy of detecting plant material in independent images from other sources. This problem becomes particularly apparent, when these indices are applied to the collection of photographic plant databases currently available.

Plant classification might be expanded to hyper spectral imaging (Okamoto, et al. 2007). Wavelet along with discriminant analyses were used to identify spectral patterns of pixel samples for a 75–80 percent classification rate of five young plant species. Typically, hyper spectral cameras are expensive.

In summary, color image classification systems utilize the red (R), green (G), and blue (B) tonal intensity components. Color is a special form of spectral reflectance, which can be derived from spectral measurements (Wyszecki and Stiles, 1982; Murch, 1984; Jain, 1989; Gonzalez and Woods, 1992; Perry and Geisler, 2002). Perceived (human) color is based on the (RGB) primary colors. Woebbecke et al. (1995) discovered that the excess green index $(2 \cdot G - R - B)$ could provide excellent near-binary segmentation of weed canopies over bare soil for canopy shape feature analysis. El-Faki et al. (2000b) studied different RGB indices, as potential weed detection classifiers, but none possibly as good as excess green. The best correct segmentation rates (CCR) found were around 62%, while some misclassification rates were less than 3%. Meyer et al. (1999, 2004) proposed an excess red index $(1.3 \cdot R - G)$, based on physiological, rod-cone proportions of red and green. This index also provides near-binary silhouettes of plants under natural lighting conditions. Marchant, et al, (2004) proposed additional procedures for dealing with machine vision and natural lighting. The utilization spectral wave bands and color components have been used arithmetically and called vegetation indices. The index Meyer and Camargo Neto (2008) is an advanced color vegetation index.

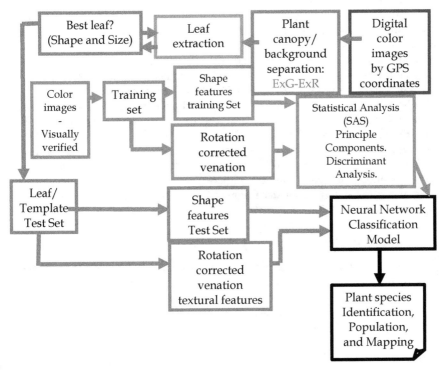

Fig. 2. Prototype Plant Species Identification and Enumeration using Leaf features.

5. Computerized single leaf extraction

Only a few methods of unsupervised leaf extraction from canopy images have been reported in the literature. Franz et al. (1991b) reported the use of curvature functions and the Fourier-Mellin correlation to identify completely visible and partially occluded sets of leaves. Leaf statistical features of mean, variance, skewness, kurtosis were computed, using spectral wavebands of red, green, blue, and near infrared. These features were used to discriminate leaf types of unifoliolate soybean, ivy, morning glory cotyledons, velvetleaf cotyledons, foxtail, first leaf of ivy, morning glory, and the first leaf of velvet leaf. Franz et al. (1995) further developed an algorithm to extract boundaries of occluded leaves using an edge detection technique to link the end points of leaf edge segments. User intervention was required at various steps of the algorithm. The fractions of individual leaves obtained were reported to be 0.91, 0.87, 0.95, and 0.71 for velvetleaf, soybean, ivy leaf morning glory, and foxtail, respectively.

To clarify this issue, occluded or partial fractions of leaves are probably not that useful for species identification. However, all canopies will exhibit whole individual leaves at the canopy apex, which can be seen in overhead photographs. Some leaves may standout by themselves (non-concealed) against the soil-residue background. Others will have vegetation from occluded leaves around them, which we will call concealed leaves. The latter would represent a difficult image processing problem, not easily solved by traditional algorithms such as edge detection, erosion, dilation, and such.

Deformable templates using active contours were used by Manh, et al. (2001) to locate boundaries of green foxtail leaves. Manh's process attempted to combine color separation and shape feature analysis into a single operation. The procedure began with identification of a leaf tip, and followed by shape analysis across the rest of the green material. However, a manually selected energy level or color was needed. Segmentation accuracy for a single species of foxtail leaves was reported to be 84%. No other species were studied.

Individual, whole, and fragments of leaves were isolated using the Gustafson-Kessel fuzzy clustering method over bare soil, corn stalks, and wheat straw color images (Hindman, 2001, Meyer et al., 2004b, Gustafson and Kessel, 1979). Zadeh intensification of the fuzzy cluster membership functions resulted in definitive green canopy areas, but not individual leaves. However, Camargo Neto, et al (2006) used the Gustafsen-Kessel fuzzy leaf cluster fragmentation method on green canopy regions of interest. He also developed a reassembling method of the green cluster fragments resulting in individual leaves using a genetic algorithm (Holland, 1975).

6. Shape feature analysis

If the process of image vegetative/background classification is to be useful, the separated plant region of interest (ROI) must provide important canopy or leaf shape feature or property information to at least discriminate between broadleaf and grass species (Woebbecke et al., 1995b; Meyer et al., 1998a; Meyer et al., 1998b).

Supervised leaf and single plant canopy shape feature analysis has been studied the most. Petry and Kuhbauch (1989) found shape parameters using five canonical indices found distinctly different for several weed species. Guyer et al. (1986, 1993) used image shape feature analysis on individual leaves to distinguish between weed species and corn. Guyer et al. (1993) using only leaf and canopy shapes, reported a 69% correct identification rate for 40 weeds and agricultural crop species. Guyer found that no single shape feature alone could distinguish corn from all other species. Franz et al. (1991 a,b) identified plants based on individual leaf shape at two growth stages using the Fourier-Mellin correlation. Woebbecke et al. (1995a, b) used basic image shape feature analysis to discriminate between broadleaf and grassy plant canopies. Woebbecke found that broadleaf and grass shape features best appeared to a vision system at early stages of growth or within a specific window of time, from 1-4 weeks after emergence. Downey, et al (2004) described a field canopy shape identification system which used a binary canopy erosion technique to discriminate between grasses and broadleaf plants. Yonekawa et al. (1996) presented a set of classical shape features for a leaf taxonomy database. Chi, et al (2002) fitted Bezier curves to different leaf boundary shapes. Mclellan and Endler (1998) compared several morphometric methods for describing complex shapes. They found that approximately 20 harmonics of the elliptic Fourier method accurately depicted shapes of *Acer saccharinum*, *Acer saccharum*, and *Acer palmatum* leaves. A leaf shape image retrieval systems was also reported by Wang, et al (2003).

Du et al (2005, 2006, 2007) proposed the Douglas-Peucker approximation algorithm for leaf shapes and the shape representation was used to form the sequence of invariant attributes.

A modified dynamic programming (MDP) algorithm for shape matching was proposed for the plant leaf recognition. Oide and Ninomiya (2000) used the Elliptic Fourier (EF) method to classify soybean varieties, using a normalized leaf shape. The EF method using a chain-coded, closed contour, invariant to scale, translation, and rotation was first introduced by Kuhl and Giardina (1982). EF has been used in recent studies to describe the shape of objects. Innes and Bates (1999) used an Elliptical Fourier descriptor to demonstrate an association between genotype and morphology of shells. Chen et al. (2000) used Elliptic Fourier descriptors to describing shape changes in the human mandible for male and female at different ages. Most methods previously investigated ignore leaf edge serration. Leaf serration or edgeness is an important morphologic feature used for identifying plant species. For example, the curvature functions developed by Franz et al. (1991b) were found generally inadequate where leaflet serration was quite pronounced. Camargo Neto, et al, 2006b applied the Elliptic Fourier shape feature analysis to extracted leaves of velvet leaf *Abutilon theophrasti*, pig weed *Amaranthus retroflexus L.*, sunflower *Helianthus annus*, and soybean *Glycine max*. A velvet leaf example is shown in Figure 3.

Hearn (2009) used a database of 2,420 leaves from 151 plant species for a plant leaf shape analysis. Using metrics derived during Fourier and Procrustes analyses, it was found that a minimum of ten leaves for each species, 100 margin points, and ten Fourier harmonics were required to develop any accuracy using the leaf shape of a species. His results indicated a success rate of 72% correct species identification for all 151 species used. This may mean that more than leaf shape is needed for classification.

7. Textural feature analysis

Color and/or leaf shape features alone may not be sufficient to consistently distinguish between young weed and crop plant species. Textural features may supply some additional botanical information, such as leaf venation, leaf pubescence, but also leaf disease and insect damage. The color or tonal detail for texture was first described by quantification of co-occurrence of tonal pairs or contrast also known as spatial tonal frequency (Haralick, 1978 and 1979). Wavelet analysis and energy have been recently suggested as a frequency based textural analysis for segmenting weeds imbedded in canopies (Chang and Kuo, 1993, Strickland and Hahn, 1997, Tang, et al, 2003). Shearer and Holmes (1990) used color co-occurrence matrix method to identify the textural features of isolated plants. Shearer and Jones (1991) proposed a texture-alone plant detection system based upon hue-saturation-intensity (HSI) images. Oka and Hinata (1989) used side view images of rice to distinguish between old and new Japanese rice cultivars. Zhang and Chaisattapagon (1995) tested a combination color, shape, and texture approach for detecting weeds in wheat fields and found that leaf –surface- coarseness indices defined by Fourier spectra may be effective in differentiating wheat from broad-leaf weeds. Meyer et al. (1999) showed that combined color, shape, and textural statistical discriminate analysis system could separate grasses from broadleaf canopies against bare soil backgrounds. Major problems for obtaining botanical textural detail involve image resolution, leaf orientation or rotation, shadows, bidirectional reflectance of leaf surfaces, and background lighting. Uneven lighting for example, could obscure venation - mesophyll leaf detail. Diffuse lighting could provide more even illumination than direct-beam lighting. Fu and Chi (2006) presented an algorithm for extracting leaf vein details from detached leaves under artificial light. Park, et al (2008) described a prototype system for classifying plants based on leaf venation features. Their

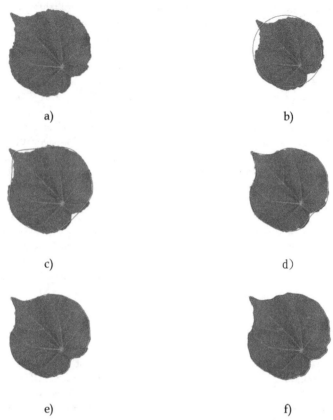

Fig. 3. Elliptic Fourier shape approximations for velvetleaf (*Abutilon theophrasti Medicus*), a) original leaf image, b) 1st EF harmonic, c) 1st + 2nd + 3rd + 4th EF, d) 1st + ... + 8th EF harmonics, e) 1st + ... + 16th EF harmonics, and f) 1st + ... + 30th EF harmonics.

method detected the differences between tree and parallel venations in leaves, and thus could be considered as an enhancement to the classification tool set.

De Oliveira Plotze (2009) combined computer vision techniques and plant taxonomy protocols, these methods are capable of identifying plant species. The biometric measurements are concentrated in leaf internal forms, specifically in the venation system. The methodology was tested with eleven species of passion fruit of the genus *Passiflora*. The features extracted from the leaves were then applied to a neural network system to develop a classification of species. The results were very accurate in correctly differentiating among species with 97% of success. Zheng and Wang (2009, 2010) presented the results of mathematical morphology used on images of single leaf samples. Mathematical morphology provides four fundamental operations of dilation, erosion, opening, and closing in image processing. Their goal was to extract only leaf veins using hue and intensity information. Camargo Neto and Meyer (2005) classified the plant species botanically sing additional leaf venation textural features acquired during the previous steps. One thing is clear, lack of care in the photography of a leaf may affect image textural properties and classification.

8. Plant species classification

Most studies in the last 20-years have addressed the classification of only two crop-weed classes or general cases of broad leaf versus grasses and in other cases, crop row versus between crop row (Tang, et al, 2003). However, to precisely classify a plant species that may be imbedded within other different species of plants in an image is a botanically challenging exercise.

Agarval, et al (2006) described an ongoing project to digitize information about plant specimens that would become available to field botanists and crop managers. They indicated that the first step required acquisition of digital images and possibly plant architectural models, along with an effective retrieval method and mobile computing mechanisms for accessing this information. At that time they had indicated progress in developing a digital archive of the collection of various plant specimens at the Smithsonian Institution.

Analytical tools are improving for classifying plant species. The artificial neural network (ANN) has been proposed for many classification activities. Plotze and Bruno (2009) have also proposed a plant taxonomy system. Yang et al., (2000, 2002, 2003) used RGB pixel intensities as inputs for a fuzzy artificial neural network (ANN) for distinguishing weeds from corn, with success rates as high as 66% for corn and 85% for weeds. To encompass the uncertainty of image classification processes, fuzzy set theory (FST) has been proposed for plant classification (Gottimukkala et al., 1999). FST provides a possibilistic alternative (different, but in many cases complementary) to the probabilistic or statistical approaches. FST embraces virtually all (except one) of the definitions, precepts, and axioms that define classical sets that supports common mathematics, (Ross, 2004). It uses variables in the form of membership functions with degrees of support for fuzziness, incorporating uncertainty (Zadeh, 1965; Mamdani, 1976; Li and Yen, 1995). Pal, et al 1981, 1994, Bezdek, 1973, 1993) summarized the use of a FST neural network for pattern recognition, generating membership functions, performing fuzzy logic (FL) operations, and then deriving inference rule sets. Jang (1993) invented the artificial neural fuzzy inference system (ANFIS) for training membership functions and rule sets that could be used for classification (Figure 4).

Fuzzy logic machine vision classification systems are intended to imitate human perception or vision and to handle uncertainty. In the weed discrimination example, expert human perception or scouting validation is required for ground truthing. Bhutani and Battou (1995) and Tizhoosh (1998, 2000) provide computational overviews and various examples of fuzzy logic applied to image processing. Incorporating unsupervised fuzzy logic clustering and image analysis into site-specific technologies has tremendous potential (Kuhl and Giardina, 1982, Gath and Geva, 1989, De and Chatterji, 1998, Babuska, 1998, Manthalkar, at al, 2003, Meyer, et al. 2004). The very nature of site-specific data collection, image analysis, decision-making, etc., is characterized by uncertainty, ambiguity, and vagueness, which may be over overcome with these techniques.

Hindman and Meyer (2000) demonstrated a prototype fuzzy inference system for plant detection. Jones et al. (2000) used remotely sensed data with FL classification to detect crop status, resulting in a fuzzy description of crop phenology based upon spectral data. Yang et al., (2003) also presented potential herbicide savings in weed control with a fuzzy logic system. Heming and Rath (2001) proposed a fuzzy weed classifier that yielded correct classification accuracies between 51 and 95%. The potential fallacy of any regression, ANN, or fuzzy ANN (ANFIS) model is that they can be designed to mimic signal errors and random noise data too well, especially with an inadequate size of the training data set. Fuzzy inference systems can also incorporate the "I do not know" result.

Fuzzy clustering refers to unsupervised partitioning of data into subclasses for pattern recognition (Ross, 2004). Babuska (1998) presented six different clustering techniques that might be used to organize tonal image data with their limitations. These included the fuzzy c-means, the Gustafson-Kessel, fuzzy maximum likelihood, fuzzy c-varieties, fuzzy c-elliptotypes, and possibilistic clustering that might be used on tonal images. Moghaddamzadeh et al. (1998) described a fuzzy nearest-neighbor, clustering method for segmenting color images. Townsend (2000) discussed methods for making comparisons of fuzzy ecological pattern recognition methods. Beichel, et al. (1999) discussed the use of an unsupervised Gath-Geva clustering method for Landsat thermatic mapper (TM) images. Classification accuracy reached a maximum value of 86 % with five clusters. Individual, whole, and fragments of leaves were isolated using the Gustafson-Kessel fuzzy clustering method over bare soil, corn stalks, and wheat straw color images (Meyer et al., 2004b). Zadeh intensification of the membership functions resulted in definitive green canopy areas.

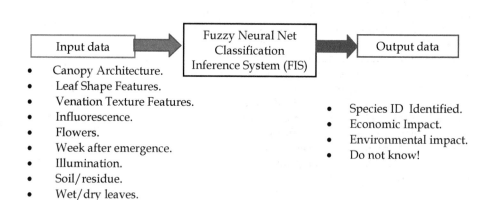

Fig. 4. Advanced Species Classifier Method– Fuzzy Logic- Neural Network using Image metrics and others.

A machine vision system with unsupervised image analysis and mapping of features was presented by Camargo Neto (2006a) and Camargo Neto, et al. (2006b). A classification system was trained using statistical discriminant analysis which was tested using individual test leaves and clusters from several plants. As many as 75 percent of exposed whole leaves were extracted, and can be further species identified at 75% or better. When such a system is improved and validated with scientific-based methods, it could dramatically assist understanding crop-weed relationships, growth, competition, and control. A machine vision system certainly should be able to identify and distinguish weed species that are 7 - 21 days old, a time when post emergence herbicides are most effective.

9. Linking machine vision with weed management systems

Predicting crop yield loss due to weed competition is one critical component of dynamic decision making for integrated weed management. Moreover, spatial variation in weed occurrence must be accounted for to accurately predict crop yield loss (Lindquist et al. 1998, 2001 a,b). The fuzzy logic machine vision classification system will be extremely useful where weeds are distinguished from crop plants and precisely mapped within a farm field. Shape feature analysis also provides a means for determining the relative surface area of weed plants relative to crop plants. Kropff and Spitters (1991) argued that the competitive strength of a species is determined by its share in leaf area at the moment when interspecific competition begins. Kropff et al. (1995) presented an equation that expresses yield loss (YL) as a function of weed and crop LAI. This approach has recently been expanded to relate yield loss to weed and crop relative volume (Conley et al. 2003) and could easily be used to relate yield loss to weed and crop relative surface area obtained from our image analysis. This kind of detail requires close-in imaging within a few meters with current high pixel rate digital cameras.

Holst, et al (2006) reviewed the progress of weed population modeling and of course the use is similar: strategic decision making for weed management. Freckleton and Stephens (2009) discussed the use if dynamic plant models for weed management. They concluded that there exist a discrepancy in the field of weed population modeling; many of the problems faced by weed ecologists require detailed quantitative predictions, but few modelers are attempting to provide such predictions. FST has also been used for modeling biological systems. Ambuel et al. (1994) used FL to develop a crop yield simulator for assessing spatial field variability for accuracy and optimizing pesticide application rates. Weed plant growth and plant population models that also describe the canopy architecture would be very helpful for weed classification.

10. Conclusions

The literature is rich in selected or component ideas for machine vision, plant species identification. Now is the time to put together a complete robust system that essentially mimics the human taxonomic, plant identification keying method. If one returns to Stubbendick, et al (2003), one can verify that the human classification process requires metrics on leaves, stems, flowers, influorescence, and a picture of the plant. Leaf shape and venation images alone may not close the classification process.

Shape analysis for image processing is very well-understood and computer algorithms are readily available. The leaf angle in the plane of the canopy is of interest (the first elliptic Fourier harmonic), and that is a critical angle for rotationally invariant leaf texture or venation analysis. Additional studies regarding leaf orientation relative to the camera lens might help to reduce classification errors. Modern digital cameras are capable of acquiring large amounts of image-pixel data. Future studies need to determine minimal digital image resolutions needed to maintain the highest species discrimination performance.

Fuzzy logic, cluster algorithms and cluster reassembly routines work well for extracting convex leaf shapes from plant canopy images. However, for more botanically diverse leaf shapes, such as species with complex leaves, lobed margins (indented), trifoliolates, etc., new fitness criteria need to be developed to accommodate these leaf shapes. Undoubtedly, integration of specific shape and textural feature analyses as fitness criteria may be a key to improvement of this process. New leaf extraction/species classification algorithm can become especially useful, if acceptance criteria can be designed to accommodate more a extensive leaf taxonomy digital library (shape and texture of single and compound leaves). Work has been extended on utilizing digital canopy architecture metrics in three dimensions which is important plant taxonomy.

Species classification and mapping has been tested using a neural-fuzzy inference model, which can be improved with inclusion of additional training information, including: stage of growth, expected canopy architecture, distance from a designated crop row, crop row spacing and direction.

Studies and discussion should be conducted to determine if older photographic plant image data bases can be used as references for new unknown digital plant images. Considerable field testing and validation are always needed for plant identification studies using machine vision.

11. References

_____, 2004. *National Roadmap For Integrated Pest Management.* http://www.csrees.usda.gov/nea/pest/pdfs/nat_ipm_roadmap.pdf.

Ambuel, J.R., Colven, T.S., Karlen, D.L., 1994. *A fuzzy logic yield simulator for prescription farming.* Trans. ASAE 37 (6), 1999–2009.

Agarwal, G., Belhumeur, P., Feiner, S., Jacobs, D., Kress, W.J., Ramamoorthi, R., Bourg, N.A., Dixit, N., Ling, H., Mahajan, D., Russell, R., Shirdhonkar, S., Sunkavalli, K. and White, S. 2006. *First steps toward an electronic field guide for plants.* Taxon 55: 597–610.

Babuska, R. 1998. Fuzzy Clustering Algorithms. *Fuzzy Modeling for Control.* Kluwer Academic Publishing , Boston 3:49-74.

Beichel, R., R. Bolter, and A. Piz, 1999. *Fuzzy Clustering of a Landsat TM Scene.* Geoscience and Remote Sensing Symposium, IGARSS '99 Proceedings. IEEE 1999 International 1597-1599.

Bezdek, J.C. 1973. *Fuzzy mathematics in pattern classification.* Ph. D. Thesis, Applied Math. Center, Cornell University, Ithaca.

Bezdek, J.C. 1993. *A review of probabilistic, fuzzy, and neural models for pattern recognition.* Journal of Intelligent and Fuzzy Systems. 1(1):1-25.

Bhutani, K.R. and A. Battou, 1995. *An application of fuzzy relations to image enhancement.* Pattern Recognition Letters. 16:901-909.

Camargo Neto, G.E. Meyer , D. D. Jones. 2006a. *Individual Leaf Extractions from Young Canopy Images using Gustafson-Kessel Clustering and a Genetic Algorithm.* Computers and Electronics in Agriculture (Elsevier) 51:65-85.

Camargo Neto, G.E. Meyer, D. D. Jones, A.K. Samal. 2006b. *Plant Species Identification using Elliptic Fourier Analysis.* Computers and Electronics in Agriculture (Elsevier), 50:121-134.

Camargo Neto, J., 2004. *A Combined Statistical -- Soft Computing Approach for Classification and Mapping Weed Species in Minimum Tillage Systems.* Unpublished PhD Dissertation, University of Nebraska, Lincoln, NE.

Camargo Neto, J., G.E. Meyer. 2005. Crop species identification using machine vision of computer extracted individual leaves. In: Chen, Y.R., Meyer, G.E., Tu S. (Eds.), *Optical Sensors and Sensing Systems for Natural Resources and Food Safety and Quality,* Proc. SPIE, Bellingham WA., 5996: 64-74.

Camargo-Neto, J., G.E. Meyer, and D.D. Jones, 2004. *Advances in Color Image Segmentation of Plants for Weed Control.* ASAE Paper Number 043060, The Society for Engineering in Agricultural, Food, and Biological Systems, St Joseph MI.

Campbell, J B, 1996. *Introduction to Remote-Sensing,* The Guilford Press, London).

Chang, T. and C.J. Kuo, 1993. *Texture Analysis and Classification with Tree-Structured Wavelet Transform.* IEEE Transactions on Image processing. 2(4): 429-441.

Chen, S. Y.Y., P.E. Lestrel, W.J.S. Kerr, and J.H. McColl, 2000. *Describing shape changes in the human mandible using elliptic Fourier functions.* European Journal of Orthodonics. 22:205-215.

Chi, Y.T., C.F. Chien, and T.T. Lin, 2002. *Leaf Shape Modeling and Analysis using Geometric Descriptors derived from Bezier Curves.* Trans. ASAE. 46(1):175-185.

Conley, S. P., D. E. Stoltenberg, C. M. Boerboom, and L. K. Binning. 2003. *Predicting soybean yield loss in giant foxtail (Setaria faberi) and common lambsquarters (Chenopodium album) communities.* Weed Science. 51:402-407

Criner, B.R., J.B. Solie, M.L. Stone, R.W.Whitney. 1999. *Field-of-View Determination for a Bindweed Detection Sensor.* Trans. ASAE. 42(5):1485-1491.

De Oliveira Plotze, R., 2009. *Automatic Leaf Structure Biometry: Computer Vision Techniques and Their Applications In Plant Taxonomy.* International Journal of Pattern Recognition and Artificial Intelligence 23(2): 247–262.

De T.K. and B.N. Chatterji, 1998. *An approach to generalized technique for image enhancement using the concept of fuzzy set.* Fuzzy Sets and Systems. 25:145-158.

Downey, D., D.K. Giles, and D.C. Slaughter, 2004. *Weeds are accurately mapped using DGPS and ground-based vision identification.* California Agriculture. 58(4):218-221.

Du, J.X., Huang D.S., Wang, X.F. and Gu, X. 2006. *Computer aided plant species identification (CAPSI) based on leaf shape matching technique.* Trans. Inst. Measurement Control 28: 275–284.

Du, J.X., Wang, X.F. and Gu, X. 2005. *Shape matching and recognition based on genetic algorithm and application to plant species identification.* Lecture Notes Computer Sci. 3644: 282–290.

Du, J.X., Wang, X.F. and Zhang, G.J. 2007. *Leaf shape based plant species recognition.* Appl. Math. Computation 185: 883–893.

El-Faki, M. S., Zhang, N., Peterson, D. E. 2000. *Weed detection using color machine vision.* Trans. ASAE. 43(6):1969-1978.

El-Faki, M. S., Zhang, N., Peterson, D. E. 2000. *Factors affecting color-based weed detection.* Trans. ASAE. 43(4):1001-1009.

Franz, E., L.D. Gaultney, and K.B. Unklesbay. 1995. *Algorithms for extraction leaf boundary information from digital images of plant foliage.* Trans. ASAE 38(2):625 - 633.

Franz, E., M.R. Gebhardt, and K.B. Unklesbay. 1991. *Shape description of completely visible and partially occluded leaves for identifying plants in digital images.* Trans. ASAE. 34(2):673-681.

Franz, E., M.R. Gebhardt, and K.B. Unklesbay. 1991. *The use of local spectral properties of leaves as an aid for identifying weed seedlings in digital images.* Trans. ASAE. 34(2):682-687.

Freckleton, R.P. and P.A Stehens, 2009. *Predictive models of weed population dynamics.* Weed Research. 49:225-232.

Fu, H. and Z. Chi. 2006. *Combined thresholding and neural network approach for vein pattern extraction from leaf images.* IEEE Proceedings Vision, Image and Signal Processing 153: 881–892.

Gath, I., and A.B. Geva, 1989. *Unsupervised optimal fuzzy clustering.* IEEE Transactions on Pattern Analysis and Machine Intelligence. 2(7):773-781.

Gausman, H.W., R.M. Menges, A.J. Richardson, H. Walter, R.R. Rodriguez, and S. Tamez. 1981. *Optical properties of seven weed species.* Weed Science. 29(1):24-26.

Gausman, H.W., W.A. Allen, C.L. Wiegand, D.E. Escobar, R.R. Rodriguez, and A.J. Richardson. 1973. *The leaf mesophylls of twenty crops, their light spectra, and optical and geometrical parameters.* USDA Technical Bulletin No. 1465-59.

Gebhards, R. And S. Christensen, 2003. *Real-time weed detection, decision making, and patch spraying in maize, sugarbeet, winter wheat and winter barley.* Weed Research (European Weed Research Society) 43:385-393.

Giltelson, A.A., Y.J. Kaufman, R. Stark, and D. Rundquist. 2002. *Novel algorithm for remote estimation of vegetation fraction.* Remote Sensing of Environment (Elsevier) 80:76-87.

Gonzalez, R.C. and R.E. Woods. 1992. *Digital Image Processing.* Addison Wesley Publishing Company, Reading, Massachusetts.

Gottimukkala, S., N. Zhang, X. Lin, and N. Wang, 1999. *An application of fuzzy logic in weed detection.* ASAE Paper No. 99-3139 (microfiche). St. Joseph, MI. pp12.

Grube, A., D. Donaldson, T. Kiely, and L. Wu, 2011. *Pesticides Industry Sales and Usage. 2006 and 2007 Market Estimates.* Biological and Economic Analysis Division. Office of Pesticide Programs. Office of Chemical Safety and Pollution Prevention. U.S. Environmental Protection Agency. Washington, DC, 33 pages.

Gustafson, E.E., and W.C. Kessel. 1979. *Fuzzy clustering with a fuzzy covariance matrix.* IEEE CDC, San Diego:761 - 766.

Guyer, D.E., G.E. Miles, L.D. Gaultney, and M.M. Schreiber. 1993. *Application of machine vision to shape analysis in leaf and plant identification.* Trans. ASAE. 36(1):163-171.

Guyer, D.E., G.E. Miles, M.M. Schreiber, O.R. Mitchell, V.C. Vanderbilt. 1986. *Machine vision and image processing for plant identification.* Trans. ASAE. 29(6):1500-1507.

Hagger, R.J., C.J. Stent, and J. Rose, 1985. *Measuring spectral differences in vegetation canopies by a reflectance ratio meter.* Weed Research. 24:59-65.

Hagger, R.J., C.J. Stent, and S. Isaac, 1983. *A prototype hand-held patch sprayer for killing weeds activated by spectral differences in crop/weed canopies.* J. Agricultural Engineering Research. 28:349-358.

Hague, T., N.D. Tillett and H. Wheeler, 2006. *Automated crop and weed monitoring in widely spaced cereals.* Precision Agric. 7:21-32.

Haralick, R.M., 1979. *Statistical and Structural Approaches to Texture.* Proc. IEEE, 67(5): 786-804.

Haralick, R.M., 1978. *Statistical and Structural Approaches to Texture.* Proceedings   4th Int. Joint Conf. Pattern Recognition. 45-69.

Hearn, D.J., 2009. *Shape analysis for the automated identification of plants from images of leaves.* Taxon. 58 (3): 962–981.

Hemming, J. and T. Rath. 2001. *Computer-vision-based weed identification under field conditions using controlled lighting.* J. of Agricultural Engineering Research. 78(3):233-243.

Hindman, T.W. 2001. *A fuzzy logic approach for plant image segmentation and species identification in color images.* PhD Dissertation, University of Nebraska, Lincoln.

Hindman, T.W. and G.E. Meyer, 2000. *Fuzzy logic inference systems for discriminating plants from soil and residue with machine vision*. In J.A. DeShazer and G.E. Meyer (ed). *Biological Quality and Precision Agriculture II*. SPIE Optical Engineering Press, Bellingham, WA (ISBN 0-8194-3868-5), pp 111-121.

Holland, J.H. 1975. *Adaptation in natural and artificial systems*. Ann Arbor, The University of Michigan, Press.

Holst, N., I.A. Rasmussen, and L. Bastiaans, 2007. *Field weed population dynamics: a review of model approaches and applications*. Weed research. 47:1-14.

Hoppner, F., F. Klawonn, R. Kruse, and T. Runkler. 2000. *Fuzzy cluster analysis methods for classification, data analysis and image recognition*. John Wiley and Sons, LTD, NY.

Hummel, J.W., and E.W. Stoller. 2002. *On-the-go weed sensing and herbicide application for the Northern Cornbelt*. American society of Agricultural Engineers. Paper no. 021021.

Hunt, E.R., M. Cavigelli, C. T. Daughtry, J. McMurtrey III and C. L. Walthall, 2005. *Evaluation of Digital Photography from ModelAircraft for Remote Sensing of Crop Biomassand Nitrogen Status*. Precision Ag (Springer) 6:359-378.

Hunt, E.R., W.D. Hively, G.W. McCarty and C.T. Daughtry, 2011. *NIR-Green-Blue High-Resolution Digital Images for Assessment of Winter Cover Crop Biomass*. GIScience and remote Sensing 48(1)86-98.

Innes D.J. and J.A. Bates, 1999. *Morphological variation of Mytilus edulis and Mytilus trossulus in eastern Newfoundland*. Marine Biology133,691^699.

Jain, A.K. 1989. *Fundamentals of Digital Image Processing*. Prentice Hall, Inc. Englewood Cliffs, New Jersey.

Jang, J. R. 1993. *ANFIS: Adaptive-network-based fuzzy inference system*. IEEE Trans. On Systems, Man, and Cybernerics 23(3):665-684.

Johnson, G. A., D. A. Mortensen, and C. A. Gotway,1993. *Spatial and temporal analysis of weed seeding populations using geostatistics*, Weed Science. 44(3), 704.

Johnson, G.A., D.A. Mortensen, and A.R. Martin, 1995. *A simulation of herbicide use based on weed spatial distribution*. Weed Res. 35:197-205.

Jones, D. and E.M. Barnes, 2000. *Fuzzy composite programming to combineremote sensing and crop models for decision support in precision crop management*. Ag Systems (Elsevier) 65:137-158.

Kincaid, D.T. and R.B. Schneider. 1983. *Quantification of leaf shape with a microcomputer and Fourier transform*. Canadian Journal of Botany. 61:2333-2342.

Kropff, M. J. and C. J. T. Spitters. 1991. *A simple model of crop loss by weed competition from early observations on relative leaf area of the weed*, Weed Res., 31:97-105.

Kropff, M. J., L. A. P. Lotz, S. E. Weaver, H. J. Bos, J. Wallinga and T. Migo. 1995. *A two parameter model for prediction of crop loss by weed competition from early observations of relative leaf area of the weeds*, Ann. Appl. Biol., 126:329-346.

Kuhl, F.P. and C.R. Giardina, 1982. *Elliptic Fourier features of a close contour*, Computer Graphics and Image processing. 18:236-258.

Lamm, R.D., D.C. Slaughter, and D.K. Giles. 2002. *Precision weed control for cotton*. Trans. ASAE 45(1):231-238.

Lee, C.-L. & Chen, S.-Y. 2006. *Classification of leaf images*. Int. J. Imaging Systems and Technology 16: 15–23.

Li, H.X., and V.C. Yen, 1995. *Fuzzy sets and fuzzy decision-making*. CRC.

Lindquist, J. L. and S. Z. Knezevic. 2001. *Quantifying crop yield response to weed population: Applications and limitations. Biotic Stress and Yield Loss.* Pages 205-232, in Peterson R. K. D. and Higley, L. G. (eds.), CRC Press, Boca Raton, FL.

Lindquist, J. L. *Mechanisms of crop loss due to weed competition. 2001. Biotic Stress and Yield Loss.* Pages 233-253, in Peterson R. K. D. and Higley, L. G. (eds.), CRC Press, Boca Raton, FL.

Lindquist, J. L., J. A. Dieleman, D. A. Mortensen, G. A. Johnson, and D. Y. Pester-Wyse. 1998. *Economic importance of managing spatially heterogeneous weed populations.* Weed Technology. 12:7-13.

Mamdani, E.H. 1976. *Advances in the linguistic sysnthesis of fuzzy controllers.* Int. J. of Man-Machine Systems. 8(6):669-678.

Manh, A.G., G. Rabatel, L. Assemat, and M.J. Aldon. 2001. *Weed leaf image segmentation by deformable templates.* Journal of Agricultural Engineering Research 80(2):139 - 146.

Manthalkar, R ; Biswas, P K ; Chatterji, B N, 2003. *Rotation and scale invariant texture features using discrete wavelet packet transform.* Pattern recognition letters. 24,(14): 2455-2463.

Mao, W., Y. Wang, and Y. Wang. 2003. *Real-time detection of between-row weeds using machine vision.* ASAE paper number 031004.

Marchant, J.A., N.D.Tillet, and C.M. Onyango, 2004. *Dealing with color changes caused by natural illumination in outdoor machine vision.* Cybernetics and Systems. 25:19-33.

Mclellan, T. and J.A. Endler, 1998. *The relative success of some methods for measuring and describing the shape of complex objects.* Syst. Biol. 47(2):264-281.

Medlin, C.R., D.R. Shaw, P.D. Gerard, and F.E. LaMastus. 2000. *Using remote sensing to detect weed infestation in Glycine max.* Weed Science 48:393 - 398.

Meyer G.E., T. Mehta, M. F. Kocher, D. A. Mortensen, and A. Samal. 1998. *Textural Imaging and Discriminate Analysis for Distinguishing Weeds for Spot Spraying.* Trans. ASAE., 41(4): 1189-1197.

Meyer, G. E. , J. Camargo Neto, D. D. Jones, T. W. Hindman. 2004. *Intensified fuzzy clusters for determining plant, soil, and residue regions of interest from color images*, Computers and Electronics in Agriculture. (Elsevier), 42(3):161-180.

Meyer, G.E. and J. Camargo Neto, 2008. *Verification of Color Vegetation Indices for Automated Crop Imaging Applications.* Computers and Electronics in Agriculture (Elsevier), 63:282-293.

Meyer, G.E., T.W. Hindman, and K. Lakshmi, 1999. Machine vision detection parameters for plant species identification, in G.E. Meyer and J.A. DeShazer (Ed.) *Precision Agriculture and Biological Quality.* Proceedings of SPIE, Bellingham, WA 3543:327-335.

Meyer, G.E., T.W. Hindman, D.D. Jones, and D.A. Mortensen. 2004. *Digital camera operation and fuzzy logic classification of plant, soil, and residue color images.* Engineering in Agriculture 20(4):519-529.

Moghaddamzadeh, A., D. Goldman, and N. Bourbakis. 1998. *A fuzzy-like approach for smoothing and edge detection in color images.* International Journal of Pattern Recognition and Artificial Intelligence 12(6):801-816.

Montserrat, J., F. Garcia-Torres, A. Garcia-Ferrer, M. Orden, and S. Atenciano, 2003. *Multi-species spatial variability and site-specific management maps in cultivated sunflower.* Weed Sci. 5:319-328.

Mortensen, D.A., G.A. Johnson, and L.J. Young. 1992. Weed distribution in agricultural fields. In *Soil Specific Crop Management*, Agronomy Society of America, 113-124.

Murch, G. M. 1984. *Physiological principles for effective use of color.* IEEE Computer Graphics and Applications. 4(11):49-54.

Oide, M., and S. Ninomiya. 2000. *Discrimination of soybean leaflet shape by neural networks with image input.* Computer and Electronics in Agriculture 29:59 - 72.

Oka, M. and K. Hinata. 1989. *Comparison of plant type between new and old rice cultivars using computer image analysis.* Japanese Journal of Crop Science. 58(2):232-239.

Okamoto, H., T. Mutata, T. Kataoka, and S. Hata, 2007. *Plant classification for weed detection using hyper spectral imaging and wavelet analysis.* Weed Biology and Management. 7:31-37.

Pal N.R. and J.C. Bezdek. 1994. *Measuring Fuzzy Uncertainty.* IEEE Transactions on Fuzzy Systems. 2(2):107-118.

Pal, S.K. and R.A. King, 1981. *Image Enhancement using smoothing with fuzzy sets.* IEEE Transactions on systems, man, and cybernetics. 11(7):494-501.

Park, J.K., E.J. Hwang, and Y.Y. Nam, 2008. Utilizing venation features for efficient leaf image retrieval Systems and Software (Elsevier) 81: 71-82.

Perez, A.J., F. Lopez, J.V. Benlloch, and S. Christiansen, 2000. *Colour and shape analysis techniques for weed detection in cereal fields.* Computers and Electronics in Agriculture. (Elsevier) 25:197-212.

Perry, J.S., and W.S. Geisler. 2002. Gaze-contingent real-time simulation of arbitrary visual fields. In: B. Rogowitz, T. Pappas (Eds.) *Human Visio Electronic Imaging.* SPIE Proceedings, San Jose, CA.

Petry, W. and W. Kuhbauch. 1989. *Automatisierte unterscheidung von unkrauten nach formparametern mit hilfe der quantitativen bild analyse.* Journal of Agronomy and Crop Science. (Berlin) 163:345-351.

Plotze, R.D., and O.M. Bruno, 2009, *Automatic leaf structure biometry: computer vision techniques and their applications in plant taxonomy.* International Journal of Pattern Recognition and Artificial Intelligence 23(2) 247–262.

Ross, T.J., 2004. *Fuzzy logic with engineering applications.* John Wiley and Sons LTD., West Sussex, England. 628 pp.

Shearer, S.A. and P.T. Jones. 1991. *Selective application of post emergence herbicides using photoelectrics.* Trans. ASAE. 34(4):1661-1666.

Shearer, S.A. and R.G. Holmes. 1990. *Plant identification using color co-occurrence matrices.* Trans. ASAE. 33(6):2037-2044.

Storlie, C.A., A. Stepanek, and G.E. Meyer. 1989. *Growth analysis of whole plants using video imagery.* Trans. ASAE. 32(6):2185-2189.

Strickland, R.N. and H.I. Hahn, 1997. *Wavelet Transform Methods for Object Detection and Recovery.* IEEE Transactions on Image Processing. 6(5):724-735.

Stubbendick, J., M.j. Coffin, and L.M. Landholt, 2003.Weeds of the Great Plains. Nebraska department of Agriculture. Lincoln, NE.

Tang, L., L. Tian, B.L.Steward, 2003. *Classification of Broadleaf and Grass Weeds using Gabor Wavelets and an Artificial Neural Network.* Trans. ASAE. 46(4):1247-1254.

Tarbell, K.A. and J.F. Reid. 1991. *A computer vision system for characterizing cotton growth and development.* Trans. ASAE. 34(5):2245-2255.

Tian, F.L., J.F. Reid, and J. Hummel. 1999. *Development of a precision sprayer for site-specific weed management.* Trans. ASAE 42(4):893-900.

Tillet, N.D., T. Hague, and S.J. Miles, 2001. *field assessment of a potential method for weed and crop mapping on the basis of crop planting geometry.* Computers and Electronics in Agriculture (Elsevier), 32(3):229-246.

Tizhoosh, H.R. 1998. *Fuzzy image processing: Potentials and state of the art.* IIZUKA 98, 5th International Conference on Soft Computing, Iizuka, Japan, 1:321-324.

Tizhoosh, H.R. 2000. Fuzzy image enhancement: An overview. In E.E. Kerre and M. Nachtegael (Editors). *Fuzzy techniques in Image processing,* Physica-Verlag (A Springer-Verlag Co.) 5:137-171.

Townsend, P.A. 2000. *A Quantitative Fuzzy Approach to Assess Mapped Vegetation classifications for Ecological Applications.* Remote Sens. Environ (Elsevier). 72:253–267

Tucker, C.J., J.H. Elgin, Jr., and J.E. McMurtrey. 1979. *Temporal spectral measurements of corn and soybean crops.* Photogrammetric Engineering and Remote Sensing. 45(5):643-653.

Wang, Z., W. Chi and D. Feng. 2003. *Shape based leaf image retrieval.* IEEE Proceedings Vision, Image and Signal Processing 150: 34–43.

Wiles, L.J., and E.E. Schweizer. 1999. *The cost of counting and identifying weed seeds and seedlings.* Weed Science 47(6):667-673.

Wiles. L. J, and E.E. Schweizer, 2002. *Spatial dependence of weed seed banks and strategies for sampling.* Weed Sci. 50:595-666.

Woebbeck, D.M., G.E. Meyer, B.K. V., and D.A. Mortensen. 1995. *Shape features for identifying young weeds using image analysis.* Trans. ASAE 38(1):271 - 281.

Woebbeck, D.M., G.E. Meyer, B.K. V., and D.A. Mortensen. 1995. *Color indices for weed identification under various soil, residue and lighting conditions.* Trans. ASAE 38:259-269.

Woebbecke, D.M., A. Al-Faraj, and G.E. Meyer. 1994. *Calibration of Large Field of View Thermal and Optical Sensors for Plant and Soil.* Trans. ASAE 37(2):669-677.

Wyszecki, G. and W.S. Stiles. 1982. *Color Science, Concepts and Methods.* John Wiley and Sons, New York, 117-248.

Yang, C., S.O. Prasher, J.A. Landry, H.S. Ramaswamy, and A. Ditommaso, 2000. *Application of artificial neural networks in image recognition and classification of crop and weeds.* Canadian Ag. Engr. 42(3):147-152.

Yang, C., S.O. Prasher, J.A. Landry, 2002. *Weed recognition in corn fields using back-propagation neural network models* Canadian Biosystems Engr. 22:

Yang, C., S.O. Prasher, J.A. Landry, and H.S. Ramaswamy, 2003. *Development of an Image Processing System and a Fuzzy Algorithm for Site-Specific Herbicide Applications.* Precision Agriculture (Kluwer). 4(1):5-18.

Yonekawa, S., N. Sakai, and O. Kitani. 1996. *Identification of idealized leaf types using simple dimensionless shape factors by image analysis.* Trans. ASAE. 39(4):525-533.

Zadeh, L.A., 1965. *Fuzzy sets, Information and Control.* 8: 338-353.

Zhang, N. and C. Chaisattapagon. 1995. *Effective criteria for weed identification in wheat fields using machine vision.* Trans. ASAE. 38(3):965-974.

Zheng, X. and X. Wang, 2009. *Fast Leaf Vein Extraction using Hue and Intensity Information.* The International Conference on Information Engineering and Computer Science. ICIECS 2009. Wuhan, China. 1-4.

Zheng, X. and X. Wang, 2010. *Leaf vein extraction using a combined operation of mathematical morphology.* The 2nd International Conference on Information Engineering and Computer Science. ICIECS 2010. Wuhan, China. 1-4.

8

Soybean Phospholipids

Daicheng Liu and Fucui Ma
College of Life Science, Shandong Normal University
China

1. Introduction

As soybean phospholipids are coproducts of soybean oil processing, the production of soybean phospholipids rises with the continuous increase of soybean oil yield. Phospholipids have been already applied widely in such fields as medicine, food, agriculture and industry etc., relating to various aspects of everyone's clothing, food, shelter and transportation. New phospholipids products will constantly sprout in large numbers with the development of science and technology.

The authors describe the structure, composition, physical and chemical properties and applications of soybean phospholipids based on the research data in hand. This chapter is focused on the processings of concentrated soybean phospholipids, powdery soybean phospholipids and modified phospholipids as well as the isolation and purification of phosphatidylcholine (PC), phosphatidylethanolamine (PE), phosphatidylinositol (PI), phosphatidylserine (PS) and phosphatidic acid (PA) in soybean phospholipids. The exploration of technologies for isolating and purifying individual molecular species of a certain phospholipids class is now one of the hot and difficult research issues in the world. The breakthrough in these technologies will enormously improve the great development of medicine (e.g. biomembrane bionics, liposomes and intracellular drug carriers etc.) and chemical industry (e.g. aggregation and dispersion of nano materials) etc.

2. The structure, composition and physical and chemical properties of soybean phospholipids

Food Chemicals Codex (FCC) definites phospholipids as follows: Food grade phospholipids are complex mixtures obtained from soybean and other plants consisting of acetone insolubles (AIs) which are mainly phosphatidylcholine (PC), phosphatidylethanolamine (PE) and phosphatidylinositol (PI).

2.1 Soybean phospholipids structure

Phospholipids mainly include glycerol phosphatides and sphingomyelin. In this chapter, we mainly discuss glycerol phosphatides. The structures of phospholipids are shown in Fig. 1.

2.2 Soybean phospholipids composition

Phospholipids constitute 0.3%-0.6% of soybean seed, or 1.5%-3.0% of crude soybean oil. The phospholipids composition is shown in Table 1. The fatty acid composition of soybean phospholipids is shown in Table 2.

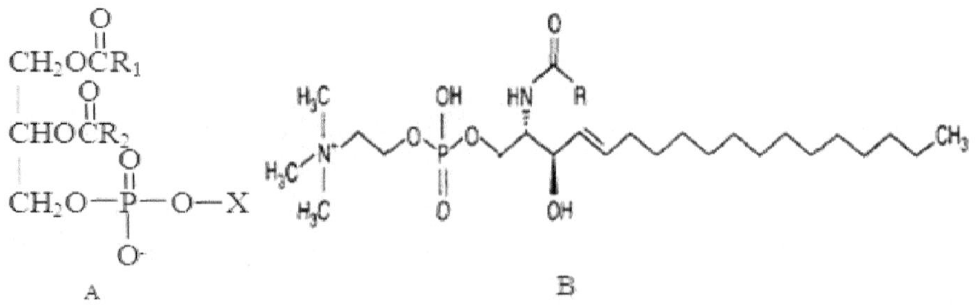

X=-CH$_2$CH$_2$N$^+$(CH$_3$)$_3$	Phosphatidylcholine PC
X=-CH$_2$CH$_2$N$^+$H$_3$	Phosphatidylethanolamine PE
X= 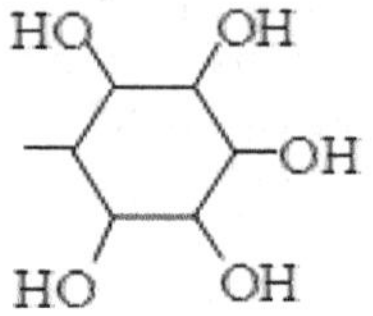 (inositol ring with HO, OH, OH, HO, OH, -OH)	Phosphatidylinositol PI
X=-CH$_2$CH(NH$_2$)COOH	Phosphatidylserine PS
X=-H	Phosphatidic acid PA

Fig. 1. Structures of phospholipids; A: Glycerol phosphatides structure; B: Sphingomyelin; R1, R2, R: Hydrocarbon chains; Point 'X' is likely composed of structures noted in the box.

Component	Abbreviation	Range(%)		
		Low	Intermediate	High
Phosphatidylcholine	PC	12.0-21.0	29.0-39.0	41.0-46.0
Phosphatidylethanolamine	PE	8.0-9.5	20.0-26.3	31.0-34.0
Phosphatidylinositol	PI	1.7-7.0	13.0-17.5	19.0-21.0
Phosphatidic acid	PA	0.2-1.5	5.0-9.0	14.0
Phosphatidylserine	PS	0.2	5.9-6.3	-
Lysophosphatidylcholine	LPC	1.5	8.5	-
Lysophosphatidylinositol	LPI	0.4-1.8	-	-
Lysophosphatidylserine	LPS	1.0	-	-
Lysophosphatidic acid	LPA	1.0	-	-
Phytoglycolipids		-	14.3-15.4	29.6

Table 1. Composition of Soybean Phospholipids (Szuhaj, 1989)

Fatty acid	Range(%)		
	Low	Intermediate	High
Myristic(C14:0)	0.3-1.9	-	-
Palmitic(C16:0)	11.7-18.9	2.5-26.7	42.7
Palmitoleic(C16:1)	7.0-8.6	-	-
Stearic(C18:0)	3.7-4.3	9.3-11.7	-
Oleic(C18:1)	6.8-9.8	17.0-25.1	39.4
Linoleic(C18:2)	17.1-20.0	37.0-40.0	55.0-60.8
Linolenic(C18:3)	1.6	4.0-6.2	9.2
Arachidic(C20:0)	1.4-2.3	-	-

Table 2. Fatty Acid Composition of Soybean Phospholipids (Szuhaj, 1989)

Soybean phospholipids are by-products of soybean oil refining process. Phospholipids composition can be affected by the oil refining processes and may decrease after frost. The lipase may contribute to phospholipids decrease during storage. Other minor compositions in soybean phospholipids include water, pigment, galactosyl glyceride, glycolipids, carbohydrates, sterols and tocopherol etc.

2.3 Physical and chemical properties
2.3.1 Physical properties
Pure phospholipid is a white solid at room temperature, odorless and colorless. The color of phospholipid may be from light yellow to brown due to refining methods, product categories and storage conditions etc. Non decolored, once decolored and twice decolored are three grades of phospholipid color which is determined by Gardner colorimeter (AOCS official method Td-La-64). The chromaticities are from 9 to 17.

Soybean phospholipids are soluble in aliphatic hydrocarbons, aromatic hydrocarbons and halogenated hydrocarbons solvents, such as ether, benzene, chloroform and petroleum ether etc., and particularly soluble in aliphatic alcohol, for example ethanol. As other non-polar surfactants, soybean phospholipid is insoluble in polar solvent, for example methyl acetate, especially acetone (solubility less than 0.03g/L at 5 degrees Celsius). Phospholipids solubility in methyl acetate and acetone increases when there is a small amount of oil in the phospholipids. PC is soluble in ethanol while PI not. The soluble and insoluble portions of PE in ethanol are about equivalent. The differences of soybean phospholipids solubilities in the above solvents may provide references for isolation, purification and quantification of phospholipids. Soybean phospholipids are soluble in animal fat and vegetable oil, mineral oil and fatty acids, but insoluble in cold animal fat and vegetable oil actually.

The hydrophilic phosphate group and alkaline and hydrophobic hydrocarbon keys in phospholipids molecules help to form a interface between water and oil which lowers the interfacial tension between water and oil and makes them stable colloidal. Soybean phospholipids exist in oil and have obvious hydrophilic colloid property. When mixed with suitable amount of water, phospholipids are isolated from oil. Particularly, in hot alkalescent water (pH>8) the phospholipids are more likely to absorb water and expand and then the colloidal solution is formed. Due to the above property, phospholipids are obtained from crude oil and are widely applied (Lu, 2004).

Phospholipids consist of fluidic and plastic phospholipids. Fluidic phospholipids have the flow property of Newtonian fluid and the fluidity of plastic phospholipids increases with the addition of fatty acids. The viscosity of phospholipids is affected by such factors as AI (acetone insoluble) content, moisture, mineral content, acid value (AV) and various additives for example plant-based surfactant. Generally, high AI or water content results in high viscosity while high AV results in low viscosity. Some bivalent minerals for example Ca^{2+} affect viscosity too.

N-hexane insolubles (HIs) make fluidic phospholipids turbid. The turbidness not only influence the appearance of the products, but also leads to precipitation in long-term storage. The phospholipids also get turbid when the water content is over 1% (Wu, 2001).

2.3.2 Chemical properties

Phospholipids are very unstable when exposed to the air or the sunshine, and color deepening and oxidative rancidity easily happen. However, phospholipids are stable in oil without water. So the oil in concentrated phospholipids can prevent oxidative rancidity and is conductive to the phospholipids storage. Phospholipids are unstable at high temperature. In oil and phospholipids processing, the color of the oil get deeper at 150 degrees Celsius and the odor of phospholipids get worse. Phospholipids decompose at over 150 degrees Celsius.

Hydrolysis of phospholipids occurs upon exposure to strong acid at high temperature. Saponification of phospholipids happens when heated in alkaline ethanol or water solution, and soaps are produced. The salts of phosphoglycerol and inositol phosphate are further heated to be hydrolyzed into glycerol, inositol and phosphoric acid. Free fatty acids and free substances of the above compound are produced after acid hydrolysis or high pressure hydrolysis.

Phospholipids can be hydrolyzed by enzymes. At least four kinds of lipases can cleave ester bonds formed by carboxylic acid and phosphoric acid attached to the glycerol molecule and some of them can only cleave unsaturated fatty acids from phospholipids and can't act on saturated fatty acids. These actions result in production of so called lysophospholipids which have strong effect of hemolysis.

Phospholipids may be modified under certain conditions. The modification reactions include hydroxylation, acetylation, hydrogenation, sulfonation, hydroxyl acetylation and enzymatic modification etc. Modified phospholipids vary in their properties and functions (Lu, 2004).

Phospholipids are regarded as the synergist of antioxidant, and they can synergize or prolong the antioxidation functioin of tocopherol. The synergism of phospholipids differs due to the differences of oil and phospholipids. Mixtures of PE, mixed tocopherol and synthetic antioxidant exhibit the highest antioxidant property (Wu, 2001).

3. Soybean phospholipids processing

3.1 Preparation of concentrated soybean phospholipids

Concentrated soybean phospholipids are products obtained by drying and dehydrating hydrated soybean crude oil foot. Industrial methods preparing concentrated soybean phospholipids include continuous and batch processings.

3.1.1 Continuous processing

The processing steps are as follows: The crude oil is heated to 80 degrees Celsius and then centrifuged and passed through the flowmeter followed by addition of 80 degrees Celsius water of 2% (w/w) of the oil.

Degumming oil and oil foot sediments are produced and they can be separated by centrifugation. The hydrated oil foot should be concentrated immediately to avoid microbial rancidity due to the high moisture and neutral oil content. Oil foot (or mixed with hydrogen peroxide or fluidity agent in advance) is pumped into the agitated film dryer. Phospholipids film is formed under gravity or centrifugal force and the pressure of incoming production materials and flow to the bottom of the vessel while moisture evaporates under high temperature and vacuum conditions. The motionless fluid product is dried at vacuum (726 mm Hg) and 100-110 degrees Celsius for 2 min and then cooled to obtain concentrated soybean phospholipids with less than 1% moisture content. The concentrating procedure should be operated under vacuum as phospholipids are thermosensitive (Ji & Li, 2005).

3.1.2 Batch processing

3.1.2.1 Preheating

Mechanically pressing crude soybean oil is preheated to 80 degrees Celsius after removal of impurities by filtration.

3.1.2.2 Hydration

The amount of water added is determined by phospholipids content in the oil and the changes of phospholipids granules formed during heating and is normally 3.5 times (w/w) the content of phospholipids. The water added is usually boiling or 0.7% hot salt solution is used. The speed of adding water is determined by the water absorption velocity of phospholipids. The faster the latter the faster the former, and vice versa. When adding water, the stirring speed must be fast and is normally 80-100rpm at the beginning and is slowed down 20-30min later when large flocculent phospholipids granules are formed and the stirring is continued for another 20-30min. Then the liquid is left standing still to settle. The supernatant of the upper phase is dehydrated to produce refined oil while the oil foot of the lower phase need to be concentrated to obtain phospholipids products.

3.1.2.3 Concentration

The hydrated phospholipids oil foot is drawn into the concentrating tank by vacuum and subjected to temperature rising and stirring. Vacuum dehydration of phospholipids occurs at about 80 degrees Celsius. When there is slight silk flash while stirring the fluidic phospholipids the moisture content is consistent with the specification. The moisture content is about 5%. Phospholipids after concentration is a brown semisolid and can be used in food, medicine and industry.

3.1.2.4 Decoloration

Decoloration of concentrated phospholipids is needed for preparation of high quality phospholipids. The amount of 30% hydrogen peroxide added to the concentrating tank is 2%-2.5% (w/w) of the concentrated phospholipids. The phospholipids are decolored in the closed tank for 1h at 50 degrees Celsius without vacuum. Then turn on the vacuum pump and heat the mixture to 70 degrees Celsius. Dehydrate until there is no water in the water knock vessel. The decolored phospholipids are light brown.

Mixed fatty acids and mixed fatty acid ethyl ester are added as fluidity agents during the vacuum concentrating procedure to improve the fluidic property of concentrated phospholipids and prevent phospholipids separating with the oil and guarantee the stability of phospholipids products.

The products obtained can flow freely at room temperature. If mixed fatty acids added is inadequate, it will not act as fluidity agent. On the contrary, excess addition of mixed fatty acids may raise the AV of phospholipids and get them rancid. The amount of mixed fatty acids added is usually 2.5%-3% (w/w) of the concentrated phospholipids. The addition of mixed fatty acid ethyl ester does not affect the AV and flavor of phospholipids and can gain high quality products but the cost is high. The amount is 3%-5% (w/w) of the concentrated phospholipids (Ji & Li, 2005).

3.2 Preparation of powdery soybean phospholipids

The applications of concentrated phospholipids are limited due to its high content of neutral oil, fatty acids and other substances and its low purity and off-flavor formation. Refining and purifying processings are needed to consistent the phospholipids products with the high purity and non off-flavor specifications of functional food material.

Methods of producing high purity phospholipids from concentrated phospholipids include solvent extraction, ultrafiltration purification, supercritical carbon dioxide extraction etc. So far, acetone solvent extraction is the most widely used method in industry.

3.2.1 Solvent extraction

3.2.1.1 Preparation of powdery phospholipids of one kind of purity from one kind of materials

The acetone solvent extraction theory is isolating phospholipids from oil by precipitation due to the fact that water and oil is soluble in acetone while phospholipids not.

30% hydrogen peroxide is pumped into the closed agitated container with the amount of 2%-3% (w/w) of the concentrated phospholipids. Concentrated phospholipids are pumped into the above container while stirring with the rotate speed of 30-40rpm. Decoloration occurs after the temperature reaches 60 degrees Celsius with a processing time of 6h. After that, the temperature is raised up to 70-75 degrees Celsius and decolor for 0.5h to decompose residual hydrogen peroxide into water. The decoloring procedure is optional due to the product requirements.

Acetone with purity above 98% is pumped through a flowmeter into the closed agitated container. Concentrated phospholipids with acetone residues of the amount of 1:10 (w/w) are pumped into the above container. Stir for another 20-30min with the speed of 80rpm. After that, the liquid is statically sedimented for 0.5h and the upper acetone extract is discharged. The above procedure is repeated three more times with each time a 5:1 ratio (w/w) of acetone to concentrated phospholipids and prolonging sedimentation time. The total amount of acetone is 25 times (w/w) that of concentrated phospholipids.

Phospholipids settle down at the fourth time is discharged and centrifuged. The diameter of the centrifuge rotor is 800mm and the rotate speed is 1200-1600rpm. Centrifuged phospholipids go directly into the lower closed agitated-container. Acetone of 2 times (w/w) the weight of the concentrated phospholipids is pumped into the same container while stirring (80rpm). The extraction procedure lasts 0.5h and then the liquid is discharged and centrifuged to produce phospholipids with 25%-50% (w/w) acetone. The phospholipids are fed into the double conic dryer with the amount of 1/3-1/2 of the whole dryer volume. The drying parameters are as follows: drying temperature 50-55 degrees Celsius, rotate speed 10rpm, vacuum -0.083--0.09 MPa, time 4-6h. Then light yellow powdery phospholipids without acetone residue are obtained and weighed for packaging.

The above method can be applied to prepare powdery phospholipids from such various raw materials as soybean, rapeseed, peanut and corn etc. as well as concentrated phospholipids prepared from hydrated oil foot and alkalized oil foot. The powdery phospholipids produced have a phospholipids content of 90%-98% due to the quality of the concentrated phospholipids (Liu & Yang et al., 2006; Liu & Feng et al., 2006; Liu, 2007).

3.2.1.2 Preparation of powdery phospholipids of various purities from one kind of materials

In acetone solvent extraction, the phospholipids purity increases with the increase of acetone amount and extraction times. The increase of phospholipids purity results in longer time needed to settle the whole phospholipids in acetone solution.

If the purity of the powdery phospholipids obtained in 3.2.2.1 is 97%-98%, half of the phospholipids will be settled in 0.5-1h in the fourth extraction while the other half in 4h. The upper phospholipids solution of acetone is discharged when the time has passed 2.5-3h and centrifuged and dried. The purity of the phospholipids produced can reach up to over 99%.

Acetone of 2 times the weight of concentrated phospholipids is added into the extraction tank with agitation. The extraction time is 0.5h and static settle time is 1.5-2h. The upper phospholipids solution of acetone is discharged and centrifuged and dried to obtain phospholipids product with purity of over 95%.

The residual liquid is discharged, centrifuged and dried to produce phospholipids product with purity of about 90% (Liu & Ma, 2011).

This method can produce phospholipids products with various purities due to the product purity obtained in 3.2.2.1 and discharging time to meet the market requirements, and make the best use of the materials.

3.2.2 Supercritical carbon dioxide extraction

Extraction temperature, pressure and time are important technological parameters of supercritical carbon dioxide extraction. Extraction yield increases with the increase of one of the parameters in a certain range while the other two conditions remain unchanged. However, there are also problems of increased cost, power consumption and unsafe factors. Generally, the extracting effect is rather good at 50 degrees Celsius and 20MPa for 5h.

Supercritical carbon dioxide extraction used to extract soybean phospholipids has significance for the industrial application and is an applicative technology with wide prospect as it has the advantages of simple, non solvent residue, safe and reliable and high purity product and it consists with the trend of current green chemical technology (Shi, 2005).

3.2.3 Ultrafiltration purification

The crude phospholipids are subjected to derosination and dissolved in solvents and then passed through ultrafiltration film with certain pore size. Components of suitable sizes pass through the membrane and are isolated.

Ultrafiltration lecithin introduced by ADM (Archer Daniels Midland Co.) which has the property of dry, easy to be mixed with other materials, high quality and high purity is produced by removing the glycerides in phospholipids by ultrafiltration. Ultrafiltration lecithin can be precisely quantified and conveniently used. In certain situation which has strict requirements for flavor ultrafiltration lecithin is precious as it has good flavor (Shi, 2005).

3.3 Preparation of modified soybean phospholipids
3.3.1 Chemical modification

3.3.1.1 Hydrogenation

After hydrogenation, the unsaturated double bonds of the phospholipids are saturated to improve the stability, oxidative stability, color and odor of the phospholipids. Hydrogenated phospholipids are mainly used in cosmetics, dyes and lubricants.

Powdery soybean phospholipids are dissolved in the mixture of dichloromethane and ethanol (3:1, v/v) with a 1:6 (w/v) ratio in the stainless steel autoclave. A 5% palladium/carbon catalyst is added into the autoclave followed by leakage checking. Then the air in the autoclave is displaced by hydrogen for several times. The reacting parameters include a temperature of 50 degrees Celsius, a pressure of 0.6MPa, a stirring speed of 300r/min, and a reacting time of 3h under constant temperature and pressure. After reaction, the temperature and pressure are reduced. The catalyst is removed and recycled by centrifuging the reaction product. 30% hydrogen peroxide with the amount of 5% (w/w) is added into the liquid portion to decolor and the solvent is removed by rotate evaporation at 55 degrees Celsius. Light yellow solid hydrogenated soybean phospholipids are obtained after vacuum drying at 70 degrees Celsius for 8h. It may be better to use pure ethanol as solvent than the mixture of dichloromethane and ethanol when hydrogenating phospholipid that is soluble in ethanol such as PC (Huang et al., 2003).

3.3.1.2 Acetylation

PE is transformed into N-acylphosphatidylethanolamine after acetylation, and its 'zwitter ion' structure is modified to obtain improvements in Hydrophile-Lipophile Balance (HLB) value, thermostability, oil in water emulsifying ability and viscosity property. Meanwhile, N-acylphosphatidylethanolamine's large solubility in acetone facilitates isolation and purification of phospholipids. Acetylation with acetic anhydride is used to produce acetylated phospholipids in industry.

Considering acetylated phospholipids are mainly applied in food processings, direct heating (noncatalytic) acetylation process is adopted to produce food grade acetylated soybean phospholipids. Acetic anhydride is added into crude phospholipids with the amount of 1%-4% (v/w) due to the PE content in phospholipids and the amino conversion rate. The process requires temperatures of 60-70 degrees Celsius and stirring reacting time of 1h-1.5h. After acetylation, the mixture is neutralized with sodium hydroxide or potassium hydroxide of certain concentration and then dried under vacuum. The specifications of acetylated phospholipids are: free amino 0.7%-1.7%, pH 6.5-8, and HLB value 5-6 (Xu et al., 2008).

3.3.1.3 Hydroxylation

The hydroxylation theory is that two hydroxyls are introduced into the double bonds of the unsaturated fatty acids of phospholipids molecules, i.e., crude phospholipids react with hydrogen peroxide with the existence of organic weak acid such as lactic acid to hydroxylate the unsaturated bonds in phospholipids and oil. The ethanolamine group of PE is also modified. The obvious hydrophilic property makes modified phospholipids more easily disperse in cold water. The degree of hydroxylation modification is controlled by the amount of hydrogen peroxide added and usually measured by the drop in iodine value (IV). The products with 10%-25% drops in IV have good water dispersibility and oil in water emulsifying property. The emulsifying property decreases and hydrophilic property increases with the increase of drop in IV, but the cost rises too.

Phospholipids hydroxylation processes include such various methods as 'lactic acid + hydrogen peroxide + phospholipids', 'acetic acid + hydrogen peroxide + phospholipids', 'peracetic acid + phospholipids' and 'basic potassium permanganate + phospholipids' etc., which belong to alkyleneortho-dihydroxylation and have various hydroxylation effects. In industrial production, the 'lactic acid + hydrogen peroxide + phospholipids' process is generally adopted as it's a mild method with no problems of the three wastes(waster gas, waster water and industrial residue) and meets the food grade requirements. 75% lactic acid and 30% hydrogen peroxide with the amount of 1%-3% and 5%-15% (v/w), respectively, are added into crude phospholipids. The reaction is carried out at 50-70 degrees Celsius with stirring for 1h-3h. The mixture is neutralized with sodium hydroxide of certain concentration and then dried under reduced pressure until a less than 1% moisture content is reached. The specifications of hydroxylated phospholipids include: drop in IV 10%-25%, pH 6.5-7.5, HLB value 9-10 (Xu et al., 2008).

3.3.1.4 Acetyl-hydroxylation

Acetyl-hydroxylation refers to acetylation of phospholipids followed by hydroxylation, i.e., double modification. Hydroxylation occurs between phospholipids and hydrogen peroxide with the help of acetic acid produced by acetylation. The procedures are as follows: acetic anhydride is added into the crude phospholipids with the amount of 1%-4% (v/w) due to the PE content in phospholipids and the amino conversion rate. The reaction is carried out at 60-70 degrees Celsius for 1h-1.5h with stirring. Then hydrogen peroxide of 5%-15% (v/w) is added. Temperatures of 60-75 degrees Celsius and stirring reacting time of 1h-3h are required. At last the mixture is neutralized with sodium hydroxide of certain concentration and then dried under reduced pressure until reach a less than 1% moisture content. The specifications of acetyl-hydroxylated phospholipids are drop in IV 10%, free amino no more than 1.65%, pH 7-8, HLB value 9-10 (Xu et al., 2008).

3.3.1.5 Hydroxyl-chlorination

100 portions of soybean phospholipids are dissolved in 300 portions of n-hexane. Sodium hypochlorite of 22.5% (w/w) of the total phospholipids is added and the pH is adjusted to 4.5 with acetic acid. The reaction is carried out at 50 degrees Celsius for 1h with stirring. The mixture is washed 3 times with each time 100 portions of water is used. The upper phospholipids solution is evaporated to recycle solvent and obtain hydroxyl-chlorinated soybean phospholipids. The emulsion stability, dispersibility and wettability are improved enormously compared with that of non-modified phospholipids (Xu et al., 2008).

3.3.1.6 Sulfonation

The most likely positions for sulfonation are the double bonds of long chain unsaturated alkanes and α -carbon near ester bonds. When sulfonation of phospholipids including PC, PE, PA and PI etc. happens, the position which is most likely to be introduced with active group is hydroxyl of PI. That is to say, sulfonation occurs on double bonds while esterification (sulfation) occurs on hydroxyls. The double bonds in products will diminish or vanish if sulfonation occurs on double bonds totally or partly. The decrease in unsaturation of sulfonated soybean phospholipids results in the drop of IV. So we can determine whether sulfonation on double bonds happens or not by measuring IV.

There have been a lot of reports on sulfonation and sulfation of phospholipids, but maturer method is sulfur trioxide gas phase continuous film sulfonation which is developed in China. The film sulfonation pipe need heat preservation jacket. The parameters of the sulfonation process are a feed temperature of 40 degrees Celsius, a sulfur trioxide/air flow rate of $1.5m^3/h$ and a protective wind flow rate of $0.25m^3/h$. Continuous sulfonation happens in film sulfonator followed by neutralization with alkali and decoloration with 5%-20% hydrogen peroxide. The sulfonated phospholipids with a 4% sulfur trioxide binding amount and 6-8 pH exhibit such properties as light color, hydrophilic property, emulsifying property and good permeability.

Sulfonation provides soybean phospholipids with special properties and raises the HLB value from 1-2 to 12-16. The physical and chemical properties of phospholipids are improved enormously to facilitate the wide applications of phospholipids as fatliquoring agent, flotation agent and emulsifying agent in leather, pharmacy and farm chemical etc. (Zhang et al., 2004).

3.3.1.7 Alkoxylation

Alkoxylation technology including ethoxylation and propoxylation etc. is a main technology producing nonionic surfactant. It is carried out by addition-condensation reaction of oxirane or epoxypropare with initiators (aliphatic alcohol and nonyl phenol etc.), and the initiator-ethoxylation or initiator-propoxylation products are obtained.

Alkoxylated phospholipids are obtained by addition-condensation reaction of alkoxylated reactant such as oxirane and phospholipids containing hydroxyl such as PE and PI. As hydrophilic oxethyl groups are introduced into the polar end of PE and PI molecules, the HLB value and hydrophilic property are increased and the oil-in-water emulsifying ability improved. As with sulfonation and hydrogenation, the alkoxylation process is very complex and the products are mainly used in non-food industry.

There are not many manufacturers producing this kind of products. R & R551 is the representative ethoxylated soybean phospholipids of ADMC. The ethoxylated phospholipids have a 12.5 HLB value.

According to patents that have been made public and reports related, the soybean phospholipids alkoxylation process is mainly as follows: addition reaction occurs between phospholipids (PE and PI) and alkoxylation reactant (oxirane, epoxypropare and glycidol etc.) and alkoxylated soybean phospholipids are produced. For example, 23.5 pounds of oil-containing soybean phospholipids are dissolved in 15 pounds of dimethylbenzene. 4.5 pounds of oxirane is added. The reaction is carried out at 100 degrees Celsius and 0MPa for 3h followed by removal of solvents. The ethoxylated soybean phospholipids which are resinous, insoluble in dimethylbenzene and soluble in water are obtained (Xu et al., 2008).

3.3.2 Enzymatic modification

Chemical modification of phospholipids improves their emulsifying and hydrophilic properties, but damages natural structure of phospholipids as well. Enzymatic modification exhibits such advantages as no need for purifying the reactant, mild reacting conditions, fast, complete, less by-products, exact action position of enzymes and easy to obtain etc. Phospholipases including phospholipase A_1, A_2, C and D can catalyze various hydrolysis of phospholipids as well as esterification and interesterification reaction with the existence of certain acyl receptor and donor to change or modify the structure of phospholipids which will gain different structures and applications. Phospholipase A_2 and D are used in industries while the other enzymes are on the experimental status (Gu et al., 1999).

3.3.2.1 Phospholipase A₁ and number 1,3-position specific phospholipase

Phospholipase A_2 can specifically hydrolyze the acyl at number Sn-1 position of natural phospholipids. But acyl at number Sn-2 position can be easily transferred onto the thermodynamically stable number Sn-1 position and this results in producing of the same products as with phospholipase A_2. The source of phospholipase A_1 is very limited.

Phospholipase with number 1,3-position specificity can selectively hydrolyze acyl at number Sn-1 position of phospholipids, and can replace phospholipase A_1. Number 1,3-position specific phospholipase can directly catalyze interesterification of phospholipids and fatty acids or oleic acid in organic solvents to produce new phospholipids. For example, Lipozyme IM20 can catalyze the intersterification of PC and fish oil fatty acids with 45% eicosapentaenoic acid (EPA). The parameters are: enzyme amount 1.5% (w/w), the optimal ratio of phospholipids to fatty acids 1:2 and the optimal organic solvent n-hexane. The binding ratio of EPA at number Sn-1 position is 17.7%. Polyunsaturated fatty acids such as EPA and docosahexaenoic acid (DHA) which are good for cardiovascular and cerebrovascular health can be attached to phospholipids and obtain better digestion and absorption properties than that with triglyceride through this kind of reaction (Gu et al., 1999).

3.3.2.2 Phospholipase A₂

Phospholipase A_2 (EC 3.1.1.4) specifically catalyzes the hydrolysis of acyl at number Sn-2 position of phospholipids to produce lysophospholipids and fatty acids. Modified soybean phospholipids exhibit obviously improved hydrophilic and emulsifying properties. They can maintain good emulsifying property under conditions of high or low temperatures or low pH or various salt concentrations. Lysophospholipids are applied in bakery food. They form complexes with amylose and retard aging of breads effectively. Lysophospholipids are two times the price of ordinary phospholipids but they have the advantages of smaller dosage, better effect, oxidative stability and antibiotic property. They are industrially produced in Japan and America. The process is as follows: phospholipids are subjected to moisture content adjustion previously and then added into the solutions with phospholipase A_2 of 0.02% (w/v) and calcium chloride of 0.3% (w/v) with stirring. The temperatures are 50-55 degrees Celsius and the reacting time is 7h-9h. The hydrolyzing degree reaches 35%-40% when the acid value (AV) is in the range of 33-30. The following procedures are required to obtain powdery lysophospholipids: concentration under reduced pressure, pressure filtration, washing with acetone, solvent removal under reduced pressure and vacuum drying (Song et al., 2007).

3.3.2.3 Phospholipase C

Phospholipase C acts on phospholipids to produce diglyceride, phosphoinositide, phosphocholine, phosphoethanolamine and phosphoric acid etc. Diglyceride is a bioactive substance which acts as the second messenger in cell signaling transmission and affects the cell metabolism. There are three kinds of specificities of microbial phospholipase C: the first one specifically hydrolyze PI into diglyceride and cyclic phosphoinositide; the second one specifically hydrolyze sphingomyelin and the third one has relatively wider specificity and takes PC as the optimal substrate (Song et al., 2007).

3.3.2.4 Phospholipase D

Phospholipase D (EC 3.1.4.4) can hydrolyze PC into phosphatidic acid and choline. In microwater system with alcohol, phospholipase D can catalyze transacylation which results in exchange of primary or secondary hydroxyl of some molecules with ethanolamine or choline groups of phospholipids and formation of new phospholipids. This character is called phospholipids' transfer characteristic or base exchange reaction of phospholipase D (Song et al., 2007).

4. Extraction and isolation of soybean phospholipids

4.1 Soybean phosphatidylcholine (SPC)
4.1.1 Organic solvent extraction
Fractions in soybean phospholipids are isolated due to their solubilities' differences in organic solvents. PC exhibits large solubility in lower alcohol (C1-C4) whereas PE and PI have small solubilities. PC- and PI-enriched products can be obtained by their solubilities' difference. When treated with lower alcohol, PC in deoiled phospholipids is soluble in alcohol leaving insoluble matter consisting mainly of PE and PI. The ratio of PC to PE increases from 1:1 (w/w) in raw material to 3:1 (w/w), and even to 12:1 (61% PC, 5% PE).

Better isolation effect on PC can be obtained by isopropanol. Mixtures of high-purity phospholipids and isopropanol with the ratios of 8:157-16:157 (12:157 is optimal) is added into the agitated- and refluxed-closed container. The extraction is conducted in thermostatic water bath or cryohydrate bath with isopropanol volume fraction of 95%-100% (100% is optimal) at -5-15 degrees Celsius (-5 degrees Celsius is optimal) for 5min-11min (5min is optimal). After the extraction, the mixture is filtered, evaporated to remove isopropanol and dried to obtain product with mass fraction of PC increased from 25.6% in raw material to 66.8%. Isopropanol extraction is the most commonly employed step to obtain high PC-containing phospholipids from deoiled phospholipids (An et al, 2001).

4.1.2 Lower alcohol with salt or acid extraction
It's also an effective method to fractionate phospholipids with the property that phospholipids can react with some salts or acids and precipitate. This method is more promising than organic solvent extraction as metal ions or acids can 'recognize' phospholipids molecules more effectively than solvents.

100g of phospholipids containing 45% PC is dissolved in 1L ethanol of 95% before addition of 4.5g of zinc chloride. The light yellow phospholipid-zinc chloride compound precipitate formed is centrifuged, decomposed with 250ml freezed acetone under nitrogen to obtain 36.7g of phospholipids containing 99.6% PC (Ni, 1995).

4.1.3 Supercritical fluid extraction
Supercritical fluid extraction (SFE) is a rapidly developed new technology in recent years. Supercritical fluids most commonly used are carbon dioxide, ammonia, ethylene, propylene and water etc. Carbon dioxide is most frequently used due to its following properties: critical temperature and pressure easy to get, stable chemical properties, non-toxic, odorless, non-corrosiveness and reusable. Supercritical carbon dioxide extraction (SCE) is a method with bright prospect as it can obtain high purity products with no solvent residues and maintain the nutritional and functional properties of the products and need simple process, single equipment and low cost (Guan et al., 2005).

Tekerikler et al. (2001)obtained phospholipids containing 91% PC after SCE of deoiled phospholipids with 10% ethanol as entrainer at 17.2MPa and 60 degrees Celsius. Increasing the pressure to 20.7MPa increased the extraction yield and PC content (95%). Increasing the temperature to 80 degrees Celsius decreased the extraction yield and PC content which was attributed to decrease of solubility and selectivity of solvents to PC as the solvents density decrease at high temperatures.

4.2 Soybean phosphatidylinositol (SPI)

PI causes concern as it is involved in the transmission of messages in the cell. PI plays and important role in maintaining normal physiological functions of central nervous system, especially in regulating calcium homeostasis. PI on cell membrane can be hydrolyzed by phospholipase C into 1,4,5-triphosphate inositol that goes into intracellular aqueous phase as the second messenger and 1,2-diacylglycerol that stays in the cell wall. These two substances synergetically induce cell reactions such as contraction, secretion, metabolism and proliferation etc.

Soybean phospholipids are rich in PI. PI is a white amorphous solid with its sodium salt a crystal and is wet-sensitive. PI is soluble in water, chloroform and benzene, slightly soluble in methanol, diethyl ether and petroleum ether, insoluble in acetone, ethanol and water. It can be easily oxidized upon exposure to the air (Deng et al., 2003).

4.2.1 Solvent method

The solvent method is conducted to isolate and purify phospholipids and increase the content of a certain constituent with single or mixture of such solvents as methanol, ethanol, isopropyl ketone, acetone, n-hexane and chloroform etc.

The deoiled soybean phospholipids are extracted with appropriate amount of ethanol. The induced ethanol-insoluble phase is vacuum dried to obtain a mixed phospholipids with more PI and less PC which are dissolved in n-hexane before adding 55% ethanol with sodium acetate. The mixture is put into the separating funnel, shaked, allowed to rest and layered. The same procedure is carried out again except that the 55% ethanol is sodium acetate free. The PI obtained is 40%-50% pure. If sodium acetate is replaced by aqueous ammonia with a 8.0 pH and the ethanol concentration increased to 90%, PI of 85% pure can be obtained with the same method.

Purer PI can be obtained by some chemical reaction methods. Soybean phospholipids containing 40% mixed phospholipids are dissolved in such organic solvents as anhydrous pyridine, acetonitrile, dimethyl formamide (DMF) and dimethylsulphoxide (DMSO) etc. Chloride dimethyl tertiary butyl silicon, chloride trimethyl silicon or allyl bromide are added into the mixture to protect hydroxyls of PI by reacting with them. Then PI is isolated from the mixture with solvents such as acetone or ethanol-acetone and hydrolyzed by alkali or acid at room temperature to remove the blocking groups and recover the hydroxyls of PI. PI obtained this way is 98% pure and applied in treating of central nervous system disorder (Deng et al., 2003).

4.2.2 Column chromatography

The phospholipids mixture is dissolved in the mixture of chloroform and methanol in the 1:1 (v/v) ratio before adding aluminium oxide. The eluate contains PC, lysophospholipids, neutral lipids and glycolipids etc. Residues are washed and extracted with the mixture of chloroform, methanol and 1% hydroxyl ammonium acetate in the 1:1:0.3 (v/v/v) ratio and the eluate is loaded on silica column of which the dimension is 30cm. Neutral lipids are eluted with chloroform; glycolipids and PE are sequentially eluted with chloroform and methanol in the 80:20 (v/v) ratio; PE is further removed with chloroform and methanol in the 20:80 (v/v) ratio; PI is finally eluted with chloroform, methanol and 25% ammonia in the 80:20:5 (v/v/v) and 65:25:5 (v/v/v) ratios. The PI-containing fraction is evaporated and dried to obtain PI of no less than 98%-99% pure (Deng et al., 2003).

Column chromatography can obtain high purity PI but the long time needed and the use of complex solvent mixture reduce its feasibility in the commercial world.

4.2.3 Enzymatic method

Phospholipids can be hydrolyzed by such phospholipase as phospholipase A_1, A_2, C and D. When treating the ethanol-treated phospholipids (containing minor PC), phospholipase selectively catalyze hydrolization of PE and PC but not PI. More special, alkaline or acid phospholipase catalyzes hydrolization of PA and some salts produced by PE hydrolization but doesn't act on PI, PC or PE. PI products used in various fields can be obtained by this method and the purity can reach up to 99%. Lypase can be used to purify PI as well, and the purifies are 60%-70% (Deng et al., 2003).

Enzymatic method has wide application prospect as it is simpler, more convenient and environmental prospective compared with solvent method and column chromatography.

4.2.4 Other methods

Ion exchange resin may be applied to isolate PI from phospholipids mixture. The resin adsorbing PI include diethylaminoethylcellulose, diethylaminoethylagarse and quaternaryammoniumethylsephadex etc. (Deng et al., 2003).

4.3 Soybean phosphatidylethanolamine (SPE)

Solid-liquid extraction is performed using powdery soybean phospholipids and ethanol. The parameters are as follows: ratio of phospholipids to ethanol 30g/L, ethanol concentration 95%-100% (100% is optimal), extracting temperatures -15-50 degrees Celsius (-15 degrees Celsius is optimal) and extracting time 5min-11min (8min is optimal). PE content increases from 19.8% in raw material to 62.8% (An et al., 2006).

Phospholipids are dissolved in isopropanol below 65 degrees Celsius to reach a final concentration in the range of 1%-4% (w/v). The mixture is cooled to 26 degrees Celsius, allowed to rest and the precipitate is filtered and dried to obtain PE of 74.7%-79.9% pure (Ni, 1995).

Zhensheng Zhong et al. (2008) removed oil and fatty acids in powdery phospholipids with acetone first, and removed PC with repeated ethanol extraction due to PC has larger solubility in aliphatic alcohol than PE and PI, and finally enriched PE with petroleum ether extraction due to PE is soluble in ether while PI not.

The powdery soybean phospholipids are extracted repeatedly with acetone to remove oil and refined phospholipids containing 35% PE are obtained. PC is removed by repeated absolute ethanol extraction with heating and stirring and alcohol insolubles are obtained. The alcohol insolubles are extracted 3 times with petroleum ether in the 1:3-1:6 (1:4 is optimal, v/v) ratios at 30-60 degrees Celsius (30 degrees Celsius is optimal). PE obtained this way is 93.5% pure and has a yield of 91.9% (Zhong & Wei, 2008).

4.4 Soybean phosphatidylserine (SPS)

Pure PS is a white waxy solid. It's soluble in most of the nonpolar solvents containing little water, insoluble in anhydrous acetone and can be extracted from histiocyte with chloroform and methanol. When PS is dissolved in water, most of the insoluble lipids form micell while very few form true solution. PS has one positive and two negative charges, resulting in a net negative charge. PS can be hydrolyzed by weak base into metal salts of fatty acids and a remained portion, and by strong alkali into fatty acids, serine and glycerol phosphate. PS is ready to oxidize upon exposure to the air, and the color gets darker from white to yellow and finally black. Natural PS practically isn't affected by alcohol while saturated PS can form interwoven catenulated gel with alcohol and dipalmiloyl-phosphatidylserine can interact with 5% alcohol at room temperature to form regular gel.

PC is dissolved in organic phase while the enzyme and serine are dissolved in aqueous phase. After preheating for a while, the two phases are combined, and reaction occurs at the interface under certain conditions. PS is obtained by isolating and extracting of the organic solvent phase and quantified by thin layer chromatography (TLC). The parameters are as follows: ratio of organic phase to aqueous phase 4:4 (v/v), PC concentration 75mg/ml, reacting temperature 40 degrees Celsius, pH of aqueous phase 4.0 and reacting time 12h. PS yield is 68.9% (Yang, 2010).

Blokland et al. (1999) compared the effects of bovine cortex phosphatidylserine (BCPS), SPS and egg phosphatidylserine (EPS) on cognitive competence of middle aged rats. The dosage given to lab mice was 15mg/kg.d. Changes of emotional behavior and cognitive competence in open field experiment, Morris water maze and two-dimension active avoidance experiment were observed. Arjan Blokland discovered that SPS and BCPS exhibit similar cognitive competence-improving effects which were higher than that of EPS. SPS might be a substitute for BCPS.

4.5 Preparation of phospholipids for injection

1 portion of powdery soybean phospholipids is mixed with 12 portions of distilled water and stirred to form colloidal dispersion liquid in boiling water bath before 1.8 times the weight of raw material of anhydrous sodium sulfate is added. The saturated sodium sulfate solution is discarded after blocky phospholipids are precipitated. Then 5 portions of distilled water and 0.8 portions of anhydrous sodium sulfate are used to repeat the salting out procedure. The salting-out soybean phospholipids are dried at reduced pressure and 70 degrees Celsius in water bath in vacuum drier, transferred into three-mouth bottle followed by addition of 8.7 times the weight of raw material of 95% ethanol and reflux extraction at 80 degrees Celsius in water bath with stirring for 1h. After cooling, the ethanol solution containing phospholipids is poured out and stored in refrigerator overnight to precipitate PC. The ethanol solution containing soybean phospholipids is poured out, heated to about 35 degrees Celsius in water bath before addition of activated aluminum oxide of 0.5 times the weight of raw material, stirred for 1h and filtered. The ratio of powdery phospholipids, ethanol and aluminum oxide is 1:8:0.5 (w/w/w).

The above ethanol solution is poured out followed by addition of activated carbon of 0.22 times the weight of raw material, stirring for 1h and filtration with sintered funnel. The filtrate is transferred into the distillation flask and subjected to reduced pressure distillation at 70 degrees Celsius in water bath under nitrogen to remove ethanol. Diethyl ether of 0.75 times the weight of raw material is added into the distillation flask to dissolve the dried soybean phospholipids. The diethyl ether solution is bottled and the bottle is airtight after filling in nitrogen and stored in refrigerator overnight before ultrafiltration. The diethyl ether is removed by reduced pressure evaporation at 40 degrees Celsius in water bath under nitrogen in evaporator. Anhydrous acetone is added into the glutinous soybean phospholipids in the evaporator. The mixture is pestled and embathed for several times to remove the residual oil and moisture. Then a powdery parenteral soybean phospholipids product is produced. It exhibits the following characteristics: AV 9.9, IV 91.29, nitrogen content 1.9%, phosphorus content 4.08% and AI 99.3% (PC content is 96.7%) (Shao et al., 2000).

5. Extraction and purification of individual molecular species of soybean phospholipids

Extraction and purification of phosphatidic acid of C_{18} fatty acids

Powdery soybean phospholipids containing 20% PA are extracted with five folds (by weight) of 95% ethanol at 45 degrees Celsius for 2h with stirring at the speed of 100rpm. After centrifugation at 700xg for 10min, the supernatant is discarded. The above extraction procedure is repeated four more times until the ethanol fraction is colorless. The solid fraction is extracted with 5 folds (by weight) of methanol with a stirring speed of 100rpm for 12h at room temperature. The methanol fraction is obtained after centrifugation at 700xg for 10min. The methanol extraction procedure is repeated three times in total. The methanol fractions are combined. If the methanol is removed by evaporation, the solid residue will contain about 50% PA.

The pH of the methanol extract is adjusted to 8-9 by 1mol/L sodium hydroxide and obvious white precipitate is observed. The supernatant is obtained by centrifugation at $700 \times g$ for 5min. If the methanol is removed, the solid residue will contain about 70% PA. The pH of the supernatant is adjusted to 5-6 by 1mol/L hydrochloric acid before n-hexane of four times the volume of the methanol solution is added and mixed. The n-hexane phase is obtained after extracting for 15min and being left standing still for 2h. The n-hexane is evaporated at 45 degrees Celsius and the residue is dissolved in methanol of ten times the volume of the n-hexane phase. 60% zinc chloride solution is added into the methanol solution until no more white precipitate is formed. The precipitate is obtained by centrifugation at $700 \times g$ for 10min, washed three times with acetone which is removed by filtration and dried under nitrogen gas steam. The solid obtained is PA of C_{18} fatty acids of about 98% pure. The yield is 1/60 of the raw materials (Liu et al., 2008).

6. Applications of phospholipids

Phospholipids are widely applied in pharmaceutical field, food, feed, agriculture, daily chemical industries and other chemical industries.

Pharmaceutical field: Phospholipids exist in all of the biomembranes and play important roles in such various physiological processes as regulating of cell osmosis and membrane enzymes, transmitting of lipoids and sterols and metabolizing of cyanocobalamin, folic acid and methionine etc. Brain tissues contain 25% of phospholipids, and the metabolic abnormality of phospholipids may lead to such diseases as cancer and Alzheimer's Disease etc. High-dosage phospholipids can effectively treat neurological disorders and other diseases of nervous system. In recent years, almost 25% of the non-food application patents of phospholipids are about their applications in pharmaceutical field, especially the applications of liposomes. Phospholipids exhibit huge potential in health care products market.

Food industry: The amount of phospholipids used in food is usually 0.1%-2% of the fat in food. Phospholipids are used in margarines, shortenings, candies, soup bases, pot foods, instant foods (e.g. milk powder), bakery products (e.g. bread, cookie, dessert, biscuit and cracknel) and processed foods of meat and seafood etc. They are also used as coatings of can, soup packaging and casing of meat such as sausages etc.

Feed industry: Phospholipids are applied in animal feed such as milk replacer for calf and feed of cattle, pig, poultry, hairy animals, pets and aquatic animal (e.g. fish and crustaceans) etc. Phospholipids are the essential additives of the eel feed as they decrease the diseases of the eel and improve their growth.

Agriculture: Phospholipids can inhibit the growth of powdery mildew on cucumber, eggplant, green pepper and strawberry. The solution of 0.1% phospholipids-sodium carbonate can effectively inhibit orange green mold, cucumber powdery mildew and rice blights. Phospholipids are used as the coating components after harvesting of fruits and vegetables to improve the storage effect. Phospholipids are additives of pesticide formulae which can improve the adhesivity and permeability of pesticide and reduce their toxicity to plants.

Daily chemical industries: Phospholipids are applied in such cosmetics as moisturizer, facial cleanser, sunblocking cream, soap, bath oil, shampoo, hair care agent, shaving cream, shave clean agent, nail polish, makeup powder, blush, rouge, eye shadow, lipstick and hairspray etc. Applying of phospholipids in detergent can improve the dirt-removing power of anionic detergents.

Other chemical industries: Phospholipids are widely used in various paints, wax, shoes polish, wood preservatives, mold spray, tape coating, printing ink, ink, toner, additives of photographic materials and polyamide coating etc. as well as in papermaking and printing. They are also widely used in cement, pitch, tar shingle, surface sealant of linoleum and putty gum etc.

7. References

An, H.; Wang, M.L. & Cheng, L.B. (2001). Fractionation of Soya Lecithin With High Phosphatidylcholine Content Using Isopropanol. *Fine Chemicals*, Vol.18, No.7, pp. 385-387, ISSN 1003-5214

An, H.; Zhang, L. & He, X.F. (2006). The Fractionation of Phosphatidylethanolamine. *Food Science*, Vol.27, No.12, pp. 343-346, ISSN 1002-6630

Blokland, A.; Honig, W.; Browns, F. & Jolles, J. (1999). Cognition-Enhancing Properties of Subchronic Phosphatidylserine (PS) Treatment in Middle-Aged Rats: Comparison of Bovine Cortex PS with Egg PS and Soybean PS. *Nutrition*, Vol.15, No.10, pp. 778-783, ISSN 0899-9007

Deng, Q.G.; Qi, L. & An, H. (2003). Study On Separating Technique of Soybean Lipositol. *Chemical Engineer*, No.6, pp. 14-16, ISSN 1002-1124

Gu, L.W.; Gao, X.R. & Zhao, J.L. (1999). Recent Advances in Enzymatic Modification of Phospholipids. *China Oils and Fats*, Vol.24, No.6, pp. 60-62, ISSN 1003-7969

Guan, R.L.; Zou, J. & Zhu, H. (2005). Recent Advances in Isolation and Purification of Soybean Lecithin. *China Food Additives*, No.5, pp. 44-48, ISSN 1006-2513

Huang, G.W.; Zhao, J.T. & Gu, K.R. (2003). Hydrogenation of Soybean Powder Lecithin. *China Oils and Fats*, Vol.28, No.2, pp. 51-54, ISSN 1003-7969

Ji, B.P. & Li, B. (2005). *Soybean Production Safety and Quality Control,* Chemical Industry Press, ISBN 7-5025-7036-5, Beijing, China

Liu, D.C. (2007). A Method of Producing Powdery Phospholipids Using Basified Concentrated Phospholipids, Patent number ZL200410024223.8

Liu, D.C.; Feng, J.J.; An, L.G.; Zhu, H.R. & Han, L.Y. (2006). A New Method of Producing Powdery Phospholipids Using Concentrated Phospholipids of Peanut, Soybean or Colza As Raw Materials, Patent number ZL200310105345.5

Liu, D.C. & Ma, F.C. (2011). A Method of Producing Powdery Phospholipids of Various Phospholipids Contents With One Kind of Concentrated Phopholipids, Patent number 201110060102.9

Liu, D.C.; Tao, Y.H.; Sun, F.Z. & Li, Y.L. (2008). Extraction and Purification of Phosphatidic Acid of C_{18} Fatty Acids from Powdered Soybean Phospholipids. *Asian Journal of Chemistry,* Vol.20, No.7, pp. 5595-5600, ISSN 0970-7077

Liu, D.C.; Yang, Y.T. & An, L.G. (2006). A New Technique of Producing Powdery Phospholipids, Patent number ZL200510042316.8

Lu, Q.Y. (2004). *Production Technology of Oil and Fat Chemical Products,* Chemical Industry Press, ISBN 7-5025-5098-4, Beijing, China

Ni, H.Z. (1995). Refining and Isolation of Phospholipids. *Jiangsu Chemical Industry,* Vol.23, No.1, pp. 12-15, ISSN 1002-1116

Shao, X.F.; Zhang, J.L.; Wang, F.L. & Chen, Q.H. (2000). Studying on Injection Phosphatide. *China Oils and Fats,* Vol.25, No.5, pp. 34-35, ISSN 1003-7969

Shi, Y.G. (2005). *Soybean Products Technology* (2nd edition), China Light Industry Press, ISBN 7-5019-4807-0, Beijing, China

Song, Y.Q.; Yu, D.Y.; Luo, S.N. & Wang, S.R. (2007). Modification of Soybean Phospholipids: A Review. *Soybean Bulletin,* Vol.4, pp. 22-29, ISSN 1009-2765

Szuhaj, B.F. (Ed.). (1989). *Lecithins: Sources, Manufacture and Uses (AOCS Monograph),* American Oil Chemists' Society, ISBN 9780935315271, Urbana, USA

Tekerikler, L. (2001). Selective Extraction of Phosphatidylcholine from Lecithin by Supercritical Carbon Dioxide/Ethanol Mixture. *Journal of the American Oil Chemists' Society (JAOCS),*Vol.78, No.2, pp. 115-118, ISSN 0003-021X

Wu, S.M. (2001). *Functional Oils and Fats,* China Light Industry Press, ISBN 7-5017-3049-X, Beijing, China

Xu, Z.S.; Wang, L.J. & Liu, B.Z. (2008). Modification of Soybean Phospholipid and Its Key Technique. *Cereals & Oils,* No.11, pp. 3-9, ISSN 1008-9578

Yang, W.D. (2010). Synthesis of Phosphatidylserine Catalyzed by Phospholipases D. *Modern Food Science and Technology,* Vol.26, No.9, pp. 994-996, ISSN 1673-9078

Zhang, H.F.; Gao, H.; Huang, Y.R.; Chen, J.Y. & Xia, X.Y. (2004). Study of Structures and Performances About the Soybean Lecithin Sulfonate (I)— Structures and Physical Chemistry Performances of Soybean Lecithin Sulfonate. *China Leather,* Vol.33, No.13, pp. 48-51, ISSN 1001-6813

Zhong, Z.S. & Wei, B. (2008). Extraction and Purification of Cephalin from Soybean Phospholipid. *Modern Food Science and Technology,* Vol.24, No.4, pp. 333-335, ISSN 1673-9078

9

Soybean Protein Fibres (SPF)

Tatjana Rijavec and Živa Zupin
University of Ljubljana
Slovenia

1. Introduction

Soybean protein fibres (SPF) are manufactured fibres, produced from regenerated soya *Glycine Max* soybean proteins in combination with synthetic polymer (polyvinyl alcohol) as a predominant component. According to textile fibre labelling (FTC, 2010), textiles from SPF can be marked as azlons from soybean. Azlons are manufactured fibres in which the fibre-forming substance is composed of regenerated naturally occurring proteins (FTC, 2011).

The first commercially successful method for producing regenerated protein fibres was developed by the Italian chemist Antonio Ferretti in 1935 (Ferretti, 1944; White, 2008). In 1936 Snia Viscosa (Milan) started with the production of the world's first commercially produced protein fibres Lanital™ which were made from milk casein (Anon., 1937). Courtaulds in Great Britain (casein fibres Caslen, Fibrolan), Enka in Netherlands, Germany and United States of America (casein fibres Aralac, R-53) soon followed with their commercial productions. Fibres were treated with formaldehyde or aluminium salts, to create cross-links between proteins in the fibre and improve fibre's wet properties. In the year 1945 Snia Viscosa replaced Lanital™ fibres with Merinova™ casein fibres (Fig. 1), which had better properties than Lanital™ fibres.

In the middle of the 20th century and until 1960, vegetable regenerated protein fibres from oilseed peanuts proteins (Ardil fibres, produced by British ICI Company) (Fig. 1) and from corn zein proteins (Vicara fibres produced by American Virginia-Carolina Corporation) were also produced among casein fibres. Fabrics made from regenerated protein fibres were soft, lustrous, resilient, with a good hand and thermal resistance. They were used as a wool or silk substitute by many European fashion designers.

Rapid development of cheaper synthetic fibres with excellent mechanical properties in the early sixties had influence on the commercial production of regenerated protein fibres that was completely discontinued in the middle of the 1960s.

Nowadays, increasing world population need additional quantities of textiles. The world fibre production increases from year to year and in 2010 there was globally produced 78 million tons of fibres, including about one million tons of wool and 0.15 million tons of silk (Kanitkar, 2010). Wool and silk are still very expensive fibres, with selling prices of about 14–23 €/kg and 28–40 €/kg, respectively (Reddy & Yang, 2007).

Today's fibre production strategy is redirected from crude oil to renewable raw materials, eco-friendly and sustainable fibres, that could be biodegraded or recycled. Important raw materials for future textile fibres production could be cheap and worldwide available agricultural by-products, like lignocellulose (from rice straw), wheat gluten (Yang et al., 2006), casein protein from milk after butterfat is removed, zein protein from corn after starch manufacture, and soybean protein after beans are pressed and oil is removed.

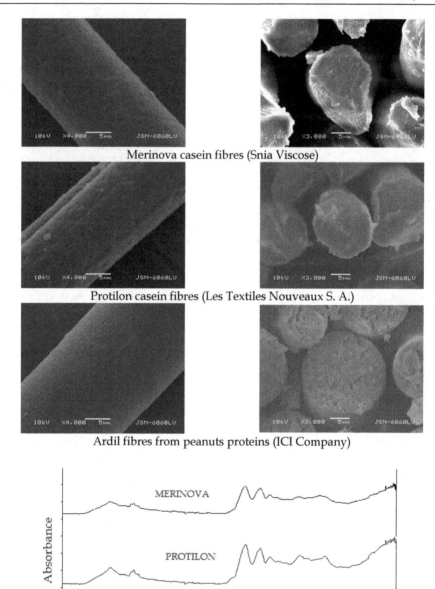

Fig. 1. Scanning electron microscope views and comparative FT-IR/ATR spectra of pure protein fibres with typical absorption peaks at 1658 cm-1 (amide I) and 1538 cm-1 (amide II).

First protein fibres had low tensile properties, especially in wet state. In order to improve mechanical properties of protein fibres, proteins were combined with synthetic polymers such as acrylonitrile or vinyl alcohol by graft copolymerization or polyblending. First such fibres, made on the patent basis of Morimoto (Morimoto et al., 1962), were produced by the Japanese Toyobo in 1969. The copolymer fibres Chinon® were made from 30% casein and 70% acrylonitrile. Acrylonitrile was grafted on protein with the addition of minor amounts of vinyl or vinylidene chloride for flame retardation. Fibre's density was 1.22 g/cm³, tensile stress in dry state 3.5-4.5 cN/dtex and moisture regain 4.5–5.5%.

Combining natural proteins and synthetic polymers to get fibres with good moisture absorbency and high tenacity led to new researches in the field of fibres at the beginning of the 21st century.

New fibres from casein proteins have been commercialized as milk protein fibres in China by Shanghai Zhengjia Milkfiber Sci& Tech Co., Ltd., under the brand name ZhengJia®. In the year 2005 a Chinese patent for producing the fibres was granted (Shanghai Z., 2011). Milk fibres are chemical casein acrylic fibres made from graft copolymer of casein and acrylonitrile. Fibres contain about 25–30% of milk proteins and 70–75% of acrylic component. The process is ecological (in 2004 it passed the Oeko-Tex Standard 100 green certification) with no formaldehyde content. Milk fibres with about 55% crystallinity have round cross section with many irregular vertical trenches and pockmarks on the surface (Wang et al., 2009). The fibres with linear density 2.22 dtex have breaking tenacity 2.5 cN/dtex and higher, breaking elongation 35.5%, elastic recovery 76.5%, moisture regain 4–5% and bacteria resistance ≥80% (Shanghai Z., 2011). Milk fibres could be dyed with reactive and acid dyes and after treated with crease-resist finishing and softening agents (Arslan, A., 2007, 2008, 2009).

New viscose filament yarn Lunacel, produced by Kurabo Industries Ltd. (Osaka, Japan), has combined properties of vegetable and animal fibres. The fibres are made from cellulose cotton linter pulp that is cross-linked with water-soluble food protein (Kurabo, 2007).

Using animal proteins as raw material for spinning fibres is very expensive. New soybean protein fibres (SPF) from soybean proteins and polyvinyl alcohol were developed in China by G. Li at Huakang R&D Center (Li, 2003, 2007). The fibres are first manufactured fibres, invented by China. The production process for new fibres was laboratory established in 1993 and commercially promoted in 2000. In 2001 the fibres were standardised and in 2003 they were launched.

The objective of this study was to investigate the contemporary SPF biodegradation in soil at controlled laboratory conditions.

2. Soybean protein fibres

2.1 Fibre forming soybean proteins

Soybeans are very reach with proteins (about 37–42% of dry bean) (Krishnan et al., 2007) in comparison to milk (3.2%), corn (10%) and peanuts (25%). Soybean proteins are used for food and feed and in many industries as adhesives, emulsions, cleansing materials, pharmaceuticals, inks, plastics and also textile fibres. Raw material for spinning textile fibres is obtained from soybean remaining flakes after the extraction of oils and other fatty substances (Li, 2004).

Amino acids content of soybean proteins is given in Fig. 2. Soybean proteins contain 18 different amino acids. There are about 23% of acidic amino acids (glutamic acid and aspartic amino acid), about 25% of alkaline amino acids (serine, arginine, lysine, tyrosine, threonine, tryptophan) and about 30% of neutral amino acids (leucine, phenylalanine, valine, alanine, isoleucine, proline, glycine). Sulphur containing amino acids are present also in soy proteins: about 1.0% of cysteine and 0.35% of methionine.

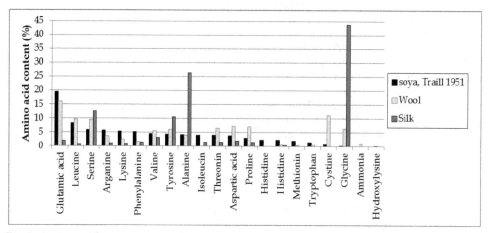

Fig. 2. Amino acids content in soybean proteins, wool keratin and silk fibroin (Brooks, 2005).

Soybean proteins consist of various groups of polypeptides with a broad range of molecular size: about 90% are salt-soluble globulins (soluble in dilute salt solutions) and the remainder is water-soluble albumins (Zhang, 2008). Very important as raw material for producing textile fibres are storage globulins with predominant β-conglycinin (30–50% of the total seed proteins) and glycinin (ca. 30% of the total seed proteins). β-conglycinin is a heterogeneous glycoprotein composed of three subunits (α', α, β) contained asparagine, glutamine, arginine and leucine amino acids. Subunits are non-covalently associated into trimeric proteins by hydrophobic interactions and hydrogen bonding without any disulphide bonds. Glycinin is a large hexamer, composed from acidic and basic polypeptides linked together by disulphide bonds (Zhang, 2008). On the basis of the sedimentation coefficient, a typical ultracentrifuge pattern of soybean proteins has four major fractions: 2S, 7S, 11S, and 15S (Zhang, 2008).

Globular proteins are composed of segments of polypeptides connected with hydrogen bonds, electrostatic interactions, disulphide bonds and hydrophobic interactions. Conformational changes of unfolding globular proteins through denaturation process (Zhang & Zeng, 2008) and reducing the inclination of denaturated proteins to form aggregates are important for spinnability of a spinning dope with proper relative viscosity. It is also important for later drawing of fibres and crystallization of proteins in fibres. Denaturation (Fig. 3) is modification of the secondary, tertiary, and quaternary structure of protein. Exposure of soybean proteins to strong alkali/acids, heat, organic solvents, detergents and urea causes the denaturation of native globular proteins, i.e. converting into unfolded polypeptide chains, which are connected with interchanging of disulphyde bonds. Extruded fibres coagulate in a precipitation acid bath and new disulphide bonds are formed. The structure of soybean proteins and changes at converting globular proteins into fibre forming proteins are given in Fig 3.

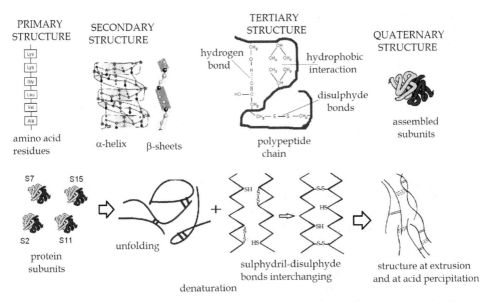

Fig. 3. Protein levels and conversion of globular proteins into fibre forming proteins (Zhang & Zeng, 2008; Kelly & Pressley, 1996).

Oils extraction with solvents used in the mid-twentieth century, was critical for the whole spinning process of soybean fibres, because the chosen temperatures, pH, urea, salts, organic solvents (hexane) and reducing agents influence on the degree of denaturation of proteins, degradation of proteins and changing of proteins colour. Protein degradation is detrimental to the production of high-strength protein fibres. Modern method of modifying soybean globular proteins is biochemical with using enzymes and auxiliary agent (Swicofil, 2011).

2.2 Pure soybean protein fibres from the mid-twentieth century

First researches for developing fibres from soybean proteins were made by the Japanese. In the year 1940 the first US patent was granted to Japanese Toshiji Kajita and Ryohei Inoue (Kajita & Inoue, 1940). The oil-free protein substance was extracted with dilute alkaline solution and precipitated by adding metallic salts. The protein was then washed in water and added by tartaric acid when the precipitate was wet. Then it was again dissolved in alkaline solution to form a spinning dope. Fibres were spun in an acid bath with organic coagulating agent (alcohol, formaldehyde, acetone etc.), where filaments hardened (Kajita & Inoue, 1946). The fibres had natural white to light tan colour. They were crimped, with high resiliency, warmth and soft feel. In comparison to wool they had lower tensile strength, especially in wet state, and lower moisture absorbency.

Patents for spinning fibres from soybean proteins were granted to the American Oscar Huppert from Glidden Company (Huppert, 1945) and Robert A. Boyer from Ford Motor Company (Boyer et al., 1945). In 1939 the American Ford Motor Company produced soybean protein fibres for their car's upholstery and seats fillings. The fibres, which have never been commercialized, had about 80% the strength of wool, higher elongation in dry and wet state than wool and didn't wet so easily as wool or casein fibres (Boyer, 1940).

Chemical and dyeing properties of pure regenerated soybean protein fibres were similar to wool.

Soybean protein fibres were also produced in Japan under the name Silkool (Myers, 1993). In 1939 the fibre production reached about 450–1,200 tons.

Low tensile strength of soybean protein fibres in wet state limited their commercial application. Fibres were used predominantly in blends with wool, cotton or synthetic fibres in woven and knitted fabrics for apparel and in upholstery, also in cars, despite of lower abrasion resistance than wool (Fletcher, 1942). The production of the mid-twentieth soybean protein fibres was ceased at the end of the World War II.

2.3 Researches on soybean protein fibres in the early twenty-first century

Huang et al. (1995) have made experimentally the textile fibres from soybean protein by re-examining the wet spinning method, described in the literature (Croston et al., 1945). The properties of the fibres made by wet spinning method from alkaline solution of soybean protein isolate and coagulated in acid bath were compared with the fibres made by dry spinning method of water solution of soybean protein isolate. Tensile properties of treated fibres were 0.77 cN/dtex at 11% relative humidity (r.h.), 0.75 cN/dtex at 65% r.h. and only 0.08 cN/dtex in wet state. They were lower than those of wool in most conditions (Huang et al., 1995). They found out that dry spinning was a suitable method for spinning soybean protein fibres because of their good solubility in water and glycerol. In the next experimental step they tried to increase tensile properties by decreasing the moisture absorption of soybean protein fibres. They used relatively nonpolar zein proteins (20, 30 and 40%), which were added to soybean protein into the spinning dope. Fibres were made by dry spinning method. The optimum soy protein-zein blended fibre was made from a suspension containing 80% of soybean protein and 20% of zein in glycerol (Zhang et al., 1997), but the tenacity was only 0.20 cN/dtex.

Another idea to improve low tensile strength and decrease shrinkage in boiling water was using water-soluble polymer, such as polyvinyl alcohol (PVA). PVA fibres are produced in similar conditions as phytoprotein fibres. Zhang (Zhang et al., 1999) experimented with bicomponent fibres from soybean protein and PVA. Fibres with a side-by-side configuration were not successful because of splitting of the components. The reason was in too large difference in swelling of the components in water. The next experiment of spinning sheath-core bicomponent fibres, with PVA component in the sheath and soybean proteins in the core, showed brittle core that couldn't be drawn. „The degradation of the soybean protein and the existing microgels in the protein spinning solution were thought to be the causes for the poor fibre drawability" (Zhang et al., 1999).

After ten years of intensive researches the Chinese scientists with Guanqi Li succeeded in producing high-tenacity soybean protein fibres from soybean protein and polyvinyl alcohol (Li, 2007). The process and fibre's properties are presented in section 2.4. Polyvinyl alcohol adds strength and acceptable wearability characteristics to the new SPF.

Biconstituent fibres from a biocompatible soy protein isolate and cellulose were produced experimentally from new aqueous solution NaOH/thiourea/urea. Strong hydrogen bonds between hydroxyl groups of cellulose and amid groups of protein were formed. Fibres with linear density of 6.2 dtex were produced with tensile strength of 1.86 cN/dtex and breaking elongation of 10.3% (Zhang et al., 2009).

High-wet strength fibres containing 5–23% of a soybean protein isolate from oiled soybean cake and 77-95% of polyvinyl alcohol were developed by scientists with Guanqi Li at

Huakang R&D Center in China (Li, 2003, 2007). A soybean protein isolate is treated with an auxiliary agent and biological enzymes to modify the structure of globular proteins. Additives break the disulphide bonds in globular proteins and convert them into linear molecules, which are stable in temperature range 55–90 °C (Mathur & Hira, 2004).

Fibres are wet spun from deaerated spinning dope composed of a soybean protein and polyvinyl alcohol dissolved in distilled water, followed by adding of borax or boric acid and mixing at temperature between 40 and 98 °C. After coagulation in a water bath with salt and alkali, as spun fibres are wet drawn, then dried, pre-heat set, heat-set at 170-185 °C, cooled, winded, stabilised by acetalysing, washed, oiled, crimped and cut into staple fibres. Production process doesn't pollute the environment. Most added agents in the process can be recovered from semi-finished fibres and used again.

The molecules of protein are laterally bonded with molecules of polyvinyl alcohol in the fibres. This enables during additional extension, orientation and crystallisation of proteins in the fibres during drawing. The morphological structure of SPF consists of less oriented sheath and well oriented microfibrilar core. The fibres have about 10% of hydrophilic groups in amorphous regions (Mathur & Hira, 2004).

Properties of soybean protein fibres taken from yarn SoySilk™ and milk protein fibres taken from yarn SilkLatte® are given in Table 1 (Brinsko, K. M., 2010).

Yarns	soybean protein fibres	milk protein fibres
available from	Southwest Trading Company	Southwest Trading Company, Tempe, AZ
cross-section and longitudinal view	bean-shaped with pronounced and elongated micro-pores inclusions	bean-shaped with small micro-pores inclusions
birefringence	0.021–0.027	0.016–0.024
melting point	250–260	235–245
chloroform, AcOH, acetone, DMF	insoluble	insoluble
formic acid	swell	swell
conc. H_2SO_4, conc. HNO_3	partially soluble	soluble, gels
characteristic peaks on FT-IR spectrum	amide I at 1640 cm^{-1} amide II at 1530 cm^{-1}	amide I at 1640 cm^{-1} amide II at 1530 cm^{-1}

Table 1. Properties of soybean protein fibres taken from yarn SoySilk™ and milk protein fibres taken from yarn SilkLatte® (Brinsko, K. M., 2010).

Adding some metallic salts into spinning dope, endows soybean fibre with far-infrared, negative ion and anti-bacterial functions. Only 3% of such fibres added into yarn can give stable and permanent antibacterial effect. Another technology from the same university is adopting $ZnSO_4$ as the dehydrating agent for soybean fibre spinning. In the course of after-processing, $ZnSO_4$ reacts with NaOH, forming ZnOH, which after drying is deoxidized into nanograde ZnO that can form covalent bond with fibre itself, taking a strong screen effect to ultraviolet radiation (Yang, 2011).

2.4 Commercial soybean protein fibres in the early twenty-first century

SPF based on the Li Guanqi patent (Li, 2007) are the first industrially produced fibres from soybean proteins in the world and they are the only soybean protein fibres present on the market today. These fibres are also the first manufactured fibres, developed by China. The production process of the new SPF was laboratory established in 1993 and commercially promoted in 2000. In 2001 the fibres were standardised and in 2003 launched.

About 1,500 tons of the fibres per year are produced under the brand name Winshow by Shanghai Winshow Soybean Fibre Industry Co., Ltd. Six manufacturing bases were established in four provinces in China for producing SPF (Shanghai, 2011). Zhejiang Jiali Protein Fiber Co., Ltd. is the owner of the soybean protein fiber international intellectual property rights and production line.

The Chinese manufacturer of soybean protein fibres Harvest SPF Textile Co., Ltd. (www.spftex.com) is a Chinese-foreign joint venture co-incorporated by China Harvest International Industry Ltd. and Zhejiang Jiali Protein Fiber Co. Ltd. (Shanghai, 2011). They are specialized in the research and development of new textile fibre raw material application technologies and application of the new-type textile materials from SPF. Fibres and yarns from soybean protein fibres are also available from Swicofil AG Textile Service (Anon, 2011), South West Traiding Company with yarn SoySilk™ (SWTC, 2011).

Since SPF resemble in their softness and shine to silk and cashmere, producers market them as "artificial cashmere", "vegetable cashmere" or "soy silk" fibres to partially decrease needs for natural silk and cashmere fibres. Cashmere goats cause damages to lands, so reducing their number has ecological benefits.

Physical and chemical properties of soybean protein fibres are given in Table 2.

PROPERTIES	SPF	Cotton	Viscose	Silk	Wool
Breaking strength (cN/dtex) in dry state	3.8–4.0	1.9–3.1	1.5–2.0	2.6–3.5	0.9–1.6
Breaking strength (cN/dtex) in wet state	2.5–3.0	2.2–3.1	0.7–1.1	1.9–2.5	0.7–1.3
Breaking elongation (%) in dry state	18–21	7–10	18–24	14–25	25-35
Initial Modulus (kg/mm²)	700–1300	850–1200	850–1150	650–1250	
Loop strength (%)	75–85	70	30–65	60–80	
Knot strength (%)	85	92–100	45–60	80–85	
Moisture regain (%)	8.6	9.0	13.0	11.0	14–16
Density (g/cm³)	1.29	1.50–1.54	1.46–1.52	1.34–1.38	1.33
Heat resistance	Yellowing and tackifing at about 120 °C (Bad)	Becoming brown after long time processing at 150 °C (Excellent)	Strength down after longtime processing at 150 °C (Good)	Keep stable when temperature <=148 °C (Good)	(Good)
Alkali resistance	At general level	Excellent	Excellent	Good	Bad
Acid resistance	Excellent	Bad	Bad	Excellent	Excellent
Ultraviolet resistance	Good	At the general level	Bad	Bad	Bad

Table 2. Comparison of physical and chemical properties of soybean protein fibres (SPF) in comparison to cotton, viscose, silk and wool (Swicofil, 2011).

A raw SPF has light yellow colour, like silk oak. Before dyeing into light colours they should be bleached with hydrogen peroxide or reduction bleached. SPF fibres can be dyed at temperatures lower than 100 °C with weak-acid dyes and substantive dyes for very few colours, because the dyeing fastness is poor. As SPF are less sensitive to high pH, they could be also dyed with reactive dyes (Mathur & Hira, 2004). SPF fibres have good light fastness and good resistance to ultraviolet radiation, which is better than that of cotton, viscose and silk. They are stable to washing even at higher temperatures, but they yellow at dry heat at 120−160°C (Anon., 2003).

Likewise regenerated cellulose bamboo fibres, SPF fibres are promoted on the market as biocompatible and health giving with natural antibacterial properties. The Chinese herbal medicine with sterilising and anti-inflammatory properties can be bonded on side chains of the proteins during the production of SPF (Yi-you, 2004) due to the bacterial resistance of SPF fibres to *Styphalococcus aureuses*, *coli bacillus* and *Candica albicans* (Swicofil, 2011). Mathur has mentioned that SPF resistance to golden and yellow *Styphalococcus aureuses* is more than 5.8 and hence they are inherently anti-bacterial fibres (Mathur & Hira, 2004).

Beside in yarns from 100% of SPF, the SPF cotton type fibres could be used in yarn mixtures with cotton, polyester, viscose and bamboo viscose. The wool type SPF should be mixed with cashmere (80/20 SPF/cashmere), lyocell, silk or wool (50/50 SPF/wool). Smooth surface of SPF has influence on low spinnability because of low friction coefficient and low cohesion force, and on pilling.

Fabrics with SPF should not be mercerised because SPF are not resistant to strong caustic soda. Woven and knitted fabrics can be used for apparel (personal underwear, T-shirts, pullovers, sweaters, evening dresses, children's clothing and sportswear) and home textiles (towels, bed linen, blankets, bathrobes, pyjamas). Since the fibres have lower abrasion resistance than wool they can be used as upholstery in automobile textiles.

A cloth made of SPF fibres exhibits good wiping properties (Reek, 2008). At least 10% of Winshow SPF of linear density 1.5 dtex and 38 mm length from Shanghai Winshow Soybeanfibre Industry Co., Ltd. of Shanghai, China is used in combination with viscose and/or other textile fibres in thermo-bonded nonwoven fibrous material.

2.5 Biodegradation of contemporary SPF

Biodegradable fibres degrade relatively quickly through biological process, which depends on many factors, such as chemical and morphological structure, temperature, pH, relative humidity and remains of auxiliary agents, which are accumulated (brought) on fibres during manufacturing and are not completely washed after finishing process (Simončič & Tomšič, 2010).

The chemical structure has influence on biodegradability with its hydrophilic nature (wettability), crystallinity of the polymer, chemical linkages in the polymer backbone, pendant groups, end groups and molecular weight distribution. Peptide bonds are susceptible to enzymatic degradation. Additional polymers may (interaction with other polymers) act as barriers to prevent migration of microorganisms, enzymes, moisture or oxygen into the polymer domain of interest (Zee, 2005).

The biodegradation process of proteins is initiated through exposure to water. Long macromolecules under hydrolytic process convert into many small molecules, which are more proper for the metabolism of microorganisms.

The mid-twentieth pure soybean protein fibres mildewed less easily than natural and casein fibres but more easily than synthetic fibres (Fletcher, 1942). Mid-twenty century soybean protein fibres were susceptible to microbiological growth. Casein fibres were readily damaged by mildew, they quickly mildewing especially in damp conditions. Changing protein molecules by chemicals and tanning (hardening) has influence on lower biodegradability of fibres (Wormell, 1954).

Very little data is yet available about biodegradability of contemporary soybean protein fibres. The fibres are promoted as biodegradable fibres in landfill (Mathur & Hira, 2004; Swicofil, 2011). Fibres from water-soluble polyvinyl alcohol are biodegradable in soil. Considering the chemical structure of SPF (Fig. 2), the soybean proteins susceptibility to biodegradation should be similar to wool and not to silk. Wool contains 80% of keratin, the rest are no-keratin proteins. Degradation of wool is mostly caused by fungus and less by bacteria. Ideal conditions for growth of microorganisms on wool fibres are temperature 30°C, relative humidity of 95% and pH from 6,5 to 8,5 (Edwards & Vigo, 2001). In the initial stage, of biodegradation of wool is hard to be noticed. When the growth of microorganisms increases, unpleasant odour appears, coloured spots can be seen on fabrics and tensile strength as result of defibrillation decreases (Edwards & Vigo, 2001, Szostak-Kotowa, 2004).

3. Experimental part

3.1 Materials
Ring spun yarns and twill 2/2 woven fabrics were used in our experiments of biodegradation of contemporary soybean protein fibres:
-   100% soybean protein yarn with linear density of 15 tex (SPF yarn) and 100% cotton yarn with linear density of 19 tex for comparison;
-   a fabric with yarn from soybean protein fibres in weft direction and cotton yarn in warp direction (SPF/CO) and a 100% cotton fabric with cotton yarn in warp and weft direction for comparison (CO).

The same cotton yarn with linear density of 28 tex were used for warp and SPF yarn with linear density of 15 tex were used for weft for all woven fabrics. The density of fabrics was 30 ends/cm and 28 picks/cm for SPF/CO and 100% cotton fabrics.

3.2 Methods
3.2.1 Method of controlled biodegradation in soil
A laboratory experiment of biodegradation (Fig. 4) of yarns and woven fabrics was made in accordance with the standardised method SIST EN ISO 11721-1. Commercial humus, rich in microorganisms, was used as a soil. During experiment the soil humidity was 60±5%, which was regularly measured by a hygrometer and maintained by spraying the soil with tap water. The temperature of the soil was 25−30 °C. Samples of yarns and woven fabrics were buried in the soil for 2, 7, 11, 16 and 21 days. After that, the samples were washed out in tap water, then immersed into ethanol for 30 minutes to stop the activity of microorganisms, and dried in the air.

Tensile properties of samples were measured on dynamometer Instron 5567 in accordance with the standard SIST EN ISO 2062 for yarns and SIST EN ISO 13934 for fabrics. For measuring tensile properties in wet state, the yarns were immersed into distilled water with detergent at room temperature for an hour. Tensile properties of yarns were analysed with the DINARA program (Bukošek, 1988). Tensile properties of fabrics were measured only in weft direction.

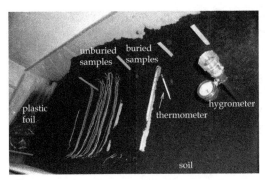

Fig. 4. Experiments of biodegradation were made in a wooden box surrounded with a foil and filled with humus soil.

3.2.2 Other methods
Fourier transform infrared spectra (FTIR/ATR) were obtained on the Spectrum GX (Perkin Elmer) with the Michelson interferometer and Spectrum 5.01 software using 16 scans at a resolution of 4 cm^{-1} in a range of wavenumber from 4000 to 500 cm^{-1}. Microphotographs were made with the Jeol JSM 6060 LV scanning electron microscope and the Nikon SMZ 800 stereomicroscope.

4. Results and discussion

4.1 Fibres properties
The SPF yarn was made from cotton type soybean fibres of 1.27 dtex with an average length of 39.5 cm. Fibres were thermoplastic with melting point at 224°C. Dry fibres absorbed 2.47% of moisture when exposed for 48 hours to the air of relative humidity 50% and temperature of 23 °C.

The cross-section shape of used soybean protein fibres was bean-shaped with diameter of 11-20 μm in longer axle and 6-7 μm in shorter axle (Fig. 5). A very smooth surface of fibres imparted high lustre to fibres. On the longitudinal view irregular grooves and wrinkles can be seen. These grooves can help to transport moisture along fibres. On the optical microscope photograph a nonhomogeneous structure with many voids is seen.

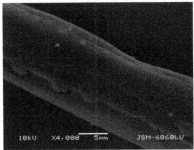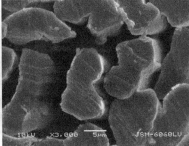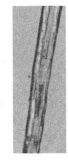

Fig. 5. Scanning electron microscope view of soybean protein fibre: top: longitudinal view at magnification of 4.000 and bottom: cross-section at magnification of 3.000. Right: optical microscope longitudinal view of soybean protein fibre.

Soybean protein fibres are composed of a mixture of two polymers, soybean proteins and polyvinyl alcohol. Protein and polyvinyl alcohol macromolecules are connected by intermolecular interactions like hydrogen bonds (Fig. 3) and van der Waals hydrophilic and hydrophobic forces.

The soybean protein fibres, used in experiment, consisted of polyvinyl alcohol and soybean proteins. The SPF FT-IR spectrum (Fig. 6) has very intensive peaks at 3301 cm^{-1}, which is typical for stretching O-H bonds, and at 1408 cm^{-1} and 1327 cm^{-1}, which corresponds to N-H stretching in amide III. FT-IR absorption spectrum of SPF is different to FT-IR spectrum of PVA fibres at peaks 1644 cm^{-1} and 1535.32 cm^{-1}.

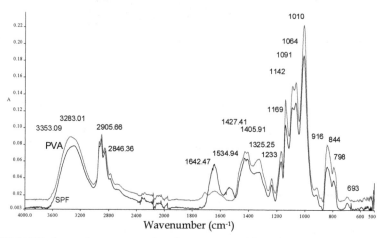

Fig. 6. FT-IR/ATR absorption spectra of soybean protein fibres (SPF) and PVA Kuralon®.

Pure soybean protein isolate has typical infrared absorption bands at 1636-1680 cm^{-1} and 1533-1559 cm^{-1} that are attributable to the –NH- bonds of amide I at 1640 cm^{-1} and at 1550 cm^{-1} for amide II in peptide bonds forming primary backbone of proteins. The absorption peak at 3294 cm^{-1} refers to the hydrogen-bond association between protein chains and moisture in protein. The absorption band at 1241-1472 cm^{-1} is attributable to the (C)O-O and C-N stretching and N-H bending (amide III) vibrations (Su et al., 2008).

At room temperature pure polyvinyl alcohol powder with –OH groups on carbon chains has a typical infrared absorption band at 2918-3565 cm^{-1}, which corresponds to –OH absorption (Su et al., 2008).

4.2 Yarns properties
4.2.1 Tensile properties of SPF yarn in dry and wet state
Water has a significant influence on tensile properties of the SPF yarn (Fig. 7, Tab. 3). After an hour in distilled water, the yarn lost its specific breaking stress for almost one third. Wet yarn had lower modulus then dry yarn in the whole deformation range and attained by 11.4% higher breaking elongation than dry yarn.

4.2.2 Biodegradation of SPF yarns
Biodegradation of SPF yarn was studied after the yarn had been buried for 2, 7, 11, 16 and 21 days in the soil with temperature about 30 °C and 65% relative humidity (Fig. 8). For the

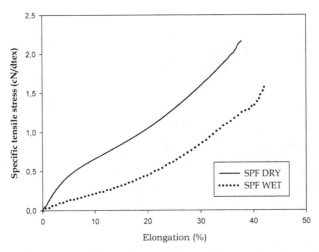

Fig. 7. Stress-elongation curves of dry and wet SPF yarns

Properties of yarns from SPF	DRY	WET	Δ (%)
Specific breaking stress (cN/dtex)	2.16	1.58	-26.9
Breaking elongation (%)	37.75	42.05	+11.4
Initial modulus (GPa)	5.08	3.25	-36.0
Specific work of rupture (mJ/kg)	39.81	22.71	-43.0

Table 3. Tensile properties of dry and wet SPF yarns and relative differences between them (Δ).

purpose of comparison, cotton yarn was buried at the same time in the soil. After 7 days cotton yarn degraded very intensively and only small remains of yarn were left in the soil. Tensile properties of biodegraded cotton yarn could be measured only after 2 days. Biodegradation of cotton showed that microorganisms in the soil were active during the experiment.

The microphotographs of SPF yarns in Fig. 8 show that the quantity of bacteria and fungus, present on the surface of soybean protein fibres, increased with time of biodegradation. After 21 days in the soil, it is hard to say that there are any physical degradations of the fibre's surface because the fibres have natural irregular grooves and wrinkles (0 days).

Soybean protein fibres in comparison to the mid-twentieth century protein fibres (Fig. 1) have essentially smoother surface and relatively lower quantity of surface grooves that could enable bacteria to penetrate into the fibres.

Specific tensile stress-elongation curves of biodegraded cotton yarns (Fig. 9 and 10) show a significant decrease of breaking force and breaking elongation after 2 days, while for the yarns from soya protein fibres, they didn't change essentially.

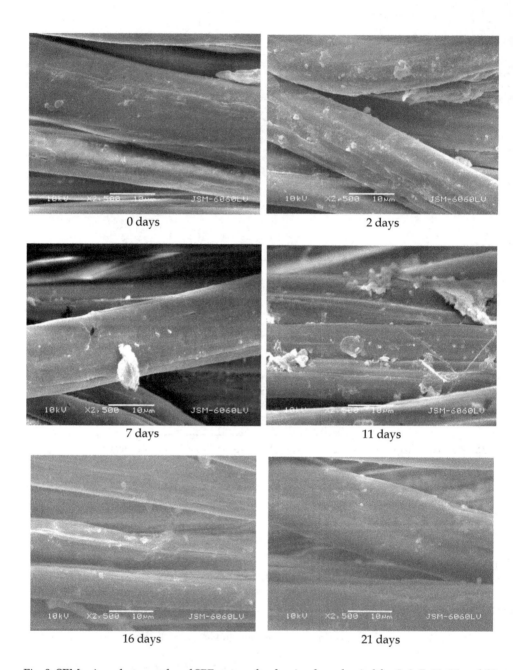

Fig. 8. SEM microphotographs of SPF yarns after having been buried for 0, 2, 7, 11, 16 and 21 days in the soil (at magnification 2500-x).

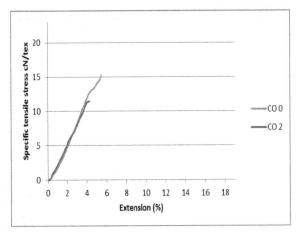

Fig. 9. Stress-elongation curves of cotton yarns after having been buried for 0 (CO 0) and 2 (CO 2) days in the soil.

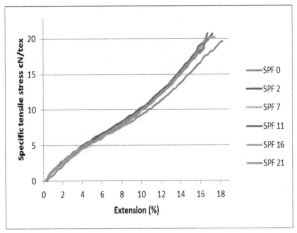

Fig. 10. Stress-elongation curves of cotton yarns after having been buried for 0 (SPF 0) to 21 (SPF 21) days in the soil.

4.3 Biodegradation of fabrics with SPF yarns in weft direction

Woven fabrics were buried at the same time as yarns into the soil at temperature 30°C and relative humidity 65% for 2, 7, 11, 16 and 21 days. Fabric samples (Fig. 11) changed the colour and became browner with many colour spots on the surface, which confirmed the existence of fungus.

Pure cotton fabrics degraded in one weak to such degree that they broke up into pieces when we tried to dig them out of the soil. After 21 days only very little remains were found in the soil. Fabrics with yarns from soybean protein fibres in weft direction were more compact in weft direction than pure cotton fabrics. But in warp direction from cotton yarns the fabrics lost their strength and were easily torn (Fig. 11).

Fig. 12 shows that after 21 days in the soil the cotton cuticle was destroyed. On the soybean protein fibres the quantity of fungus and bacteria increased, but the surface of fibres was not damaged.

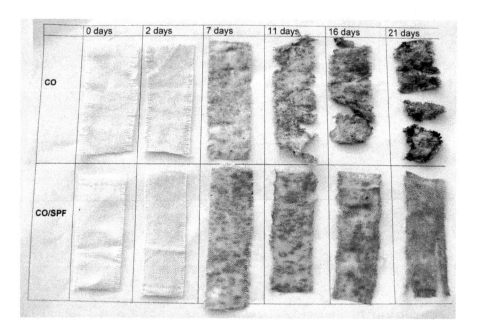

Fig. 11. Fabric samples after having been buried in the soil for 0 to 21 days. (Photo: Marica Starešinič)

The longer was the time of being buried in the soil, the greater was the loss of tensile strength of cotton fabrics: by 12% after 2 days and by 62% after 7 days of being buried in the soil. Breaking elongation decreased also rapidly: from 23% of the unburied fabric to only 8% after 7 days of being buried in the soil (Fig. 13).

Degradation of cotton yarns in warp direction affected tensile strength of SPF/CO fabrics in weft direction (Fig. 14). Fabrics buried in the soil for 7 days lost 12% of their tensile force, but after 21 days of being buried in the soil their breaking force decreased by additional 13% in comparison to unburied fabrics. Breaking elongation in weft direction of SPF/CO fabrics did not change significantly.

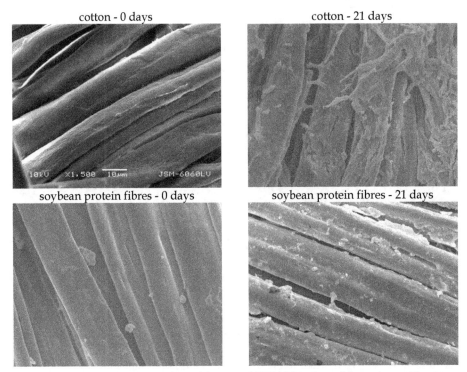

cotton - 0 days cotton - 21 days

soybean protein fibres - 0 days soybean protein fibres - 21 days

Fig. 12. SEM microphotographs of CO fabrics and SPF/CO woven fabrics (magnification 1500-x).

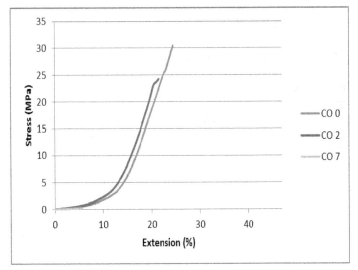

Fig. 13. Stress-elongation curves for cotton woven fabrics.

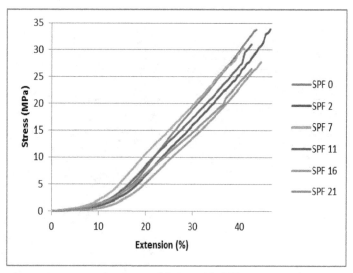

Fig. 14. Stress-elongation curves for woven fabrics with SPF yarn in weft.

5. Conclusion

The mid-twentieth century regenerated soybean protein fibres were made from pure soy proteins treated with formaldehyde or aluminium salts. Because of low tensile strength in wet state they were, like all other mid-twentieth protein fibres, noncompetitive to synthetic fibres in the 1970th.

Due to increasing prices of petroleum and a growing concern about the environmental damage arising from a slow degradation and poor biodegradability of synthetic fibres, researchers began to search for new possibilities of developing fibres from renewable raw materials, also from soybean proteins.

The fact that proteins are renewable and biodegradable materials has attracted considerable attention of many researchers in the area of textile fibres in the last two decades to re-examine the production of fibres from soybean proteins and casein. Soybean proteins have a greater potential for use as textile fibres because of their lower cost than casein proteins derived from milk.

In all experiments made until now, a soybean protein isolate (SPI) has been used, which is a highly purified protein (>90% w/w), obtained after extracting oils and fats from protein cakes. The residues after purification of protein for producing fibres can be also used as foodstuff (Yi-you, 2004).

In the last two decades researches have focused on different spinning methods (Huang et al., 1995), on new fibres from soybean proteins and polyvinyl alcohol (Zhang et. al., 1999, Li, 2007) or zein proteins (Zhang et al., 1997), on new economical biochemical processes that modify physical structure of soya proteins, and on new solvents (Zhang et al., 2009).

The experimental soybean protein fibres were made from two macromolecular components combined together into:

- *biconstituent fibres*, where a spinning dope was prepared from a homogeneous mixture of two solutions – a soybean protein water solution and a water solution of synthetic polymer polyvinyl alcohol or cellulose or zein proteins. Single fibres made from such spinning dopes had homogenous structure.

- *bicomponent fibres*, where the fibre's core was made from a soybean protein and the fibre's sheath from polyvinyl alcohol.

Polyvinyl alcohol was used, because it is a water-soluble polymer, it dissolves at similar conditions as proteins and when added to proteins, it increases the fibre's strength. Polyvinyl alcohol is also biodegradable in the soil (Brooks, 2005).

The combination of cotton yarns and the yarns from soybean proteins in woven fabrics imparts comfort, soft hand and good moisture absorption properties to undergarments, outerwear, infants' wear, towels and beddings. Biodegradation of contemporary soybean protein fibres in early phase, up to 21 days in the soil at 30 °C and 65% relative humidity, is a slow, hardly perceivable process.

6. References

Anon. (2003). Innovations China develops soyabeanf fibre, *Textiles Magazine*, Vol. 30, No. 2, p. 4, ISSN 1367-1308

Anon. (1937). Artificial Wool Production in Italy, *Nature*, Vol. 140, pp. 1090, ISSN 0028-0836

Arslan, A. (2007). Milk protein fibres (Part 1), Chemical Fibre International, Vol. 57, 108-109. ISSN 1434-3584

Arslan, A. (2008). Milk protein fibres (Part II), Melliand International, vol. 14, 20-21. , ISSN 0947-9163

Arslan, A. (2009). Milk protein fibres (Part III), Melliand International, vol. 14 (2009), 92-93. , ISSN 0947-9163

Boyer, R. A. (1940). Soybean protein fibers. Experimental production, *Industrial and Engineering Chemistry*, Vol. 32, p. 1549-1549, April 27, 2011, Available from: <http://pubs.acs.org/doi/abs/10.1021/ie50372a004>

Boyer, R. A., Robinette, C. F. & Atkinson, W. T. (1945). *Artificial fibres and manufacture thereof*, US Patent 2,377,854

Brinsko, K. M. (2010). Optical characterization of some modern "eco-friendly" fibers, *Journal of Forensic Sciences*, Vol. 55, No. 4, pp. 915-923, ISSN 1556-4029, April 15, 2011, Available from: <interscience.wiley.com>

Brooks, M. M. (2005). Soya bean protein fibres – past, present and future, In: *Biodegradable and sustainable fibres*, R. S. Blackburn, pp. 398-440, Woodhead Publishing Series in Textiles, No. 47, ISBN 1-85573-916-X (Woodhead Publishing), Cambridge

Bukošek, V. (1988). *Program »DINARA«: program izračuna in vrednotenja mehanskih in visokoelastičnih lastnosti vlaken iz diagrama specifična napetost /raztezek*, NTF Oddelek za tekstilstvo, Ljubljana

Croston, C. B., Evans, C. D. & Smith, A. K. (1945). Zein fibers - Preparation by wet spinning, *Industrial & engineering chemistry*, Vol. 37, No. 12, pp. 1194-1198, ISSN 1541-5724

Edwards, J.V. & Vigo, T.L. (2001). *Bioactive fibres and polymers*, Americal Chemical Society, pp. 421-430, ISBN 0-8412-3714-X, Washington, DC

Federal Trade Commission [FTC] 2011). *Rules and Regulations Under the Textile Fiber Products Identification Act, 16 CFR Part 303.§303.7(g) Generic names and definitions for manufactured fibers*, USA Federal Trade Commission, Mar. 05, 2011, Available from: <http://www.ftc.gov/os/statutes/textile/rr-textl.htm>

Federal Trade Commission [FTC], (2010). Textile fiber labeling, April 10, 2011, Available from:
<http://www.nordstromsupplier.com/NPG/PDFs/Product%20Integrity/Labelin g%20Requirements/Textile%20Fiber%20Labeling.pdf>

Ferretti, A. (1944). *Process for manufacturing artificial textile fibers from* casein, US Patent 2,338,917

Fletcher, H. A. (1942). Synthetic Fibers and Textiles, *Kansas Bulletin* 300, p. 8-10

Huang, H. C. , Hammond E. G. , Reitmeier C. A. & Myers D. J. (1995). Properties of fibers produced from soy protein isolate by extrusion and wet-spinning, *Journal of the American Oil Chemists' Society*, Vol. 72, No. 12, p. 1453-1460, ISSN 0003-021X (Print) 1558-9331 (Online)

Huppert, O. (1945). *Process of manufacture of synthetic wool from soya bean protein*, US Patent 2,377,885

Kajita, T. & Inoue, R. (1940). *Process for manufacturing artificial fiber from protein contained in soya bean*, US Patent 2,191,194

Kajita, T. & Inoue, R. (1946). *Process for manufacturing artificial fiber from protein contained in soybean*, US Patent 2,394,309

Kanitkar, U. P. (2010). Man Made Cellulose Fibre Industry over-view. North South Textile Summit, May 2010, Available from: <http://nsts.in/images/UPKANITKAR-GRASIM-Presentation-Man-Made-Fibre-20052010.pdf>

Kelley, J. J. & Pressley, R. (1966). Studies with soybean protein and fiber formation, *Cereal Chemistry*, Vol. 43, pp. 195-206, ISSN 0009-0352

Krishnan, H. B., Natarajan, S. S., Mahmoud, A. A. & Nelson, R. L. (2007). Identification of glycinin and beta-conglycinin subunits that contribute to increased protein content of high-protein soybean lines, *J. Agric. Food Chem.*, Vol. 55, p. 1839-1845, ISSN 1520-5118

Kurabo, 2007. New viscose filament yarn Lunacel, *Chemical Fibers International*, Vol. 57, pp. 315, ISSN 1434-3584

Li, G. (2003). *Phytoprotein synthetic fibre and the method of making the same*, WO/2003/056076

Li, G. (2007). *Phytoprotein synthetic fibre and method of manufacture thereof*, US Patent 7,271,217

Mathur, M. and Hira, M. (2004). Speciality fibres – I: soybean protein fibre, *Man-made Textiles in India*, Vol. 32, No. 10, pp. 365-369, ISSN 0377-7537

Morimoto, S., Ishihara, M., Yamamoto, A., Hamada, K., Imai, K. & Otsuka, M. (1962). *Graft copolymer containing spinnable solution and method for preparing and spinning thereof*, U.S. Pat. 3,104,154

Myers, D. (1993). Past, present and potential uses of soy proteins in non-food industrial applications, *Proceedings of the World Conference on Oilseed Technology and Utilization*, ISBN 0-935315-45-4, Budapest, March 10, 2011, Available from: <http://books.google.si/books?id=Fh2rNOiEIWQC&pg=PA278&lpg=PA278&dq=Past,+present+and+potential+uses+of+soy+proteins+in+non-food+industrial+applications,+Proceedings+of+the+World+Conference+on+Oilseed+Technology+and+Utilization&source=bl&ots=-Z7SEywKa5&sig=Torn8hK4rwCcap0fypN9bYxXDZ0&hl=sl&ei=Gpu9TefjG8TFswawp5z6BQ&sa=X&oi=book_result&ct=result&resnum=1&ved=0CBUQ6AEwAA#v=onepage&q&f=false>

Reddy, N. & Yang, Y. (2007). Novel Protein Fibers from Wheat Gluten, *Biomacromolecules*, Vol. 8, pp. 638-643, ISSN 1525-7797

Reek, A. (2008). *Wipe materials comprising regenerated plant-protein fibres*, PCT/US2009/059406 WIPO

Shanghai Winshow Soybean Fibre Industry. Co., Ltd. (2011). About us, April 10, 2011, Available from: <http://swsficl.en.china.cn>

Shanghai Zhengjia milkfiber sci & tech Co ., ltd., (2011). Zhengjia milk fibre. January 20, 2011, Available from: < http://www.milkfashion.com>

Simončič, B. & Tomšič, B. (2010). *Biorazgradnja tekstilnih vlaken in njihova protimikrobna zaščita*, Naravoslovnotehniška fakulteta, Oddelek za tekstilstvo, ISBN 978-961-6045-81-0, Ljubljana

South West Traiding Company [SWTC], (2011). April 20, 2011, Available from: <www.soysilk.com>

Sue, J.-F., Huang, Z., Yang, C.-M. & Yuan, X.-Y. (2008). Properties of Soy Protein Isolate/Poly(vinyl alcohol) Blend "Green" Films: Compatibility, Mechanical Properties, and Thermal Stability, *Journal of Applied Polymer Science*, Vol. 110, pp. 3706-3716, ISSN 1097-4628

Szostak-Kotowa, J. (2004). Biodegradation of textiles, *International biodeterioration & biodegradation*, Vol. 53, pp. 156-170, ISSN 0964-8305

Swicofil AG, (2011). Soybean protein fibres, April 27, 2011, Available from: <http://www.swicofil.com/soybeanproteinfiber.html>

Wang, N., Ruan, C., Yu, Y., Zheng, Y. & Yu, J. (2009). Composition and structure of acrylonitrile based casein fibers, *Chemical Fibres International*, Vol. 59, No.2, pp. 88-89, ISSN 1434-3584

White, M. (2008). *U.S. Alien Property Custodian patent documents: A legacy prior art collection from World War II – Part 2, statistics*, World Patent Information, Vol. 30, No. 1, pp. 34-42, ISSN 0172-2190

Wormell, R. L. (1954). *New fibres from proteins*, London, Butterworths Scientific Publications, pp. 145

Yang, V. (2011). Soybean fibre application stepping into a new stage in China. April 23, 2011, Available from: < http://www.fibre2fashion.com/industry-article/textile-industry-articles/soybean-fiber-application-stepping-into-a-new-stage-in-china/soybean-fiber-application-stepping-into-a-new-stage-in-china1.asp >

Yang, Y. et al. (2006). *Process for the production of high quality fibers from wheat proteins and products made from wheat protein fibers*, WO 2006/138039

Yi-you, L. (2004). The soybean protein fibre – a healthy & comfortable fibre for the 21st century, *Fibres & Textiles in Eastern Europe*, Vol. 12, No. 2, pp. 8-9, ISSN 1230-3666

Zee, M. van der (2005). Biodegradability of polymers – mechanisms and evaluation methods. In: *Handbook of biodegradable polymers*, Catia Bastioli , pp. 1-32, Rapra Technology, ISBN 978-1-85957-389-1, Shawbury, Shrewsbury, Shropshire

Zhang, M., Reitmeier, C. R., Hammond, E. G., & Myers, D. J. (1997). Production of textile fibers from zein and a soy protein-zein blend. *Cereal Chemistry*, Vol. 74, pp. 594-598, ISSN 0009-0352

Zhang, Y., Ghasemzadeh, S., Kotliar, A. M., Kumar, S., Presnell, S. & Williams, L. D. (1999). Fibers from soybean protein and poly(vinyl alcohol), *Journal of Applied Polymer Science*, Vol. 71, pp. 11-19, ISSN 0021-8995

Zhang, L., & Zeng, M. (2008). Proteins as sources of materials, In: *Monomers, Polymers and Composites from Renewable Resources*, M. N. Belgacem and A. Gandini, pp. 479-493, Elsevier, ISBN 9780080453163, Oxford, Boston

Zhang, L. (2008). *Physicochemical, morphological, and adhesion properties of sodium bisulfite modified soy protein components – B. S. Thesis*. Kansas, Manhattan: Kansas State University. April 20, 2011, Available from:
<http://krex.k-state.edu/dspace/bitstream/2097/1707/1/Lu%20Zhang2008.pdf>

Zhang, S., Zhao, S., Tian, Y., Li, F.-X. & Yu, J.-Y. (2009). Preparation of cellulose/soy protein isolate blend biofibers via direct dissolving approach, *Chemical Fibers International*, Vol. 59, No. 2, pp. 106-107. ISSN 1434-3584

10

Extraction and Analysis of Inositols and Other Carbohydrates from Soybean Plant Tissues

J.A. Campbell[1], S.C. Goheen[1] and P. Donald[2]
[1]Battelle, Pacific Northwest National Laboratory
Chemical and Biological Signature Sciences
Richland, WA
[2]USDA/ARS, Crop Genetics Research Unit
Jackson, TN
U.S.A.

1. Introduction

An outstanding characteristic of soybean plants is their ability to produce large amounts of the carbohydrate pinitol. Pinitol and the closely related inositols are currently undergoing widespread investigation for their biological and nutritional value. These and all the carbohydrates are typically extracted and analyzed together. Therefore, this review includes a general discussion about the extraction and analysis of carbohydrates in plants as well as a more in depth examination of the biosynthesis and use of compounds related to pinitol. The multiple roles of these substances in plants and animals, and their synergism have not been fully realized. This review discusses not only the extraction and analysis, but also the diverse roles of the inositols with an emphasis on inositols from the soybean plant.

2. Carbohydrate production and nitrogen fixation

Carbohydrates are produced in plants by photosynthesis. Zhu et al. (2010) reviewed photosynthesis in relation to improving crop yield. Agronomically, there has been little benefit in breeding for increased photosynthesis indicating that the relationship of photosynthesis to yield is still not well understood (Farquhar & Sharkey, 1982; Pessarakli, 2005). The relative growth rate of shoots was shown to be correlated to the soluble carbohydrate level in the plant, but shoot growth was also impacted by plant stress (Masle *et al*, 1990). One commonly studied plant stress in relation to carbohydrate production is drought stress. There is confusion regarding the regulation of carbohydrate synthesis when plants are under drought stress. Drought stress in addition to reducing shoot growth, increases root growth (Sharp & Davies, 1979).

Approximately 70 million tons of fixed nitrogen or about 50 % of the total nitrogen that enters the terrestrial ecosystem comes from biological nitrogen fixation (Brockwell *et al.*, 1995; Tate, 1995). The relationship of carbohydrate availability to photosynthesis, phloem sap supply and N_2 fixation in legumes is complex and knowledge is incomplete (Udvardi & Day, 1997).

Carbohydrates are the main energy source for humans. Carbohydrates are classified according to the number of monomers they contain as monosaccharides (simple sugars), oligosaccharides, or polysaccharides. Carbohydrate metabolism in plants has been reviewed (Colowick & Kaplan, 1951; Ochoa & Stern, 1952; and Horecker & Mehler, 1955). Carbohydrate levels in soybean seed are highest at growth stage R 5.5, or when the seed is half-developed (Wilson, 2004). A significant portion of the carbohydrate produced by photosynthesis is respired in the plant roots. (Lambers et al., 1996).

3. Simple sugars

The most common simple sugars are glucose and fructose. Disaccharides consist of two covalently bound sugar molecules. Sucrose, for example, is a disaccharide consisting of glucose and fructose. Sugars have a role in energy, carbon transport molecules, hormone-like signaling factors, and as the source for building proteins, polysaccharides, oils and woody materials (Halford et al., 2010). Plant genotype and environment greatly affect the levels found in plants (Halford et al., 2010).

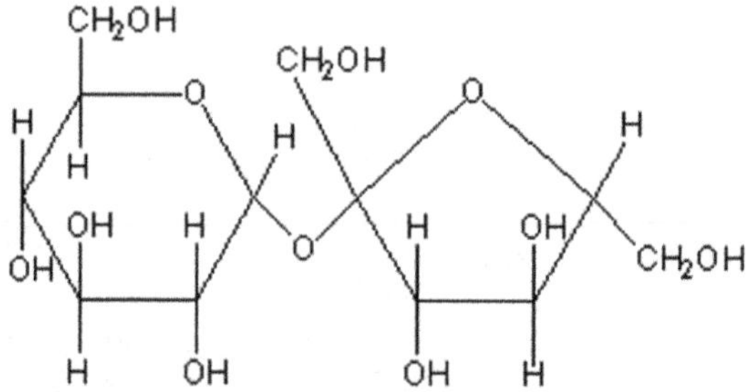

Sucrose

4. Complex carbohydrates

Complex carbohydrates (polysaccharides) are polymers of the simple sugars . Starch is the principal polysaccharide used by plants to store glucose.

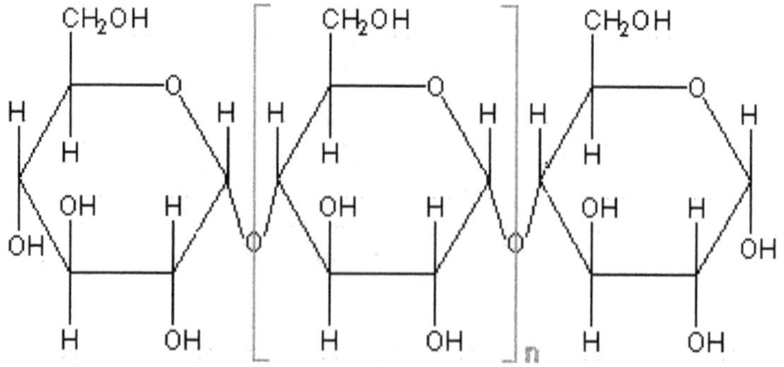

(n is the number of repeating glucose units and ranges in the 1,000's)

Starch

Zeeman *et al.*, 2010 reviewed the role of starches in plants. Starch breakdown commonly occurs when seeds germinate. Starch is also involved in malting (Halford *et al.*, 2010). Glycogen, also a polymer of glucose, is the polysaccharide used by animals to store energy. Another important polysaccharide is cellulose. Cellulose is used as a structural molecule to add support to leaves, stems, and other parts of plants. Although cellulose can't be used as an energy source in most animals, it provides essential fiber in the diet. Cell wall polysaccharides vary with plant groups and can include cellulose, xyloglucan, arabinoxylan, and pectin. In plants they make up the primary biomass and contribute to fiber in the human diet. This area has been reviewed by Scheller & Ulvskov, 2010; Fontes & Gilbert, 2010.

5. Extraction and cleanup

The methods used for isolating carbohydrates depend on the carbohydrate type, matrix, and purpose or type of analysis. However, some extraction procedures are commonly used for isolating carbohydrates from other classes of compounds in plants and foods. As an example, foods are usually dried under vacuum to prevent thermal degradation, ground to a fine powder to enhance extraction efficiency, and then remove the fats using appropriate solvent extraction.

A commonly used method for extracting low molecular weight carbohydrates from foods is to boil a sample with a 70-80% alcohol solution (Hall 2003, Asp 1993, Smith 1973.). Monosaccharides and oligosaccharides are soluble in alcohol solutions; however, most proteins, polysaccharides and dietary fiber are insoluble. The soluble components can then be separated from the insoluble components by filtering, soluble portion passes through the filter and the insoluble part retained by the filter. The two fractions can then be dried using lyophilization or nitrogen blow down techniques. In addition, monosaccharides and oligosaccharides and various other small molecules (e.g. organic acids, amino acids) may be present in the alcoholic extract. It is usually necessary to remove those components prior to carrying out a carbohydrate analysis, for example, with clarifying agents or by elution through one or more ion-exchange resins.

Water extracts of many foods contain substances that are colored or produce turbidity, and may interfere with analyses of carbohydrates; as a result, clarifiers may be needed. The most commonly used clarifying agents are heavy metals (e.g. lead acetate) which form insoluble complexes with interfering substances that can't be removed by either filtration or centrifugation. Ion-exchange is another method for removing interfering components prior to analysis. Many monosaccharides and polysaccharides are polar non-charged molecules and can therefore be separated from charged molecules by passing samples through an ion-exchange column. By using a combination of cationic and anionic resins it may be possible to remove most charged contaminants. Non-polar molecules can be removed by eluting through a column with a non-polar or hydrophobic stationary phase. Proteins, amino acids, organic acids, and hydrophobic compounds can be potentially removed from the carbohydrates in this manner prior to analyses.

Before analysis of the carbohydrates, residual alcohol (or other organic solvents) can be removed, if necessary, from the solution by evaporating under nitrogen or under vacuum using a rotary evaporator. For aqueous solutions, the sample can be concentrated using lyophilization.

Solid phase extraction (SPE) has also been reported for the cleanup and quantification of sugars and organic acids in herbal dry extracts. A three step SPE sequence was used for the separation of sugars from the other components. A hydrophobic cartridge was used as the first cartridge followed ion and cation exchange cartridges (Schiller et al., 2002).

6. Analysis

Once the carbohydrate fraction has been isolated from other components of the plant, either the total carbohydrate content can be determined, or individual carbohydrates can be isolated, identified and quantified. The analysis of carbohydrates can be performed using any of several different methods. Two of these techniques include gas chromatography (GC) and liquid chromatography (LC). There are also spectral methods available including nuclear magnetic resonance (NMR), infrared (IR) and Raman spectroscopy. In this review, our focus is on the chromatographic and mass spectrometric methods.

7. Derivatization for GC or GC/MS analyses

The most prevalent method used for analyzing carbohydrates is probably GC and GC coupled with mass spectrometry (MS) due to the high resolution of GC and definitive nature of MS. Since carbohydrates are nonvolatile, it is necessary to hydrolyze the sugars and then derivatize them to increase their volatility so they can elute through a GC column for analysis. Methods involving the formation of methylated glycosides, acetates, acetals, trimethylsilyl ethers, and more volatile alditol acetate derivatives of monosaccharides have been widely used (McInnes et al., 1958; Bishop & Cooper 1960; Bishop 1964; Lehrfeld 1981; Blakeney et al., 1983). More recently, trimethylsilyl (TMS) derivatives of carbohydrates have been used principally due to their relative ease of preparation and increased volatility. (Sweeley et al. 1963; Sullivan & Schewe 1977; Honda et al., 1979; Li et al., 1983; Twilley 1984). Different structural forms of carbohydrates can complicate their chromatograms due to the production of several (as many as 5) peaks for each monosaccharide. Formation of the corresponding oxime TMS-derivative reduces the number of potential peaks (Decker & Schweer 1982; Al-Hazmi & Stauffer 1986; Long & Chism 1987). Dmitriev et al. (1971) prepared the aldononitrile acetate derivatives with the oxime intermediate. Churms (1990) found the derivatization process was not affected by the presence of water in the reaction mixture, helping to minimize processing steps. Methods for the separation of neutral sugars in gums have also been reported using similar methods (Al-Hazmi & Stauffer, 1986).

Silylation is a versatile technique to increase the volatility of various analytes, including carbohydrates, making them amenable to GC and GC/MS analyses. There are several practical considerations that should be addressed prior to derivatization of a sample by this method. One major disadvantage of silylation derivatives is that they are susceptible to hydrolytic attack by any moisture present in the sample, resulting in incomplete silylation. However, the trimethylsilylation of aqueous samples of hydroxyl compounds has been achieved using a large excess of derivatizing reagent (Valdez 1985). Evershetd (1993) discussed another problem associated with silylation of carbohydrates, the existence of multiple reaction products, resulting in complicated chromatograms. The multiple products result from the formation of anomers and interconversion between pyranose and furanose rings. Interconversion of the anomers occurs via the open chain form of the sugar, while mutarotation results from the opening and closing of the ring. The interconversions can be minimized by the use of rapid and mild derivatization conditions. If silylation is the method of choice for derivatization, it may be desirable to protect the keto group of the monosaccharides prior to silylation in order to prevent the formation of enol-TMS ethers. These derivatives are unstable and complicate the analyses by giving rise to multiple products that can't be prepared quantitatively (Halket 1993).

In most instances, the silylating reagent is an adequate solvent. However, sometimes an additional solvent is required in the reaction. The selection of that solvent is critical to the success of the derivatization process. Any active hydrogens, including those present in the solvent, may be silylated. Pyridine has been found to be an ideal solvent for silylation reactions due to the increased solubility of the carbohydrates and their derivatives in that solvent (Evershed 1993). Heating slightly is often utilized to aid in efficient silylation,

One of the earliest reagents used for silylation was hexamethyldisilazane (HMDS). Usually, there is no need for additional solvents when HMDS is used. Recently, Ruiz-Matute et al. (2010) reviewed derivatization techniques of carbohydrates for GC and GC/MS analyses. Included in the discussion were derivatization of common sugars through the formation of ethers and esters, oximes, alditol acetates, aldononitriles, and dithoacetals (Evershed 1993). Another silylating reagent is trimethylsilylimadazole (TMSI). Garland et al. (2009) analyzed soybean roots for pinitol using GC/MS (see Figs. 1-3). Roots were extracted in methanol and derivatized using TMSI. In this example a DB-5 capillary column was used in the splitless mode. The column eluents were analyzed by a double-focusing, four-sector mass spectrometer in the electron-ionization mode. Accurate mass measurements were also performed to determine the elemental composition of the parent and fragment ions. Under these conditions, a pinitol standard produced a single peak in the total ion chromatogram with a retention time of 9.18 min as shown in Fig. 1. Although several peaks appeared, pinitol's peak at 9.18 min was well-resolved.

The mass spectrum of TMSI-derivatized pinitol in Fig. 2 shows the major ion fragments detected from this, the most common carbohydrate in soybeans (Garland et al, 2009). In this example, the base ion is m/z 260. A comparison of the extracted ion plots of the soybean extract is shown in Fig. 3. A vertical, solid black line was added to each at the retention time of derivatized pinitol as determined from the standard. In the extracted ion plot of the soybean root, Figure 4 shows the total ion chromatogram of a TMSI-derivatized sugar beet extract. In this example no significant peaks appeared at the retention time of pinitol. The sugar beet root extract also showed no substantial peaks with the m/z 260 mass fragment.

The concentration of pinitol in soybean roots was approximately 4% of the soybean root's dry mass using a dry/fresh weight ratio of 54.5 mg DW/g FW (which is similar to 73.6 mg DW/g FW reported for alfalfa by Fougere, et al. (1991). The methanol extraction method appears to be effective for removing pinitol from the root tissue of soybean plants. The extent of extraction at the cost of time was encountered as well by Streeter and Strimbu's simultaneous extraction and derivatization method (Streeter & Strimbu 1998). Although they were able to reduce processing time, they were unable to extract as much pinitol from fibrous plant tissues in pyridine in 1 h when compared to complete extraction with ethanol for 24 h before derivatization (Streeter & Strimbu 1998).

Another benefit to using methanol extraction and TMSI derivatization is the relative simplicity of the resulting chromatograms. Eleven peaks were observed in the soybean extract chromatogram in Fig. 1, with pinitol clearly defined near 9.18 min. This compares with only 6 major peaks from sugar beet (Fig. 4) and 10 from snap bean roots (Fig. 5). The simplicity of the chromatograms is an indicator that pinitol and a small amount of other compounds are present in the methanol extract, which reduces the likelihood of coelution or some other interfering matrix effect with pinitol. This also provides support for the possible use of methanol extraction as a first step in the purification of pinitol from soybean root tissue.

The mass spectrum of the derivatized pinitol shown in Figure 2 is very similar to that reported previously (Savidge & Forster, 2001). Identification of pinitol by mass spectrometry is made exceedingly easy by the presence of a high-intensity m/z 260 fragment ion. The fragment ion at m/z 260 appears to be a unique ion associated with pinitol and the other O-methylinositols compared with the other sugars observed using this analytical procedure. This allows for a high probability of quantitative results even in the event of another analyte coeluting with pinitol. The elemental composition obtained from accurate mass measurements for m/z = 260 was determined to be $C_{11}H_{24}O_3Si_2$, which was matched within 4.3 millimass units (mmu).

We have also extracted roots using 80% ethanol rather than methanol. This led to the extraction of a greater variety of inositols and O-methylinositols from several plant roots (unpublished data).

Permethylation is another derivatizing method for the analysis of carbohydrates. The methods using permethylation initially provided relatively long retention times. Some of the reactions to form permethylated derivatives include the use of methyl iodide/silver oxide (Gee & Walker 1962; Walker et al., 1962; Kircher, 1960) methylsulfinylcarbanion/methyl iodide (Hakomori 1964; Corey & Chaykovsky 1962; Moor & Waight 1975), and potassium/liquid ammonia/methyl iodide (Muskat 1934a; Muskat 1934b). Permethylation has also become very popular in the LC/MS analysis of carbohydrates.

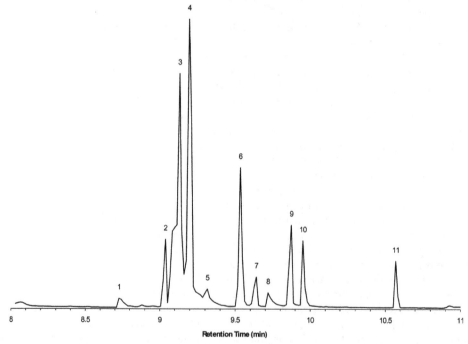

Fig. 1. Total ion chromatogram of an extract of soybean roots. Peak 4 was determined to be pinitol. From Garland, *et al* (2009).

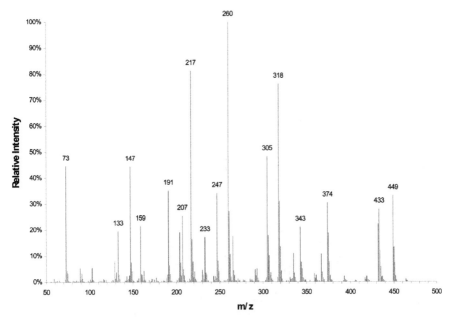

Fig. 2. Mass spectrum of TMSI-derivatized pinitol. From Garland, *et al.* (2009).

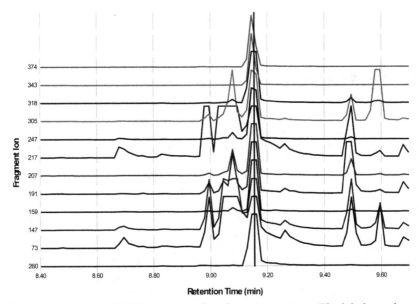

Fig. 3. Extracted ion plot of TMSI-derivatized soybean root extract. The labels on the vertical axis indicate the fragment mass of each extracted ion chromatogram. The chromatograms were spaced for easier representation. All peaks are on the same scale relative to their baselines. A vertical black line was inserted at the retention time of pinitol for reference. From Garland, *et al.* (2009).

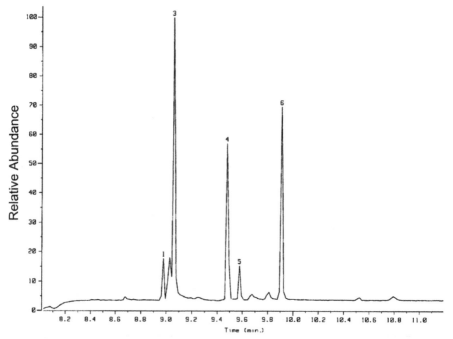

Fig. 4. Total ion chromatogram of derivatized sugar beet extract. Conditions were those of the chromatogram in Figure 1. Pinitol (retention time 9.2 min.) was not detected, as confirmed by MS analysis (Garland, *et al.*, 2009).

8. Other analytical techniques

Another technique for the separation and analysis of carbohydrates is liquid chromatography (LC). The column used in LC to provide the separation depends on whether the carbohydrates have been derivatized or not. Underivatized carbohydrates are commonly separated using ion exchange resins with water as an eluent and refractive index (RI) for detection. Refractive index detectors are, however, typically low in sensitivity, so samples need to be concentrated for quantitative analyses. The concentration of the carbohydrate must be in the percent range, and the RI detector can only be used with isocratic elution (Martens & Frankenberger 1990).

Other alternative detectors including both UV/visible absorbance and fluorescence require either pre-column or pre-detection derivatization of sugars, due to the fact that carbohydrates do not have a chromophore. Evaporative light scattering (ELS) is a detection technique used in high performance chromatography (HPLC) and supercritical fluid chromatography (SFC). It has been used for the analysis of carbohydrates and can act as a qualitative or quantitative detector (Wei & Ding 2000 ; Karlsson et al., 2005). The ELS is limited to solutes of low volatility. With the ELS, the column effluent is passed through a nebulizer and then into a heated drift tube; the solvent is evaporated leaving behind a particulate or aerosol form of the target compound. Light striking the dried particles that exit the drift tube is scattered and the photons are detected by a photodiode or photomultiplier tube at a fixed angle from the incident light. (LaPosse & Herbtreteau 2002).

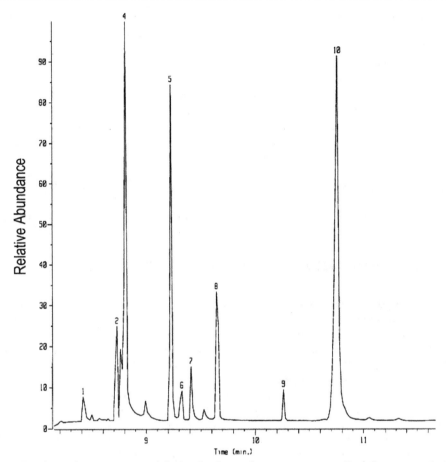

Fig. 5. Total ion chromatogram of derivatized snap bean root extract. Peak 5 was at a similar retention time to that of pinitol in Fig 1 (9.2 min.), but MS analyses were unable to detect pinitol in snap bean root extract (Garland, *et al.*, 2009).

Another detector commonly used is a pulsed amphoteric detector (Lee 1996; Johnson et al., 1993).

One derivatization procedure for carbohydrates to provide a chromophore for LC analysis involves a reaction with p-nitrobenzoyl chloride and pyridine. The reaction replaces the active hydrogens with a nitrobenzoyl group. The method was applicable to mono-, di-, and trisaccharides except fructose (Nachtmann & Budna 1977; Nachtmann 1976). Many of the derivatization reactions for carbohydrates are discussed by Knapp (1979). In addition, other derivatization techniques have been discussed (Meulendijk & Underberg 1990).

Mass spectrometry can also be coupled with LC. Examples are LC/MS and capillary electrophoresis/MS. Many of the LC techniques allow carbohydrates to be analyzed without prior derivatization as is necessary in GC and GC/MS analyses.

It should be noted that there is not one LC column that has been reported to separate every carbohydrate. Togami et al. (1991) discussed the separation of carbohydrates using cation-exchange columns. Richmond et al. (1991) separated carbohydrates in dairy products. Henderson and Berry (2009) have utilized Zorbax columns for the separation of carbohydrates in Stevia sweetener. Romano (2007) discussed carbohydrate analysis in food products emphasizing column chemistries and detection. Several vendors offer LC columns for carbohydrate separation. Wilcox et al. (2001) also discussed several column types used for carbohydrate separation. Hydrophilic interaction chromatography (HILIC) has also been reported as a method for analyzing ionic or polar compounds, particularly biomolecules and drug metabolites (http://www.laboratoryequipment.com/article-is-hilic-in-your-future-ct92.aspx). Simple carbohydrate separations can also be performed on functionalized silica or resin-based columns (http://www.labnews.co.uk/feature_archive. php/4000/5/just-juice).

The separation of mono- and oligosaccharides are also performed using capillary electrophoresis. Different formats are capillary zone electrophoresis (CZE), capillary isoelectric focusing (CIEF), capillary isotachophoresis (CITP), and micellar electrokinetic chromatography (MEKC). These techniques are summarized in a review by Thibault and Honda (2003).

9. Liquid chromatography/mass spectrometry (LC/MS) and other MS techniques

Efficient separation methods such as high performance liquid chromatography (HPLC) and capillary electrophoresis combined with detection methods (e.g. mass spectrometry) that supply structural or compositional information is a preferred tool for the analysis of biomolecules, particularly carbohydrates. Liquid chromatography/mass spectrometry with both electrospray (ESI) and atmospheric pressure ionization (APCI) has spurred a major interest in the analysis of carbohydrates.

In ESI , the liquid containing the analyte(s) of interest is dispersed into a fine aerosol. Because the ion formation involves extensive solvent evaporation, the typical solvents for electrospray ionization are prepared by mixing water with volatile organic compounds (e.g. methanol, acetonitrile). To decrease the initial droplet size, compounds that increase the conductivity (e.g. acetic acid) are customarily added to the solution. Large-flow electrosprays can benefit from additional nebulization by an inert gas such as nitrogen. The aerosol is sampled into the first vacuum stage of a mass spectrometer through a capillary, which can be heated to aid further solvent evaporation from the charged droplets. The ions observed by mass spectrometry may be quasimolecular ions created by the addition of a hydrogen ion and denoted $[M + H]^+$, or of another cation such as sodium ion, $[M + Na]^+$, or the removal of a proton, $[M - H]^-$. Multiply-charged ions such as $[M + nH]^{n+}$ are often observed (Gaskell 1997). As examples, Fountain and Grumbach (2009) used negative ion electrospray mass spectrometry for the analysis of fructose, glucose, sucrose, and lactose. Taormina et al. (2007) and Mauri et al. (2002) used flow injection techniques with mass spectrometry. Fugimoto et al. (2005) used rubidium in the mobile phase as a complexing agent for both nuclear magnetic resonance and electrospray mass spectrometry analysis. Taylor et al. (2005) utilized ESI/MS to study fragmentation patterns of carbohydrates. Schlichtherle-Cerny et al. (2003) utilized a HILIC column coupled with ESI/MS for the analysis of amino acids, peptides, glycoconjugates, and organic acids in foods without prior derivatization.

In APCI, typically the mobile phase containing eluting analyte is heated to relatively high temperatures (above 400 C), sprayed with high flow rates of nitrogen and the entire aerosol cloud is subjected to a corona discharge that creates ions. Often APCI can be performed in a modified ESI source. The ionization occurs in the gas phase, unlike ESI, where the ionization occurs in the liquid phase. A potential advantage of APCI is that it is possible to use a nonpolar solvent as a mobile phase solution, instead of a polar solvent, because the solvent and molecules of interest are converted to a gaseous state before reaching the corona discharge pin. Typically, APCI is a harder ionization technique than ESI, i.e. it generates more fragment ions relative to the parent ion.(Kostianinen et al., 2003). Kumaguai (2001) used atmospheric pressure chemical ionization mass spectrometry for the analysis of sugars and sugar alcohols without derivatization but did use methylene chloride or chloroform that was added post column to increase the sensitivity. The ions detected included $(M+Cl)^-$. Shimadzu application note also used solvent addition post column to improve sensitivity. This application also used APCI in the negative ion mode. Keski-Hynnila et al. (2004) compared APCI, atmospheric pressure photoionization, and electrospray in the analysis of phase II metabolites.

Other types of mass spectrometers used for analysis of carbohydrates include quadrupole time-of-flight (QTOF) mass spectrometers which allow both accurate mass (elemental composition) and MS/MS studies to be performed. Another mass spectrometer very useful for the analysis of carbohydrates is the ion trap (IT) MS. Ion trap technology has been described in (March & Todd 2005a, 2005b), and its major advantage includes the capability of MS^n which can provide additional structural information. Examples of glycoprotein analysis using IT have been described by (Stumpo & Reinhold 2010; Jiao et al., 2010; Reinhold et al., 1990).

Another technique that has been utilized for the analysis of carbohydrates is matrix assisted laser desorption/time-of-flight mass spectrometry (MALDI/TOFMS)(Harvey 1999, 2009)). In MALDI, the sample to be analyzed is mixed with a matrix, which in turns absorbs heat energy from irradiation with a nitrogen laser light. For example, dihydroxybenzoic acid (DHB) or ferulic acid which are commonly used as a matrices have a carboxyl group on a benzene ring. The DHB absorbs the energy and acts as a proton donor (Zenobi & Knochenmuss 1998). Time-of-flight mass spectrometry allows the majority of the ions generated throughout the mass range to be collected by the detector. MALDI has been primarily used to obtain spectra of very large polymers, biomolecules, and a variety of thermally labile materials (Hillenkamp et al., 1991, Nelson et al., 1990. We have also used MALDI/TOF for the analysis of smaller molecules (e.g. <500 amu) (Goheen *et al.*, 1997; Campbell *et al.*, 2001).

10. The inositols

Inositols (Fig. 6) are polyols of cyclohexane with the empirical formula $C_6H_{12}O_6$. There are potentially 9 stereoisomers of inositol but only five are naturally occurring (structure shown below). They are *myo*-inositol, *chiro*-inositol, *scyllo*-inositol, *muco*-inositol, and *neo*-inositol. Of these, *myo*-inositol is the precursor of the other four. *myo*-Inositol is synthesized from glucose.

The synthesis of *myo*-inositol uses the enzyme L-*myo*-inositol 1-phosphate synthase to catalyze the reaction which produces L-*myo*-inositol-1-phosphate from D-glucose 6-phosphate (Hoffmann-Ostenhof and Pittner, 1982). The L-*myo*-inositol-1-phosphate is then dephosphorylated through inositol monophosphate to produce *myo*-inositol (Loewus & Murthy, 2000). The enzyme that catalyzes this step is L-*myo*-inositol 1-phosphate synthase

(Stieglitz et al, 2005). The four other inositol isomers are derived from *myo*-inositol (Loewus and Murthy, 2000). The sequoyitol can then be epimerized to D-pinitol (See Fig. 7) which is demethylated to D-*chiro*-inositol using NADP-specific D-pinitol dehydrogenase (Stieglitz et al, 2005).

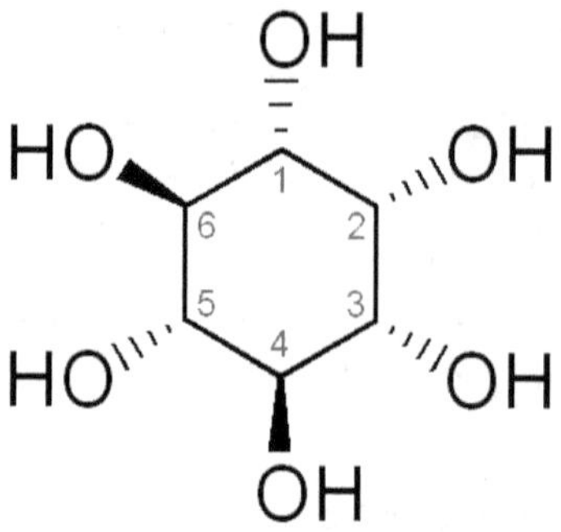

Fig. 6. Inositol.

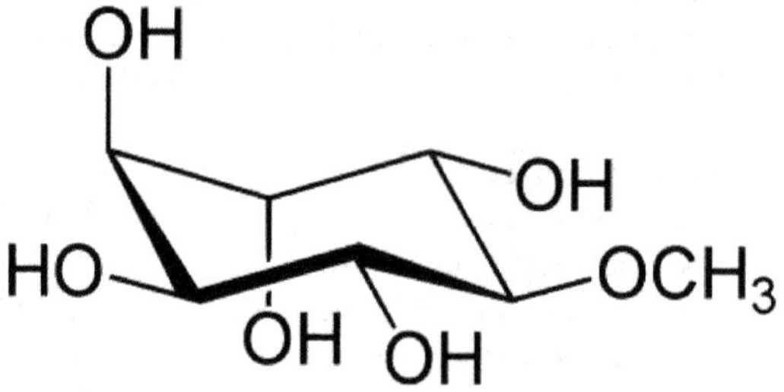

Fig. 7. Pinitol.

In addition to the five stereoisomers of inositol, the *O*-methylinositols can also be synthesized from *myo*-inositol. Of these, ononitol and pinitol are common to soybeans. Ononitol is a precursor to pinitol in soybeans (Loewus and Murthy, 2000; Chiera et al. 2006). Of the *O*-methylinositols, pinitol is most abundant in soybeans.

myo-Inositol is probably the most studied of all the inositols because it is the most commonly available. It has a very important function as it is required in the formation of Lecithin, which protects cells from oxidation and is an important factor in the building of cell membranes. Inositol, also has a metabolic effect in preventing too much fat to be stored in the liver, which is why it is called a lipotropic and is a vital part in maintaining good health. Inositols have been found in many plants both foodstuff and other plants at varying evolutionary stages (Clements & Darnell, 1980; Chiera *et al.*, 2006; Guo & Oosterhuis, 1997; Henry, 1976; Johansen *et al.*, 1996; Johnson and Sussex, 1995; Johnson & Wang, 1996; Lind *et al.* 1998; Loewus *et al.*, 1984; Manchanda and Garg, 2008; Ogunyemi *et al.*, 1978; Phillips, et al 1982; Sheveleva *et al.*, 1997; Streeter *et al.*, 2001). Different soybean plant parts contain different levels of inositols as do soybean plants in vegetative verses reproductive growth stages (Phillips and Smith, 1974). Comparison of total inositols among plants should be examined carefully because each plant may produce different proportions of the various inositols (Larson & Raboy, 1999). Research with pinitol in soybean documents that this cyclitol is a major consitutent of soybean (Phillips & Smith, 1974; Streeter, 1980; Phillips, et al. 1982; Dougherty & Smith, 1982). Because pinitol diffuses faster than carbohydrates during imbibition, it is theorized that loss of pinitol from soybean seed encourages the growth of Bradyrhizobium (Rhizobium) species in the soil needed for nitrogen fixation (Nordin, 1984). Accumulation of ononitol and pintol in soybean and other plants under drought conditions has been documented (Streeter *et al.*, 2001; Guo & Oosterhuis, 1997; Manchanda & Garg, 2008; Sheveleva, et al, 1997).

Inositols are very important in general plant growth, seed storage, nitrogen fixation and protection of plants during stress. Inositol metabolism and its role in photosynthesis, plant health, and subsequent potential increase in yield is complex but new discoveries in this area may lead to future yield improvements. The role of inositols in nitrogen fixation is also complex and not currently fully understood. Inositols play an important role in phosphorus movement in the environment. Efforts are being made to alter the phytate content of soybean so animals can use the phosphorus and also reduce the amount that is excreted as manure. There are implications here not only for animal health but also for the preservation and sustainability of watersheds.

Phytate, *myo*-inositol hexakisphosphate, is found in almost all plant and animal cells and serves as an important phosphate reserve in plants (Irvine & Schell, 2001). Exposure of soybean cell suspension to *Psuedomonas syringae* pv *glycinea* indicated that whether a virulent or avirulent strain is used, the plant starts defense systems at the expense of housekeeping cell functions (Logemann *et al.*, 1995; Shigaki & Bhattacharyya, 2000). Part of this defense reaction involves cellular cyosolic inositol and the IP3 pathway. This pathway is involved in cell division, growth and elongations and there is evidence that this pathway is inhibited when the plant is exposed to pathogens (Perera *et al.*, 1999; Shigaki & Bhattacharyya, 2000). Selection of plants with reduced phytate levels raised the question of these plants' response to stress in the form of diseases. Murphy et al., 2008 found that disruption of phytate biosynthesis resulted in increased susceptibility in *Arabidopsis thaliana* to virus (potato virus Y), fungal (*Botrytis cinerea*) and bacterial (*Psueodomonas syringae*) diseases. The role of phytate in basal resistance to plant pathogens was previously unknown. Klink et al. (2009) found 1-phosphatidylinositol phosphodiesterase-related genes expressed when soybean plants are exposed to *Heterodera glycines*, a pathogen of soybean. The findings of inositols in plant defense are important findings and the next step is to determine whether the defense reaction is a general reaction or specific to different types of attacks.

Transgenic plants that release extracellular phytase from their roots have a significantly increased ability to acquire phosphorus from inositol phosphates from growth medium; however, there is less evidence that phosphorus nutrition of plants can be improved in plants grown in soil (George, *et al.*, 2004). Phytate and phytic acid represent the major form of phosphorus in animal feed derived from plants. Phosphorus in seeds and tubers is stored primarily as phytate (*myo*-inositol exakisphosphate), which is poorly digested by non-ruminant animals such as swine, poultry and fish (Saghai Maroof *et al.*, 2009; Kim *et al.*, 2006). The lack of the hydrolytic enzymes necessary for phytate to be utilized by these animals requires supplemental phosphate. Plant breeding efforts involve plant selections for improved phosphorus usage by animals and different feed additives resulting in less environmental pollution.

Inositol is synthesized sparingly in the body but is present in many foods. The inositols are essential nutrients for plants (Loewus and Murthy, 2000) and animals (Holub, 1986). Concentrations of the inositols and their metabolites can be much higher in some plant species than in mammalian tissue. For example, in soybeans, the concentration of pinitol alone approaches 30 mg/g (Streeter & Strimbu 1998; Garland *et al.*, 2009) whereas in human blood, the levels of free *myo*-inositol is 3000 times lower (1 mg/100 mL). Levels of pinitol in blood is not widely known, but are anticipated to be orders of magnitude less than *myo*-inositol.

One form of inositol, inositol hexaniacinate, has been used to support circulatory health because it functions like niacin in the body. The major dietary forms of myo-inositol are inositol hexaphosphate or phytic acid, which is widely found in cereals and legumes and associated with dietary fiber, and myo-inositol-containing phospholipids from animal and plant sources.

Inositol is involved in the glucuronic acid and pentose phosphate pathways. Inositol exists as the fiber component phytic acid, which has been investigated for its anti-cancer properties. Inositol is primarily used in the treatment of liver problems, depression, panic disorder, and diabetes (Narayanan, 1987). Used with choline, it also aids in the breakdown of fats, helps in the reduction of blood cholesterol, and helps to prevent thinning hair (Walker, 2010). It promotes the export of fat from the liver. Inositol is required for the proper function of several brain neurotransmitters. Inositol may improve nerve conduction velocities in diabetics with peripheral neuropathy. Inositol may help protect against atherosclerosis and hair loss. There has also been the suggestion that it may help to reverse some nerve damage caused by diabetes (Gregersen et al. 1978; *Ibid*,1983). Inositol has also been tried for other psychological and nerve-related conditions including the treatment of side effects of the medicine lithium. Inositol also has a prominent calming effect on the central nervous system, so it is sometimes helpful to those with insomnia. Inositol may also be involved in depression.

Under pinitol deficiency, detrimental health conditions may exist such as higher blood sugar in diabetics (Geethan and Prince, 2008). Myo-inositol deficiency can lead to depression and other mental disorders (Levine et al, 1995; Benjamin et al, 1995; Fux, et al, 1996). Also, polycystic ovary syndrome (PCOS) has been reported to be related to a deficiency in dietary inositol (Gerli, et al. 2003; Ibid, 2007). Correlations with depression and similar disorders may be related to the abundance of inositol phospholipids in brain and other nervous system tissues. However, the relationship between pinitol and blood sugar levels is more likely correlated with the similarities in structure between the 0-methyl inositol and glucose.

There is no recommended daily allowance for inositol, but the normal human dietary intake is about 1 gram per day. Inositol is available from both plant and animal sources. Natural sources of inositol include soybeans, wheat germ, brewer's yeast, bananas, liver, brown rice, oak flakes, nuts, unrefined molasses, vegetables, and raisins. Most dietary inositol is in the form of phytate, a naturally occurring plant fiber.

Dietary effects of pinitol and ononitol are still in the earlier stages of discovery. It has recently been shown that pinitol lowers blood glucose levels in type II diabetics while significantly decreasing total cholesterol, LDL-cholesterol and the LDL/HDL-cholesterol ratio (Kim et al 2005). The dietary benefits or hazards of the other metabolites of isomers of inositol (other than myo-inositol) are under active investigation.

11. Conclusions and future directions

It is clear from this review that there are many different tools to study the carbohydrates in soybean plants. Results from any of the various analytical methods can be compared as long as they have been tested with adequate standards. The outstanding carbohydrate found in soybeans is pinitol, part of the inositol family. There has been considerable research into the value of the inositols, but most of the emphasis has been on myo-inositol, probably because it is widely available. However, for the soybean industry, it would be valuable to better understand the role of pinitol in health and nutrition. There are good indications that pinitol may have unique nutritional value, and key roles in soybean plant biology. Future directions should include the use of effective analytical methods to perform more research into the roles of pinitol and related inositols in various fields of nutrition, medicine, and plant biology.

12. Acknowledgements

We acknowledge Shaun Garland and Luther McDonald for their technical support.

13. List of abbreviations

GC	gas chromatography
LC	liquid chromatography
MS	mass spectrometry
RI	refractive index
UV	ultraviolet
APCI	atmospheric pressure chemical ionization
IT	ion trap
ELS	evaporative light scattering
ESI	electrospray ionization
SPE	solid phase extraction
MS/MS	mass spectrometry/mass spectrometry
LC/MS	liquid chromatography/mass spectrometry
GC/MS	gas chromatography/mass spectrometry
MALDI/TOFMS	matrix assisted laser desorption/time-of-flight mass spectrometry
QTOF	quadrupole time-of-flight
TMS	trimethylsilyl
TMSI	trimethylsilyl imidazole
HMDS	hexamethyldisilazane

14. References

Anderson, L. & Wolter, K.E. (1966). *Cyclitols in plants: Biochemistry and physiology*. Annual Review Plant Physiology 17:209-222.

Al-Hazmi, M.I. & Stauffer, K.R. (1986). Gas chromatographic determination of hydrolyzed sugars in commercial gums. *Journal of Food Science* 51:1091-1092, 1097.

Anthony, R.M.; Nimmerjahn, F.; Ashline, D. J.; Reinhold, V. N.; Paulson, J. C. & Ravetch, J. V. (2008). *Recapitulation of Intravenous Ig Anti-Inflammatory Activity with a Recombinant IgG Fe*. Science, 320: 373-376.

Asp, N-G. (1993). Nutritional importance and classification of food carbohydrates , In: *Plant Polymeric Carbohydrates*, Meuser, F., Manners, D.J., & Seibel, (Ed.), W. Royal Soc. Chem., Cambridge, U.K., pp. 121-126.

Benjamin, J.; Levine, J.; Fux, M.; Aviv, A. & Belmaker, R.H. (1995). Double-blind, placebo-controlled, crossover trial of inositol treatment for panic disorder. *American Journal of Psychiatry*. 152: 1084-1086.

Berridge, M.J.; Helsop J.P.; Irvine, R.F. & Brown K.D. (1984). Inositol trisphosphate formation and calcium mobilization in Swiss 3T3 cells in response to platelet-derived growth factor. *Biochem. J*. 222: 195-201.

Binder, R. G. & Haddon, W. F. (1984). *Analysis of O-methylinositols by gas-liquid chromatography-mass spectrometry Carbohydr*. Res.129:21-32.

Bisho , C.T. (1964). *Gas-Liquid chromatography of carbohydrate derivatives*. Advances in Carbohydrate Chemistry 19:95-147.

Bishop, C.T. & Cooper, F.P. (1960). Separation of carbohydrate derivatives by gas-liquid partition chromatography. Candian J. Chem., 38:388-398.

Blakeney , A.B.; Harris, P.J.; Henry, J. & Stone, B.A. (1983). A simple and rapid preparation of alditol acetates for monosaccharide analysis. Carbohydrate Research, 113:291-299.

Bohnert, H.J.; Nelson, D.E. & Jensenay, R.G. (1995). The Plant Cell, *Adaptations to Environmental Stresses*, Vol. 7, 1099-1111.

Brockwell, J.; Bottomley, P.J. & Thies, J.E. (1995). Manipulation of rhizobia microflora for improving legume productivity and soil fertility: a critical assessment. *Plant Soil*. 174: 143-180.

Campbell, J. A.; Hess, W.P.; Lohman, J.R. & Goheen, S.C. (2001). Analysis of Hanford-related organics using matrix-assisted laser desorption ionization time-of-flight mass spectrometry. J. Radioanalytical and Nucl. Chem., 250: 247-253.

Campling, J.D. & Nixon, D.A. (1954). The inositol content of foetal blood and foetal fluids. J. Physiol. (1954). 126, 71-80

Chiera, J.M.; Streeter, J.G. & Finer, J.J. (2006). Ononitol and pinitol production in transgenic soybean containing the inositol methyl transferase gene from *Mesembryanthemum crystallinum*. Plant Sci. 171: 647-654.

Churms, S.C. (1990). Recent developments in the chromatographic analysis of carbohydrates. *Journal of Chromatography*. 500:555-583.

Churms, S.C. (1996). Recent progress in carbohydrate separation by high-performance liquid chromatography based on size exclusion. *Journal of Chromatography*, A., 720:151-166.

Clements, R.S., Jr. & Darnell, R. (1980).myo-Inositol content of common foods: development of a high-*myo*-inositol diet. *J.Am. Clin. Nutr*. 33:1954-1967.

Colowick, S.P. & Kaplan, N.O. (1951). Carbohydrate metabolism. *Annual Review of Biochemistr*, 20:513-558.

Corey, E. J. & Chaykovsky, M. (1962). Methylsulfinylcarbanion. *J. Am. Chem. Soc*. 84: 866-867.

Cote, G.G. & Crain, R.C. (1993). Biochemistry of phosphoinositides. *Annual Review of Plant Physiology*. 44:333-356.

Daughaday, W.H.; Larner, J. & Hartnett, C, (1955). The synthesis of inositol in the immature rat and chick embryo. *J. Biol. Chem*. 212:869-875.

Decker, P. & Schweer, H. (1982). Gas-liquid chromatography on OV-225 of tetroses and aldopentoses as their O-methoxime and O-n-butoxime pertrifluoroacetyl derivatives and of C3-C6 alditol pertrifluoroacetates. *J. of Chromatography*, 236:369-373.

Denison, R.F.; Hunt, S. & Layzell D.B. (1992). Nitrogenase activity, nodule respiration and O_2 permeability following detopping of alfalfa and birdsfoot trefoil. *Plant Physiol.* 98:894-900.

Dmitriev, B.A.; Backinowsky, L.V.; Chizhov, O.S.; Zolotzrev, B.M. & Kochetkow, N.K. (1971). Gas-liquid chromatography and mass spectrometry of aldononitrile acetates and partially methylated aldononitrile acetates. Carbohydrate Research 19:432-435.

Dogruel, D. & Williams, P. (1995). Detection of human IgM at m/z 1 MDa. Rapid Comm Mass Spectrom. 9:625

Eagle, H.; Oyama, V.I.; Levy, M. & Freeman, A.E. (1957) . myo-Inositol as an essential growth factor for normal and malignant human cells in tissue culture. *J. Biol. Chem.* 266: 191-205

Ensminger A.; Ensminger, M.; Konlande, J. & Robson J. (Ed.). (1995). *The Concise Encyclopedia Of Foods & Nutrition*, Boca Raton, London, Tokyo: CRC Press, pp.580-581.

Evershed, R.P. (1993). Advances in Silylation. In: *Handbook of derivatives for chromatography*, K. Blau & J.M. Halket, (Ed.), Chichester, U.K.: Wiley, p.53, p.59.

Farquhar, G.D. & Sharkey, T.D. (1982). Stomatal conductance and photosynthesis. Annu. Rev. Plant Physiol. 33:317-345.

Fontes, C.M.G.A. &. Gilbert, H.J. (2010). Cellulosomes:Highly efficient nanomachines designed to deconstruct plant cell wall complex carbohydrates. Annual Review Biochem. 79:655-681.

Fougere, F.; LeRudulier, D. & Streeter, J.G. (1991). Effects of salt stress on amino acid, organic acid, and carbohydrate composition of roots, bacteroids, and cytosol of alfalfa (*medicago sativa L.*). Plant Physiol. 96: 1228-1236.

Fountain, K.J.; Hudalla, C.; Grumbach, E.S. & McCabe, D. (2009). Analysis of carbohydrates by ultraperformance liquid chromatography and mass spectrometry. LCGC, ASIA Pacific, Volume 12, issue 4.

Fugimoto, T.; Sakari, S.; Tsutsuri, A.; Furihata, K.; Machinami, T. & Tashiro, M. (2005). Observation of 1,6-anhdro-beta-maltose and 1,6-anhdro-beta-D-glucopyranose complexed with rubidium by NMR spectroscopy and electrospray ionization mass spectrometry. Anal. Sci., 21: 1245-1247.

Fux, M.; Levine, J.; Aviv, A. & Belmaker, R.H. (1996). Inositol treatment of obsessive-compulsive disorder. American Journal of Psychiatry, 153:1219-1221.

Garland, S.; S. Goheen, S.; Donald, P.; McDonald, L & Campbell, J. (2009). Application of derivatization gas chromatography/mass spectrometry for the identification and quantitation of pinitol in plant roots. Anal. Letters 42:2096-2105.

Gaskell, S. (1997). Electrospray: Principles and Practices. *Journal of Mass Spectrometry*, 32:677-688.

Gee, M. & Walker H.G. (1962). Gas-liquid chromatography of some methylated mono-, di-, and trisaccharides. Anal. Chem. 34:650-653.

Geethan, P.K. & Prince, P.S. (2008). Antihyperlipidemic effect of D-pinitol on streptozotocin-induced diabetic Wistar rats, J. Biochem. Mol. Toxicol. 22:220-224.

George, T.S.; Richardson, A.E.; Hadobas, P. & Simpson R.J. (2004). Characterization of transgenic *Trifolium subterraneum* L. which expresses *phyA* and releases extracellular phytase: growth and P nutrition in laboratory media and soil. Plant Cell Environ. 27:1351-1361.

George, T.S.; Simpson, R.J.; Hadobas, P.A. & Richardson, A.E. (2005). Expression of a fungal phytase gene in *Nicotiana tabacum* improves phosphorus nutrition of plants grown in amended soils. Plant Biotechnol. J. 3:129-140.

Gerli, S.; Mignosa, M. & Di Renzo, G.C. (2003). Effects of inositol on ovarian function and metabolic factors in women with PCOS: a randomized double blind placebo-controlled trial. European Review for Medical and Pharmacological Sciences. 7:151-159.

Gerli, S.; Papaleo, E.; Ferrari, A. & Di Renzo, G.C. (2007). Randomized, double blind placebo-controlled trial: effects of Myo-inositol on ovarian function and metabolic factors in women with PCOS. European Review for Medical and Pharmacological Sciences, 11(5):347-354.

Goheen, S. C.; Wahl, K.L.; Campbell, J.A. & Hess, W.P.(1997). Mass spectrometry of low molecular mass solids by matrix-assisted laser desorption/ionization. J. Mass Spectr. 32: 820-828.

Gomes, C. I.; Obendorf, R.L. & Horbowicz, M. (2005). myo-Inositol, D-chiro-Inositol, and D-Pinitol synthesis, transport, and galactoside formation in soybean explants. Crop Sci. 45:1312-1319.

Gregersen, G.; Bersting, H.; Theil, P & Servo, C. (1978). Myoinositol and function of peripheral nerves in human diabetes. Acta Neurol. Scand. 58:241-248.

Gregersen, G.; Bartelsen, B. & Harbo, H. (1983). Oral supplementation of myo-Inositol: effects of peripheral nerves in human diabetics and on the concentration in plasma, erythrocytes, urine, and muscle tissue in human diabetics and normals. Acta Neurol. Scand. 67:164-172.

Guo, C. & Oosterhuis, D.M. (1997). Effect of water-deficit stress and genotypes on pinitol occurrence in soybean plants. Environ. Exper. Bot. 37:147-152.

Hakomori, S. (1964). A rapid permethylation of glycolipid, and polysaccharide catalyzed by methylsulfinyl carbanion in dimethyl sulfoxide. J. Biochem 55:205-208.

Halket, J.M. (1993). Derivatives for GC-MS. In: Handbook of derivatives for chromatography, ed. K. Blau and J.M. Halket, Chichester, U.K., Wiley, 304.

Halford, N.G.; Curtis, T.Y.; Muttucumaru, N.; Postles, J. & Mottran, D.S. (2010). Sugars in crop plants. Ann Appl Biol 158:1-25.

Hall, M.B. (2003). Challenges with nonfiber carbohydrate methods. J. Anim. Sci. 81:3226-3232

Halliday, J.W. & Anderson, L. (1955). The synthesis of myo-inositol in the rat. Journal of biological chemistry 217:797-802.

Hanneman, A. R.; Cesar, J.; Ashline, D. E. & Reinhold, V. N. (2006) Isomer and Glycomer Complexities of Core GlcNAcs in *Caenorhabditis e/egans*.Glycobiology, 16: 874 - 890.

Hansen, B. & Ortmeyer, H. (1996). Inositols-Potential roles for insulin action in diabetes: Evidence form insulin-resistant nonhuman primates. Lessons from Animal Diabetes U. Ed. E. Shafrie, pp. 333-348.

Harvey, D.J. (1999). Matrix-assisted laser desorption/ionization mass spectrometry of carbohydrates. Mass Spectrom. Rev. 18:349-450.

Harvey, D.J. (2010). Analysis of carbohydrates and glycoconjugates by matrix-assisted laser desorption/ionization mass spectrometry: an update for 2003-2004. Mass Spectrom. Rev. 28:273-361.

Hasegawa, P.M.; Bressan, R.A.; Zhu, J.K. & Bohnert, H.J. (2000). Plant cellular and molecular responses to high salinity, Annu. Rev. Plant Physiol. Plant Mol. Biol. 51:463–99.

Henderson, J.W., Jr.; & Berry, J. (2007). Isocratic Stevia seetner analysis using selective ZORBAX columns. Agilent Application Note. Food.

Henry, E.W. (1976). Determination of inositol in inbred and hybrid corn (Zea mays) seedlings. Journal of Experimental Botany 27:259-262.

Hillenkamp, F.; Karas, M.; Beavis, R.C. & Chait, B.T. (1991). Matrix-assisted laser desorption/ionization of biopolymers. Anal. Chem. 63:1193A-1203A

Hoffmann-Ostenhof, O. & Pittner, F. (1982). The biosynthesis of myo-inositol and its isomers. Can. J. Chem.60:1863-1871.

Holm, P.J.; Booth, M.H.; Schmidt, T.; Greve, H. & Callesen, H. (1999). Theriogenology 52: 683-700. "High bovine blastocyst development in a static in vitro production system using sofaa medium supplemented with sodium citrate and myo-inositol with or without serum-proteins"

Holub, B. J. (1986). Metabolism and function of myo-inositol phospholipids. Ann. Rev. Nutr. 6: 563-597.

Honda, S.; Kakehi, K. & Okada, K. (1979). A convenient method for the gas chromatographic analysis of hexosamines in the presence of neutral monosaccharides and uronic acids. Journal of Chromatography. 176:367-273.

Horecker, B.L. & Mehler, A.H. (1955). Carbohydrate metabolism. Annual Review of Biochemistry 24:207-274.

Irvine, R.F., & Schell, M.J. (2001). Back in the water: the return of the inositol phosphates. Mol. Cell. Biol. 2:327-338.

Ismail, M.N.; Stone, E.L.; Panico, M.; Lee, S.H.; Luu, Y.; Ramirez, K.; Ho, S.B.; Minory M.; Haslam J.D.; Stuart, M. & Dell, A. (2010). High Sensitivity O-glycomic Analysis of Mice Deficient in Core 2131,6-N-acetylglucosaminyl transferases. Glycobiology. 21:82-98.

Jaindl, M. & Popp, M. (2006). Cyclitols protect glutamine synthetase and malate dehydrogenase against the induced deactivation and thermal denaturation. Biochem and Biophysical Research Communications 345:761-765.

Jiao, J,; Zhang, H. & Reinhold, V. (2010). High Performance IT-MS" Sequencing of Glycans. Int. J. Mass Spectrometry. 303:109-117.

Johansen, S. L.; A. Sivasothy, A.; Dowd, M.K.; Reilly, P.J. & Hammond, E.G. (1996). Low-molecular weight organic compositions of acid waters from vegetable oil soapstocks. Journal of the American Oil Chemists' Society 73:1275-1286.

Johnson, D.C.; Dobberpuhl, D.; Roberts, R. & Vandeberg, P. (1993). Pulsed amperometric detection of carbohydrates, amines, and sufur species in ion chromatography-the current state of research. *J. of Chromatography* 640:79-96

Johnson, M. D., & I. M. Sussex, I.M. (1995). 1-L-*myo*-inosital 1-phosphate synthase from *Arabidopsis thaliana*. Plant Physio 107:613-619.

Johnson, M.D. & Wang, X. (1996). Differentially expressed forms of 1-L-*myo*-inosital 1-phosphate synthase (EC) in *Phaseolus vulgaris*. *J Biol Chem* 271:17215-17218.

Karlsson, G.; Winge, S, & Sandberg, H. (2005). Separation of monosaccharides by hydrophilic interaction chromatography with evaporative light scattering detection. *J. of Chromatogr.* A, 1092:246-249.

Keski-Hynnila, H.; Kurkela, M.; Elovarra, E.; Antonio, L.; Magdalou, J.; Luukanen, L,; Taskinen, J. & Kostianen, R. Comparison of electrospray, atmospheric pressure chemical ionization, and atomospheric pressure photoionization in the identification of apormophine, dobutamide, and entacapone phase II metabolites in biological samples. *Anal. Chem.* 74:3449.

Kim, J.I.; Kim, J.C.; Kang, M.J.; Lee, M.S.; Kim, J.J. & and Cha, I.J. (2005). Effects of pinitol isolated from soybeans on glycaemic control and cardiovascular risk factors in Korean patients with type II diabetes mellitus: a randomized controlled study.*Eur. J. Clin Nutrition* 59: 456-458.

Kim, T.; Mullaney, E.J.; Porres, J.M.; Roneker, K.R.; Crowe, S.; Rice, S.; Ko, T.; Ullah, A.H.J.; Caly, C.G.; Welch, R & Lei, X.G. (2006). Shifting the pH profile of *Aspergillus niger* PhyA phytase to match the stomach pH enhances its effectiveness as an animal feed additive. *Applied and Environmental Microbiology.* 72:4397-4403.

Kircher, H. W. (1960). Gas-liquid partition chromatography of methylated sugars. *Anal. Chem.* 32: 1103-1106.

Klink, V.P., Hosseini, P.; Matsye, P.; Alkharouf, N.W. & B.F. Matthews, B.F. (2009). A gene expression analysis of synctia laser microdissected from the roots of the *Glycine max* (soybean) genotype PI 548402 (Peking) undergoing a resistant reaction after infection by *Heterodera glycines* (soybean cyst nematode). *Plant Molecular Biology* 71:525-567.

Knapp, D.R. (1979). Carbohydrates In: *Handbook of Analytical Derivatization Reations*. John Wiley and Sons, New York, New York, 1979, Chapter 13, pages 539-598.

Kostianinen, R.; Kotiaho, T.; Kuuranne, T. & Auriola, S. (2003). Liquid chromatography/atmospheric pressure-ionization- mass spectrometry in drug metabolism. *J. Mass Spectrometry* 38:357-372.

Korak, D.A. (1984). Novel biosynthesis of D-pinitol in *Simmondsia chinensis.*"*Phytochemistry* 23: 65-66. "

Kumaguai, H. (2001). Application of liquid chromatography/mass spectrometry to the analysis of sugars and sugar-alcohol. Application note. 5988-4236EN. Agilent Technolgies.

Lambers, H., Stulen, I. & Van Der Werf, A. (1996). Carbon use in root respiration as affected by elevated atmospheric CO2. *Plant and Soil.* 187:251-263.

LaFosse, M. & Herbreteau, B. (2002). In: *Carbohydrate Analysis by Modern chromatography and electrophoresis, J. of Chromatography Library*, Z. El Rossi (Ed.), Elesevier Science, Chapter 30, 1101-1134.

Larson, S & Raboy, V. (1999). Linkage mapping of maize and barley *myo*-inositol phosphate synthase DNA sequences: Correspondence with a low phytic acid mutation. *Theoretical Applied Genetics* 99:27-36

Lau, K.; Partridge; E. A.; Grigorian, A.; Silvescu, C. I.; Reinhold, V. N.; Demetriou;Michael; Dennis.James W. (2007). Conditional interpretation of extracellular cues by N-glycan multiplicityand Golgi processing. Cell, 129: 123-134.

Layzell, D.B., Hunt, S.; Moloney, A.H.M.; Fernando, S.M. & Diaz del Castillo. L. (1990). Physiological, metabolic and developmental implications of O2 regulations in legume nodule. Pp 21-32. In: ? P.M. Gresshoff, (Ed.), Nitrogen fixation: Achievements and objectives. Chapman and Hall, New York.

Lee, L.S. & Morris, N.J. (1963). Isolation and identification of pinitol from peanut flour, *Agr. Food Chem.* 11: 321-322

Lee, Y.C. (1996). Carbohydrate analyses with high-performance anion-exchange chromatography. *J. Chromatogr A*. 720:137-149.

Lehrfield, J. (1981). Differential gas-liquid chromatography for the determination of uronic acids in carbohydrate mixtures. *Anal. Biochem.* 115:410-418.

Levine J.; Barak, Y,; Kofman, O.; & Belmaker, R.H. (1995). Follow-up and relapse analysis of an inositol study of depression. *The Israel Journal of Psychiatry and Related Sciences.* 32: 14-21.

Li, B.W.; Schuhmann, P.J.;. &. Holden, J.M.(1983). Determination of sugars in yoghurt by gas-liquid chromatography. Journal of Agricultural and Food Chemistry.l 31:985-989.

Lind, Y.; Engman, J.; L. Jorhem, L. & Wicklund, G.A. (1988). Accumulation of cadmium from wheat bran, sugar-beet fibre, carrots and cadmium chloride in the liver and kidneys of mice. *British Journal of Nutrition* 80:205-211.

Logemann, E.; Wu, S.; Schröder, J.; Schmelzer, E.;. Somssich, I.E. & K. Hahlbrock, K. (1995). Gene activation by UV light, fungal elicitor or fungal infection in *Petroselinum crispum* is correlated with repression of cell cycle-related genes. Plant J. 8:865-876.

Loewus, M. W.; Bedgar, D.L. & Loewus, F.A. (1984.) 1-L-*myo*-Inositol 1-phosphate synthase from pollen of *Lilium longifolium*. *J Biol Chem* 259: 7644-7647.

Loewus, F.A. & Murthy, P.P.N. (2000). *myo*-Inositol metabolism in plants. *Plant Science* 150:1-19.

Long, A.R. & Chism III, G.W. (1987). A rapid direct extraction-derivatization method for determining sugars in fruit tissue. *Journal of Food Science* 52:150-154.

McInnes, A.G.; Ball, D.H.; Cooper, F.P. & Bishop, C.T. (1958). Separation of carbohydrate derivatives by gas-liquid partition chromatography. *J. of Chromatog* 1:556-557.

Manchanda, G. & Garg, N. (2008). Salinity and its effects on the functional biology of legumes. Acta Physiol. Plant. 30:595-618.

Maniatis, S.; Hui, Z. & Reinhold, V. N. (2010). Rapid De-O-Glycosylation Concomitant with Peptide Labeling Using Microwave Radiation and anAlkyl Amine Base. Anal. Chem.,82(7):2421-2425.

March, R.E. & Todd, J.F.J. (2005a). A historical review of the development of the quadrupole ion trap. In: *Quadrupole Ion Trap Mass Spectrometry*, 2nd edition, John Wiley and Sons, Inc., Hoboken, N.J., Chapter 1.

March, R.E. & Todd, J.F.J. (2005b). Linear quadrupole ion trap mass spectrometer. In: *Quadrupole Ion Trap Mass Spectrometry*, 2nd edition, John Wiley and Sons, Inc, Hoboken, N.J., Chapter 5.

Martens, D.A. & Frankenberger, W.T. (1990). Determination of saccharides by high performance anion-exchange chromatography with pulsed amperometric detection. *Chromatographia* 29:7-12.

Masle, J.; Farquhar, G.D. & Gifford, R.M. (1990). Growth and carbon economy of wheat seedlings as affected by soil resistance to penetration and ambient partial pressure of CO_2. Aust. J. Plant Physiol. 17:465-487.

Mauri, P.; Minoggio, M.; Somonetti, P.; Gardana, C. & Pietta, P. (2002). Analysis of saccharides in beer samples by flow injection with electrospray mass spectrometry. Rapid Commun. Mass Spectrom., 16:743-748

Munnik, T.; Irvine, R.F. & Musgrave, A. (1998). Phospholipid signalling in plants, Biochim. Biophys. Acta 1389 222–272.

Murphy, A.M.; Otto, B.; Brearley, C.A.; Carr, J.P. & Hanke, D.E. (2008). A role for inositol hexakisphosphate in the maintenance of basal resistance to plant pathogens. Plant Journal 56:638-652.

Nachtmann, F. & Budna, K.W. (1977). Sensitive determination of derivatized carbohydrates by high-performance liquid chromatography. J. Chromatography 136:279-287.

Nachtmann, F. (1976). Some properties of 4-nitrobenzoates of saccharides and glycosides; application to high-pressure liquid chromatography. Z. Anal. Chem.282:209-213.

Narayanan, C. (1987). Pinitol-A new anti-diabetic compound from the leaves of *Bouganivillea*. Current Science, 56:139-141.

Nelson, R.W.; Meulenkijk, J.A.P. & Underberg, W.J.W. (1990). Ultraviolet-Visible Derivatization. In: *Detection –Oriented Derivatization Techniques in Liquid Chromatography*, H. Lingeman & W.J.M. Underberg, (Ed.), Marcel Dekker, Inc, New York, New York, pp. 268-270.

Nordin, P. (1984). Preferential leaching of pinitol from soybeans during imbibition. Plant Physiol 76:313-315.

Ochoa, S. & Stern, J.R. (1952). Carbohydrate metabolism. Annual Review of Biochemistry 21:547-602.

Ogunyemi, O.; Pittner, F., & Hoffmann-Ostenhof, O. (1978). Studies on the biosynthesis of cyclitols. XXXVI. Purification of *myo*-inositol-I-phosphate synthase of the duckweed *Lemna gibba* to homogeneity by affinity chromatography on NAD Sepharose. Molecular and catalytic properties of the enzyme. Hoppe-Seyler's Z Physiol Chem 359: 613-616.

Perera, I. Y.; I. Heilmann, &. Boss, W.F. (1999). Transient and sustained increases in inositol 1,4,5-trisphosphate precede the differential growth response in gravistimulated maize pulvini. Proc. Natl. Acad. Sci. USA 96:5838-5843.

Pessarakli, M. (2005). Handbook of photosynthesis M. Pressarakli ed, 2nd edition. Taylor & Francis Group, CRC, Boca Raton, FL.

Phillips, D.V. & A. Smith, A. (1974). Soluble carbohydrates in soybean. Canadian Journal of Botany 52:2447-2452.

Phillips, D.V.; Dougherty, D.E. & Smith, A.E. (1982). Cyclitols in soybean. Journal of Agricultural & Food Chemistry 30:456-458.

Phillips, D.V. & Smith, A. (1974) Influence of sequential prolonged periods of dark and light on pinitol concentration in clover and soybean tissue. Physiol. Plant. 54: 31-33.

Raboy, V. (2007). Seed phosphorus and the development of low-phytate crops. In: Turner, B.L., Richardson, A.E., Mullaney, E.J. (Eds.)., Inositol phosphates linking agriculture and the environment. CABI, Oxfordshire.

Reinhold, V.N.; Ashline, D.J.; Zhang, H. (2010).Unravellng the Structural Details of the Glycoproteome by Ion Trap Mass Spectrometry, in *Practical Aspects of Trapped Ion Mass Spectrometry. Vol. 4: Theory and Instrumentation.* Edited by Raymond E. March and John F.J. Todd.Chapter 23,706-736. CCR Press, Boca Raton, FL.

Richmond, M.L.;. Barfuss, D.L.; Harte, B.R.; Gray, J.I. & Stine, C.M. (1982). Separation of carbohydrates in dairy products by high performance liquid chromatography. J. Dairy Sci. 65:1394-1400.

Romano, J. (2007). Carbohydrate analysis: column chemistries and detection. Presented at the Carbohydrates in Feeds Methodology Forum, AOAC 2007 Annual Meeting, Anaheim, CA, September 18, 2007.

Ruiz-Matute, A.I.; Hernandez-Hernandez, O.; Rodriguez-Sanchez, S.; Sanz, M.L. & Martinez-Castro, I. (2010). Derivatization of carbohydrates for GC and GC-MS analyses. J. of Chromatography B Available online 8 December 2010

Saghai Maroof, M.A.; Glover, N.M.; Biyashev, R.M.; Buss, G.R. &. Grabau, E.A. (2009). Genetic basis of the low-phytate trait in the soybean line CX1834. Crop Sci: 49: 69-76.

Sartini, C.M.; Spinabelli U.; C. Riponi, C. & Galassi, S. (2001). Determination of carboxylic acids, carbohydrates, glycerol, ethanol, and 5-HMF in beer by high-performance liquid chromatography and UV-refractive index double detection. J. Chromatogr. Sci., June, 39(6):235-238.

Sanz, M.L., Villamiel, M. & Martinez-Castro, I. (2004). Inositols and carbohydrates in different fresh fruit juices. Food Chemistry 87: 325-328.

Sasaki, K. & Taylor, I.F.P. (1986). *myo*-Inositol synthesis from [1-3H]Glucose in *Phaseolus vulgaris* L. during early stages of germination. Plant Physiology 81:493-496.

Savidge, R. A. & Forster, H. (2001). Coniferyl alcohol metabolism in conifers II. Coniferyl alcohol and dihydroconiferyl alcohol biosynthesis. Phytochem. 57: 1095-1103.

Scheller, H.V. & Ulvskov, P. (2010). Hemicelluloses. Annual Review of Plant Biol. 61:263-289.

Schiller, M. H.; von der Heydt; Marz, F. & Schmidt, P.C. (2002). Quantification of sugars and organic acids in hygroscopic pharmaceutical herbal dry extracts. J. of Chrom A., 968: 101-111.

Schlichtherle-Cerny, H.; Affolter, M. & Cerny, C. (2003). Hydrophilic interaction liquid chromatography coupled to electrospray mass spectrometry of small polar compounds in food analysis. Anal. Chem. , 75:2349-2354.

Scott, R.P.W. (2011). Liquid Chromatography, Chrom Ed. Series, http://www.chromatography-online.org/HPLC/Refractive-Index/rs33.html, accessed March 8, 2011.).

Shah, M.M., Clauss, T.R.; Clauss, S.C.; Campbell, J.A.; Goheen, S.C. & Bennett, K.A. (1998). Determination of Molecular Weight Distribution of Insoluble Carbohydrates from corn fiber using HPLC and Light Scattering Detector. Presented at the Corn Utilization Conference, June 1-3, 1998, St. Louis, Mo.

Sharp, R.E. & Davies, W.J. (1979). Solute regulation and growth by roots and shoots of water-stressed maize plants. Planta 147:43-49.

Sheveleva, E.; Chmara, W.; Bohnert, H.J. & Jensen, R.G. (1997). Increased salt and drought tolerance by D-ononitol production in transgenic *Nicotiana tabacum* L. Plant Physiol. 115:1211-1219.

Shigaki, T. & Bhattacharyya, M.K. (2000). Decreased inositol 1,4,5-trisphosphate content in pathogen-challenged soybean cells. MPMI 13:563-567.

Smith, D. (1973). The nonstructural carbohydrates. In: Chemistry and Biochemistry of herbage, Vol. 1, eds. G.W. Butler, R.W. Bailey, Academic Press, London.

Somerville, C. (2006). Cellulose synthesis in higher plants. Annual Review of Cell and Developmental Biology. 22:53-78.

Stieglitz, K.A.; Yang, H.; Roberts, M.F. & B. Stec, B. (2005). Reaching for mechanistic concensus across life kingdoms: Structure and insights into catalysis of the *myo*-inositol-1-phosphate synthase (mIPS) from Archaeoglobus fulgidus. *Biochemistry* 44: 213-224.

Streeter, J. G. (1980). Carbohydrates in soybean nodules.II. Distribution of compounds in seedlings during the onset of nitrogen fixation. *Plant Physiology*. 66:471-476.

Streeter, J.G. & Strimbu, C. (1998). Simjultaneous extraction and derivatization of carbohydrates from green plant tissues for analysis by gas-liquid chromatograpy. Anal. Biochem. 259:253-257..E,. Strimbu

Streeter, J. G.; Lohnes, D.G. &. Fioritto, R.J. (2001). Patterns of pinitol accumulation in soybean plants and relationships to drought tolerance. *Plant, Cell and Environment* 24:429-438.

Stumpo, K.A.R.& Reinhold, V. (2010). The N-Glycome of Human Plasma. *Journal of Proteome Research*, 9(8):4823-4830.

Sullivan, J.E. & Schewe, L.R. (1977). Pareparation and gas chromatography of highly volatile trifluoroacetylated carbohydrates using N-methylbis[trifluoroacetamide]. *J. of Chromatographic Science* 15:196-197.

Sweeley, C.C.; R. Bentley, R.; Makita, M. & Wells, W.W. (1963). Gas-liquid chromtograpy of trimethylsilyl derivatives of sugars and related substances. J. of Am. Chem. Soc. 85:2497-2507.

Taormina, C.R.; Baca, J.T.; Asher, S.A.; Grabowski, J.J. & Finegold, D.N. (2007). Analysis of tear glucose concentration with electrospray ionization mass spectrometry. *J. Am. Soc. Mass Spectrom.*, 18:32-336.

Tate, R.L. (1995). Soil microbiology (symbiotic nitrogen fixation), pp. 307-333. John Wiley and Sons, Inc., New York, N.Y.

Taylor, V.F.; March, R.E.; Longerich, H.P. & Stadey, C.J. (2005). A mass spectrometric study of glucose, sucrose, and fructose using an inductively coupled plasma and electrospray ionization. Int. *J. Mass Spectrom.*, 243:71-84.

Thibault, P. & Honda, S. (Eds.). (2003). *Capillary electrophoresis of carbohydrates. Methods in Molecular Biology.* Humana Press, Totowa, N.J., 318 pages.

Togami, D.W.; Poulsen,B.J.; Batalao, C.W. & Rolls, W.A. (1991). Separation of carbohydrates and carbohydrate derivatives with cation-exchange columns at high pH. Biotechniques 10(5):650-655.

Twilly , J.W. (1984). The analysis of exudate glant gums In their artistic applications: An interim report. Archaeological Chemistry, Vol. 3, J.Lambert, (Ed.), American Chemical Society, Advances in Chemistry series, no. 205, Washington, D.C., American Chemical Society. 357-394.

Udvardi, M.K. & Day, D.A. (1997). Metabolite transport across symbiotic membranes of legume nodules. Annual Review of Plant Physiology and Plant Molecular Biology 48:493-523.

Vallance, S.L.; Singer, B.W.; Hitchen, S.M. & Townsend, J.H. (1998). The development and initial application of a gas chromatographic method for the characterization of gum media. *Journal of American Institute for Conservation.* 37:294-311.

Vernon, D. M. & Bohnert, H.J. (1992). A novel methyl transferase induced by osmotic stress in the facultative halophyte *Mesembuyanthemum crystalliunum.* EMBO J 11:2077-2085.

Vessey, J.K.; Walsh, K.B. & Layzell, D.B.(1988a). Oxygen limitation of N_2 fixation in stem girdled and nitrate-treated soybean. Physiol Plant. 73:113-121.

Vessey, J.K.; Walsh, K.B. & Layzell, D.B. (1988b). Can a limitation in phloem supply to nodules account for the inhibitory effect of nitrate on nitrogenase activity in soybean? Physiol. Plant. 74:137-146.

Walker, W. H.(2010). Is the "comb over" dying? A mouse model for male patter baldness (Androgenic Alopecia). Endocrinology 151: 1981-1983.

Walker, H. G.; Gee, M. & McCready, R.M. (1962). Complete methylation of reducing carbohydrates. J. Org. Chem. 27: 2100-2102.

Wei, Y. & Ding, M. (2000). Analysis of carbohydrates in drinks by high-performance liquid chromatography with a dynamically modified amino column and evaporative light scattering detection. J. of Chromatography A. 904:113-117.

Weisshaupl, V.; Scheiner, O.; Dworsky, P.; Hoffmann-Ostenhof, O. & Allgem, Z. (1976). Mikrobiol. 16: 81-86.

Wilcox, M.; Hefley, J.R. & Walsh, J.W. (2001). HPLC determination of carbohydrates: A comparison of six different columns with ELS detection. Presented at the Carbohydrate Session 59A, 2001 IFT Annual Meeting,. New Orleans, La.

Williams, J.H.H. & Farrar, J.F. (1990). Control of barley root respiration. Physiolo. Plant. 79:259-266.

Wilson, R.F. (2004). Seed composition Pp 621-677. In H.R. Boerma and J. E. Specht, co-editors, Soybeans: Improvement, production and uses. Third Edition Number 16 Agronomy Monograph. American Society of Agronomy, Inc., Crop Science Society of America, Inc., and Soil Science Society of America, Inc.. Madison, WI, USA.

Zeeman, S.C.; Smith, S.M. & Smith, A.M. (2007). The diurnal metabolism of leaf starch. *Biochemical Journal* 401:13-28.

Zeeman, S.C.; Kossmann, J. & Smith, A.M. (2010). Starch:its metabolism, evolution and biotechnological modification in plants. Annual Review Plant Biology 61:209-234.

Zenobi, R.; & Knochenmuss, R. (1998). Ion formation in MALDI mass spectrometry. Mass Spectrom. Rev. 17:337-366.

Zhu, S.G; Long, S.P. & Ort, D.R. (2010). Improving photosynthetic efficiency for greater yield. Annual Review Plant Biol. 61:235-261.

11

Extraction and Enzymatic Modification of Functional Lipids from Soybean Oil Deodorizer Distillate

Carlos F. Torres, Guzmán Torrelo and Guillermo Reglero
Departamento de Producción y Caracterización de Nuevos Alimentos
Instituto de Investigación en Ciencias de la Alimentación (CIAL), CSIC-UAM
Madrid
Spain

1. Introduction

Crude vegetable oils contain triacylglycerols as major component and various minor components such as diacylglycerols, monoacylglycerols, free fatty acids, phospholipids, tocopherols, sterols, squalene, color pigments, waxes, aldehydes, ketones, triterpene alcohols and metals that may affect the quality of the final product. The minor components are removed partially or entirely by either physical or chemical refining in order to make the vegetable oils suitable for human consumption. Deodorization is the last major processing step in the refining of edible oils. It has the responsibility for removing both the undesirable ingredients occurring in natural fats and oils and those which may be imparted by prior unit processes such as caustic refining, bleaching, hydrogenation, or even storage conditions. It is this unit process that finally establishes the oil characteristics of "flavor and odor," which are those most readily recognized by the consumer (Gavin, 1978).

Deodorizer distillate is a by-product of deodorization, which is the last major step in vegetable oil refining process. It is a complex mixture of free fatty acids, mono-, di- and triacyglycerols, sterols and their esters, tocopherols, hydrocarbons, pesticides, and breakdown products of fatty acids, aldehydes, ketones and acylglycerol species (Ramamurthi & McCurdy, 1993). Deodorizer distillate is an excellent source of valuable compounds such as phytosterols, tocopherols and squalene, which can be recovered and further used as food additives, in pharmaceutical industry and cosmetics. Their commercial value however, is mainly dependent on their tocopherol content (Fernandes & Cabral, 2007). Although in recent years, efforts from industry resulted in a significant number of reports, describing better and improved methods for phytosterol recovery and purification. Such surge is closely related to growing market for phytosterols, particularly given the widespread dissemination of functional foods (Fernandes & Cabral, 2007).

Numerous procedures have been described to isolate bioactive compounds from soybean oil deodorizer distillate to improve the value and the quality of this by-product. All these procedures can be grouped in three generic categories: crystallization and precipitation, chemical and enzymatic modification, and extraction and fractionation.

One important bioactive compound concentrated in intermediate byproducts and waste streams during the refining of soybean oil is squalene. Recently, steroidal hydrocarbons and squalene in soybean oil deodorizer distillate have been isolated and identified. Separation, purification, and chemical characterization of hydrocarbons fraction in soybean oil deodorizer distillate will help researchers finding a better utilization of these byproduct (Gunawan *et al.*, 2008b, Kasim *et al.*, 2009). Both squalene and steroidal hydrocarbons have been purified from soybean oil deodorizer distillate by means of modified soxhlet extractions with hexane to obtain two main fractions: one fraction rich in fatty acid steryl esters and squalene and a second fraction rich in tocopherols, free phytosterols, free fatty acids (FFAs) and acylglycerols. Then hydrocarbons are isolated via silica gel column chromatography. Similarly, other procedures described in the literature for isolation of sterols and tocopherols in soybean oil deodorizer distillate are based on the utilization of organic solvents (Lin *et al.*, 2004).

Recently greener technologies for isolation, purification and fractionation of bioactive compounds from soybean oil deodorizer distillate (SODD) have been developed. These methodologies can work in combination or independently to improve the fractionation of soybean oil deodorizer distillate. Hence, lipase-catalyzed methyl or ethyl esterification of SODD to transform free fatty acids into their corresponding fatty acid methyl or ethyl esters coupled to molecular distillation and/or supercritical fluid extraction has been described as a strategy to improve the separation between tocopherols, sterols and free fatty acids (Torres *et al.*, 2009). Alternatively, enzymatic esterification of the sterols with the fatty acids already present in the deodorizer sludge makes the separation of tocopherols and sterols simpler using short-path distillation or supercritical fluid extraction (Shimada *et al.*, 2000). However, short path distillation (Ito *et al.*, 2006), supercritical fluid extraction (Chang *et al.*, 2000), and enzymatic modifications (Torres *et al.*, 2007) have been also utilized independently on soybean oil deodorizer distillate. Therefore, these three technologies will be analyzed and discussed on the present chapter to evaluate the feasibility of these procedures for the valorization of side-stream products obtained during refining of soybean oil.

In addition, oxidation of sterols during refining steps such as heating, degumming, neutralization, bleaching, and deodorization, and during storage and handling should be also considered (Verleyen *et al.*, 2002b). However, limited information is available on the levels of sterols in the by-product fractions collected from chemical and physical refining processes (Verleyen *et al.*, 2001c) and the biological effects of phytosterol oxidation products on animal and human health need more investigation, as feed quality is crucial for animal health and welfare, and ultimately human health. It has been reported that the formation of sterol oxidation products is affected not only by the chemical nature of the sterols but also by their quantity (Dutta *et al.*, 2006). Positive correlations between total sterols and total phytosterol oxidation products in the by-products collected from refining processes have been found (Ubhayasekera & Dutta, 2009). Therefore, other aspects concerning soybean oil deodorizer distillate such as nonfood applications, direct consumption, and oxidative quality will be also analyzed and discussed in the present chapter.

2. Soybean oil deodorizer distillate production and characterization

The vegetable oils correspond about 70% of demand of natural oils and fatty acids consumed in the world. The soybean oil corresponds from 20 to 30% of the vegetable oils world market (Bockisch, 1998) and its production involves several steps that are necessary to render the soybean oil suitable for human consumption. These production steps have been broadly characterized as 1) soybean preparation, 2) oil extraction, and 3) oil refining.

Soybean preparation generally includes the steps of cleaning, drying, cracking, and dehulling.

Oil extraction basically consists of separating the oil from the remainder of the soybean, known as soybean meal. The great majority of commercial soybean extraction processes use a solvent to separate the oil from the meal. In the solvent extraction process, the beans are flaked to provide a large surface area. A solvent, commonly hexane, is then pumped through the soybean flakes, dissolving the oil in the hexane. The hexane is then separated from the oil and recycled.

The crude oil resulting from the extraction process must then be subjected to additional treatments, collectively called "refining", to remove various materials in order for the oil to be suitable for consumption. These materials include hydratable and non-hydratable phospholipids, free fatty acids, and various color and flavor components.

Crude soybean oil contains phosphorous compounds called hydratable phospholipids, and small amounts of calcium and magnesium that complex with a portion of the phospholipids to form non-hydratable phospholipids. Hydratable phospholipids are normally removed by a process known as "degumming", in which the oil is agitated or otherwise intimately combined with water to precipitate gums from the oil. The gums are then removed by centrifugation.

These precipitated gums can be used as a feed additive, or evaporated to remove moisture, the end product is called lecithin. Lecithin has various end uses such as food emulsifier. The degummed oil is dried under vacuum to remove any water. Removal of non-hydratable phospholipids is considerably more difficult and expensive, requiring further chemical treatment, typically chemical refining, to break the chemical bonds between the calcium or magnesium ions and the phospholipids, followed with extensive bleaching of the oil.

In most processes, free fatty acids are then removed from the oil by a process known as caustic refining, also called chemical or alkali refining, in which the oil is mixed with a caustic material, such as sodium or potassium hydroxide, which undergoes a saponification reaction with the acids, forming soaps that are then removed by centrifugation. In this case, the non-hydratable phosphotide are removed along with the free fatty acids. In addition, a significant quantity of the oil is captured by the soaps, adversely affecting oil yield.

Free fatty acid removal by a process known as physical refining has been used for oils that are low in non-hydratable phospholipids, such as lauric oils, particularly palm oil. In physical refining, the oil is vacuum distilled at high temperatures, e.g., from about 230 °C to about 260 °C to separate more volatile components from the oil. This process is used to remove various flavor components, and will also remove free fatty acids. However, the process has not been viable for removing free fatty acids from oils such as soybean oil, which contains higher levels of non-hydratable phospholipids (more than 20 ppm based of phosphorous content). The high temperatures required for physical refining tend to break down the non-hydratable phospholipids that are present in the soybean oil, producing chemical compounds that cause an unacceptable flavor and color.

Conventional refining processes also involve some bleaching of the soybean oil to remove color pigments (i. e., carotenoids, chlorophyll) that adversely affect the color of the oil. Bleaching process employs the use of adsorbents such as acid-activated clays.

Finally, deodorization is the last process step used to improve the taste, odor, color and stability of the oil by means of removing undesirable substances. The goal of deodorization is to obtain a final product, finished oil that has a bland flavor, a maximum FFA content of 0.05% and a zero peroxide value. All commercial deodorization, whether in continuous, semicontinuous or batch units, is essentially a form of physical distilling by steam, in which the oil is subjected to high temperatures (210 °C - 280 °C) under a high vacuum (1 - 6 mm Hg) for a short period of time, which is sufficient to remove FFA and other volatile flavor-causing compounds. During the process, peroxide decomposition products, color bodies and their decomposition products are eliminated and the content of sterols, sterol esters and tocopherols is also reduced.

The modern commercial deodorizers are equipped with a packed column that has three sections: vapour scrubbing section, stripping section and heat bleaching section.

Bleached oil is pre-heated by outgoing deodorized oil and sprayed into the Deaerator where dissolved air and moisture are reduced to a minimum. The oil is then heated to full temperature by hot deodorized oil in the Deodorizing Economizer and high pressure steam in the Final Heater. A portion of the free fatty acids in the oil will be flashed off as the oil temperature increases.

The hot oil enters the Packed Column, which is filled with special structured packing so that the oil is distributed into a thin film and is evenly agitated by stripping steam flowing counter currently from the bottom of the column. As a result, free fatty acids and other remaining volatile impurities in the oil are evaporated and removed with the steam. The residence time in the column is only a few minutes.

Next, the stripped oil enters the heat bleaching section where it flows through the channels of a series of vertically stacked compartments (trays) while agitated by stripping steam. The prolonged thermal action breaks down color bodies (carotenes) and other heat sensitive compounds are volatilized and removed, or rendered inactive, resulting in a lighter oil color. Also, the amount of remaining free fatty acids in the oil is reduced to an absolute minimum.

The deodorized oil is pre-cooled by deaerated oil and then sprayed into the Post Deodorizer where the final "off-flavor" compounds are removed.

Fatty acids and other materials, evaporated from the oil, are condensed by contact with recycled and cooled distillate in the Vapor Scrubbing section. The soybean oil deodorizer distillate is circulated by the Distillate Pump via the Distillate Cooler where it is cooled by cooling water. Accumulated distillate is discharged from the Scrubber to storage.

SODD is a by-product of deodorization and is a complex mixture of free fatty acids, mono-, di- and triacyglycerols, sterols and their esters, tocopherols, hydrocarbons, pesticides, and breakdown products of fatty acids, aldehydes, ketones and acylglycerol species (Ramamurthi & McCurdy, 1993, Verleyen, 2001c). SODD corresponds between 0.1 and 0.4% of crude soybean oil. Deodorizer distillate is an excellent source of valuable compounds such as phytosterols and tocopherols, corresponding approximately 10% and 20% respectively (Czuppon et al., 2003), which can be recovered and further used as food additives, in pharmaceutical industry and cosmetics (Lin & Koseoglu, 2003). Their commercial value however, is mainly dependent on their tocopherol content, depending on the market demand for this ingredient (Dumont & Narine, 2007).

SODD contains high levels of free fatty acids and acylglycerols (Chu et al., 2002). Fatty acids represent 25-75% and acylglycerols about 3-56% of Vegetal Oil Deodorizer Distillates (VODD), depending on the raw material being refined and on the type and conditions of the refining process (Ramamurthi & McCurdy, 1993). The free fatty acids from deodorizer distillate are mostly used as additives for animal food, fluidizing agents for lecithin or as medium-grade soaps. Such fatty acids also can be used as precursors in a wide variety of molecular synthesis schemes such as the production of dibasic acids of different chain lengths (Gangopadhyay *et al.*, 2007). Alternatively, deodorizer distillate have non-food applications, such as biodiesel or can be used mixed with the fuel oil to fire the steam boilers (Svensson, 1976).

Tocopherols (vitamin E) are natural antioxidants found in vegetable oils and contribute significantly to their oxidative stability. Due to the high tocopherol content of crude oils, they can be stored long periods of time if they are protected from air, moisture and high temperature (Norris, 1979) without any significant deterioration. However, the concentration of natural tocopherols in soybean oil is too high for optimum oxidative stability and flavor, because they can act as pro-oxidants by peroxide formation (Jung & Min, 1990, Warner, 2005).

Moreover, tocopherols exert several beneficial activities, such as protective role of vitamin A, β-carotene and essential fatty acids (Ferrari *et al.*, 1996). Tocopherols also prevent diseases like cancer (Kline *et al.*, 2007), cardiovascular and cataracts (Block & Langseth, 1994, Munteanu & Zingg, 2007, Rimm *et al.*, 1993). They are used in food, cosmetics and pharmaceutical industries (Chu, 2002) and a mixture of α, β, γ and δ isomers containing 60 wt% tocopherols is widely used as additive to many kinds of foods (Shimada, 2000).

On other hand, in recent years a significant number of reports, patents, and scientific publications describing improved methods for phytosterol recovery and purification have been developed. This phenomenon is closely related to the growing market for phytosterols, particularly given the widespread dissemination of functional foods (Fernandes & Cabral, 2007).

Phytosterols are useful hypocholesterolemic agents since a daily intake of 2-3 g lowers LDL cholesterol concentrations by 10-15 % as found in various populations (Kritchevsky & Chen, 2005, Quílez *et al.*, 2003). The proposed mechanism is that plant sterols reduce the micellar solubility of cholesterol and consequently lower intestinal absorption of both exogenous and endogenous cholesterol (de Jong *et al.*, 2003, Trautwein *et al.*, 2003), but also experimental investigations suggest that sterols may act modulating lipid and protein metabolism (Mulligan *et al.*, 2003, Plat & Mensink, 2005). In addition to their cholesterol lowering effect, plant sterols may possess anti-cancer (Awad *et al.*, 2003), antiatherosclerosis (Moghadasian *et al.*, 1999, Moghadasian *et al.*, 1997), anti-inflammation (Bouic, 2001) and antioxidation activities (van Rensburg *et al.*, 2000). Phytosterol compounds exhibit virtually no side effects and they have shown no evidence of *in vitro* mutagenic activity or subchronic toxicity in animals (Rozner, 2006). These compounds have been extensively used as a food ingredient in the functional food industry.

Moreover, phytosterols are valuable precursors in the production of hormones (Donova, 2007). They are used in manufacturing progesterone, corticoids, estrogens, contraceptives, diuretics, male hormones and vitamin D. They are, also, used in cosmetics (Balazs, 1987, Fernandes & Cabral, 2007).

Another important bioactive compound concentrated in intermediate byproducts and waste streams during the refining of soybean oil, is squalene, a hydrocarbon that has been used in applications such as natural moisturizer in cosmetics and biochemical precursor in the synthesis of steroids. Recently, steroidal hydrocarbons and squalene in soybean oil deodorizer distillate have been isolated and identified. Separation, purification, and chemical characterization of hydrocarbons fraction in soybean oil deodorizer distillate will help researchers finding a better utilization of these byproduct (Gunawan, 2008b, Kasim, 2009).

The refining process induces changes in the structure and concentration of tocopherols, sterols (free and bound) and squalene.

Of these various components, most attention is given to the tocopherols. Jung and coworkers (Jung et al., 1989) and Ferrari (Ferrari, 1996) have studied the tocopherol content at all stages of processing for all isomers in the finished oil. The tocopherol content decreases during each step of processing and may be markedly reduced during deodorization, as the tocopherols are volatile under these conditions. The processing removed between 30-60% of tocopherols in crude soybean oil. Even though total tocopherol content decreased during processing, the relative compositions of tocopherols in soybean oils were constant during processing.

Sterol content present in soybean oil also tend to be diminished in processing and the magnitude of such decrease is about the same as the tocopherols (Ferrari, 1996). It has been shown that the absorption of sterols is increased extensively with increased amounts of bleaching clay. The lipid extract from the bleaching clay had high concentrations of sterols in unchanged form.

Squalene content also decreases during processing (Nergiz & Çelikkale, 2010), but not drastically until deodorization, when it is partially volatilized. Total losses during all the stages of refining were found to be 31 % as compared to its content in crude soybean oil.

Numerous procedures have been described to isolate bioactive compounds from soybean oil deodorizer distillate to improve the value and the quality of this by-product. All these procedures can be grouped in four generic categories: classic method such as crystallization and precipitation, molecular distillation, supercritical fluid extraction and chemical and enzymatic modification.

3. Classic methods to obtain functional lipids from SODD

In the past, recovering tocopherols and sterols from deodorizer distillates and related mixtures has been proved to be complicated and expensive. One difficulty associated with isolating one or more distillate fractions enriched in fatty acids, tocopherols, and/or sterols from deodorizer distillates is that the molecular weights and volatilities of sterols are similar to those of tocopherols (Ghosh & Bhattacharyya, 1996). For this reason, it is difficult to recover concentrates of tocopherols and phytosterols with good yield and high quality (Lin, 2002). In addition, in order to separate the squalene present in the distillate, the main challenge is to isolate them from each other, especially in the case of the following pairs of components: tocopherol–squalene, tocopherol–fatty acids, tocopherol–sterol and sterol–squalene.

Another difficulty is that deodorizer distillate can undergo thermal degradation if it is processed for extended periods at the temperatures at which sterols and tocopherols vaporize, such temperature conditions which can cause fatty acids to convert into undesirable trans isomeric forms and may cause the degradation of tocopherols (Chu, 2002).

Classical methods for recovering tocopherols and sterols include solvent extraction, chemical treatment, crystallization, complexation, and molecular distillation (Rohr & Trujillo-Quijano, 2005). The separation process involves a series of chemical and physical techniques which are used alone or in combination. In general, most processes are designed to remove either fatty acids or sterols in the initial step, followed by tocopherol concentration by other methods.

Crystallization has frequently been used to purify sterols from SODD, either following or preceding other separation methods. Brown (Brown & Smith, 1964) reported a phytosterols product prepared by a continuous two-stage liquid-liquid extraction (LLE) with a solvent pair of methanol and hexane, and then followed by crystallization using acetone as a solvent at 4 °C for 24 h. By this approach, 73% sterol concentrate was obtained from SODD containing 6.5% sterol. Sheabar and Neeman (Sheabar & Neeman, 1987) have shown the preparation of a tocopherol concentrate through removal of sterol from SODD by a two-stage crystallization at -20 °C with hexane and acetone as crystallization solvents. Attempts have been made to isolate tocopherols from SODD by supercritical fluid extraction technology with crystallization as pretreatment to first remove sterols (Lee et al., 1991). SODD was esterified with methanol using HCl as catalyst, then a solvent pair of hexane-methanol was used to obtain tocopherols-sterols concentrate from which sterols were recovered by crystallization at -20 °C with acetone as a solvent (Brown & Smith, 1964). The results were similar to those mentioned above. Nevertheless, the information of total yield of sterols was not provided in these publications.

Crystallization seems successful as a simple and efficient process to remove and concentrate sterols and tocopherols from SODD. This process has the advantage of not causing tocopherol oxidation, because the low temperature utilized, and it does not use high pressure. While there is much information in the literature on the recovery of sterols from SODD by crystallization, little attempt is made pertaining to its optimal conditions such as solvent type, crystallization temperature and time. Lin and Koseoglu (Lin & Koseoglu, 2003) have shown crystallization of sterols from SODD without any pretreatment is practical. The best results were achieved by crystallization at -20 °C for 24 h using a solvent mixture of acetone-methanol (4:1, v/v) at a solvent-to-SODD ratio of 3:1 (v/w), followed by centrifugation, filtration, and twice washing of the wet cake. Over 90% of the original tocopherols and squalene, were retained in the filtrate fraction, while 80% of the original sterols were crystallized in the cake fraction. Khatoon and coworkers (Khatoon et al., 2010) developed a method for the preparation of phytosterols from SODD by crystallisation using hexane and water. Direct crystallisation yielded a phytosterol fraction with lower recovery of 13.2-17.8% while treatment with alkali to remove FFA and the glycerides followed by organic solvent extraction yielded unsaponifiable matter containing phytosterols with a recovery of 74.6%. Later the unsaponifiable matter was purified by double crystallisation into a mixture of phytosterols of 87% purity. Moreira and Baltanás (Moreira & Baltanás, 2004) studied the impact of the principal process variables (solvents and cosolvents, cooling rate, crystallization temperature, and ripening time) on the quality and yield of the recovered phytosterols, but in this case by using a sunflower oil deodorizer distillate "enriched" (i.e., preconcentrated). In this study, a sterols recovery as high as 84% (with 36% purity) was achieved by using a single-stage batch crystallization of hexane/ethanol mixture (ratio of 4:1, v/v) at -5 °C.

On the other hand, a modified industrial process was developed by Xu and coworkers (Xu et al., 2005) to recovery and purify valuable compounds from SODD. In this process, tocopherols and fatty acids methyl esters (FAMEs) was obtained from SODD after a process with methyl esterification by sulfuric acid catalyst, transesterification by alkaline catalyst, crystallization of sterols and molecular distillation. The waste residue of SODD was obtained after the molecular distillation and it mainly contains steryl esters, acylglycerols, and hydrocarbons.

In turn, Yang and coworkers (Yang et al., 2009) developed a catalytic and crystallization process to recover phytosterols from waste residue of SODD (WRSODD). A catalyst was employed to decompose WRSODD so as to transform steryl esters into phytosterols. The mixed solvent that generated the best crystallization results was acetone and ethanol (4:1, v/v). The yield and the purity of recovered phytosterols were 22.95 wt. % and 92-97 %, respectively.

Nevertheless, crystallization has the disadvantage of the solvents available at present are not sufficiently selective to obtain, through the current processes, a reasonable separation between the unsaponifiable components and free fatty acids. Due to this, it is often necessary to use more than one solvent, which in turn complicates and increases tremendously the cost of recovery and recycling of these solvent mixtures. Furthermore, solvents or solvent mixtures are used in very large proportions, when compared to the quantity of the material submitted for extraction, and the solvents need additional processes for their removal and/or recycling in the extraction and pre-concentration process of the valuable products. The foregoing reasons make solvent based-processes, expensive, unattractive and less environmentally friendly, resulting in a scarce and expensive final product.

Saponification is also a common practice to concentrate tocopherols and sterols since it produces alkali metal soap which, due to its insolubility in the solvent used in the process, can be separated from the dissolved tocopherols, thereby permitting recovery of the tocopherols in a form relative free from fatty acids and glycerides. The processes themselves are costly, however, and tocopherols are produced in low yield. The sterols are then isolated from the resulting concentrate mixture by crystallization (Brown & Meag, 1963, Kijima et al., 1964, Kim & Rhee, 1982).

Of the saponification processes, the lime saponification process is the most widely used. Hickman, U.S. Patent No. 2.349.270 (Hickman, 1944), discloses that deodorizer distillate can be treated with calcium hydroxide, traditionally called slaked lime, to saponify the fatty acids, followed by extraction of the unsaponifiable fraction (tocopherols and sterols) with acetone, in which the saponification products are insoluble. The extract is then washed and concentrated, as for example by solvent distillation, and then cooled to crystallize sterols which are removed by filtration, leaving a high purity tocopherol fraction. The fatty acid soaps formed by the process can be acidulated and converted into free fatty acids. Andrews, U.S. Patent No. 2.263.550 (Andrews, 1941), discloses saponification of deodorizer distillates with sodium hydroxide, followed by metathesis (a molecular process involving the exchange of bonds between the two reacting chemical species, in this case a ion exchange) with calcium chloride to convert the sodium soaps to calcium soaps (not water soluble), from which the tocopherols and other unsaponifiable matter are then extracted with acetone.

The disadvantage of each of these processes is that the calcium soap is formed in a wide particle size distribution, ranging from fine particles to lumps. The result is a soap mass which is lumpy in form and from which the unsaponificable matter is difficult to extract. To permit the extraction to take place, the soap mass must be ground into particulate form, a process which entails a substantial capital investment. Even then, solvent consumption is high and the recovery of tocopoherols and other useful unsaponificable matter such as sterols is low.

Grinding is avoided in the process disclosed by Brown and coworkers (Brown & Meag, 1963), which uses calcium silicate as a powdering agent in combination with acetone to facilitate the separation of the soluble tocopherols and sterols from the insoluble soap mass. Unfortunately, this process requires a large amount of powdering agent which remains in the soap mass, and the effectiveness of the powdering agent is diminished if the moisture content of the soap mass is too high.

Although saponification is effective to remove free fatty acids and acylglycerols, it involves the use of a large amount of alkali which is harmful to tocopherols, thus leading to low yields. Recently, molecular distillation combined with crystallization was more attractive to separately concentrate tocopherols and sterols (Gapor et al., 1989, Hunt et al., 1997, Kijima, 1964, Kim & Rhee, 1982, Smith Frank, 1967). To increase the separation efficiency, esterification and/or transesterification are usually carried out prior to molecular distillation. Free fatty acids and acylglycerol are converted to fatty acid methyl esters, which are more easily removed by vacuum distillation due to their higher vapor pressure than those of the corresponding free fatty acids and acylglycerol. However, this step made the whole process more complicated and labor-intensive when compared with the saponification process. Another drawback of molecular distillation is that it is energy consuming to maintain high vacuum all of the time during operation. Consequently, the final product is also expensive.

4. Enzymatic modification

Enzymatic reactions are based on the selective biotransformation of determined compounds in order to modify their chemical or physical properties. Hence, the utilization of enzymes, for instance, makes easier the separation of tocopherols from SODD by converting sterols to steryl esters, acylglycerols to free fatty acids and free fatty acids to fatty acid methyl or ethyl esters (FAMEs or FAEEs). Then, it is easier to separate the new product mixture by distillation or supercritical fluid extraction. From published literature, it can be point out that the main difficulties of the enzymatic processes are the numerous parameters involved such as moisture content, enzyme concentration, time, temperature, ratio of the reactants, stability, recovery and reutilization of the enzyme preparation, among others (Ramamurthi et al., 1991), (Ramamurthi & McCurdy, 1993).

The conversion of FFAs to FAMEs or FAEEs is an important step in the concentration and purification of tocopherols. If this step is omitted, the separation of FFA and tocopherols by distillation cannot be achieved due to their similar boiling points (Shimada, 2000). Furthermore, if methanol is used for the biotransformation of FFA to FAMEs, concomitant sterol esterification with fatty acids is inhibited. To avoid this problem, a lipase can be used in a two stages procedure: first to carry out hydrolysis of acylglycerols and then to promote the esterification of sterols with free fatty acids. The different components are then successfully separated by short path distillation or supercritical fluid extraction since their boiling points are now sufficiently different.

In the literature, many enzymatic procedures for the preparation of sterol esters are described, but most of them require organic solvents, water and molecular sieves or other drying agents (Haraldsson, 1992), (Shimada et al., 1999), (Jonzo et al., 1997), (Hedström et al., 1992). Although these strategies gave good conversion rates for the formation of sterol esters, the use of such multiphasic systems may complicate the final purification of the products in the case of larger scale productions. However, the enzymatic preparation of fatty acid esters of sterols, stanols and steroids in high yield by esterification and transesterification of fatty acids and other carboxylic acid esters, in vacuum at moderate temperature using immobilized lipases have been also reported (Weber *et al.*, 2001). In this case neither organic solvent, nor water or any drying reagent such as molecular sieves, are used. This and others studies (Shimada, 2000) showed that in the process of esterification of sterols with free fatty acids, the best results are obtained with *Candida rugosa* lipase and *Pseudomonas* sp. However, enzymatic conversion of FFAs to FAMEs or FAEEs is carry out frequently in the presence of *Candida antactica* lipase or *Alcalygenes* sp. (Torres, 2007), (Nagao *et al.*, 2005).

In the following paragraphs some examples of methodologies using enzymes in the pre-treatment of SODD are described. Most of them will be further developed in following sections:

Shimada and coworkers (Shimada, 2000) converted sterols from SODD to fatty acid sterol esters and completely hydrolyzed acylglycerols by applying lipase reactions (*Candida rugosa* or *Pseudomonas* sp., at 35 °C for 24 h) to the purification of tocopherols and sterols, resulting in an efficient fractionation of tocopherols and sterols as fatty acids steryl esters (FASEs) by short-path distillation. This process included the drawback that FFA and tocopherols were not efficiently fractionated because the boiling points of the two substances were close. This problem could be solved by conversion of the FFA to their corresponding methyl esters. An attempt to develop a reaction system in which the methyl esterification of FFA proceeded simultaneously with the conversion of sterols to FASEs and the hydrolysis of acylglycerols has been also reported (Watanabe *et al.*, 2004).

Nagao and coworkers (Nagao, 2005) and Watanabe and coworkers (Watanabe, 2004) have applied a procedure based on using a lipase to promote the simultaneous esterification of sterols with free fatty acids and hydrolysis of acylglycerols before the esterification of the free fatty acids with methanol. These authors use *Candida rugosa* lipase for the purification of tocopherol in SODD. Watanabe and coworkers reported 80% conversion of the initial sterols to FASEs, complete hydrolysis of the acylglycerols, and a 78% decrease in the initial FFA content by methyl esterification in 40 h. Tocopherols did not change throughout the process. Distillation of the reaction mixture purified tocopherols to 76.4% (recovery, 89.6%) and sterols to 97.2% as FASEs (recovery, 86.3%). Nagao and coworkers reported a more effective sterols esterification, with a degree of esterification reached 95%. The second-step reaction was then conducted at 30 °C for 20 h with *Alcaligenes* sp. lipase. 95% FFAs were converted to FAME, and steryl esters synthesized by the first-step reaction were not reconverted to free sterols. Finally, tocopherols and steryl esters were purified from the reaction mixture by short-path distillation. Tocopherols were purified to 72% (yield, 88%) and steryl esters were purified to 97% (yield, 97%). One of the main disadvantages of this method is that the remaining free fatty acids are not completely separated from the tocopherols.

Lipase-catalyzed esterification of sterols and ethyl esterification simultaneously, are governed by the concentration of water present. The degree to which esterification of sterols occurs relative to ethyl esterification requires to attain a balance not always easy to achieve because the presence of an excess of water favours hydrolysis, whereas esterification predominates when a very limited amount of water is present (Marangoni & Rousseau, 1995). By appropriate choice of reaction conditions, however, it is possible to separate the sterol esterification and ethyl esterification in time or space. It is then possible to optimize each of these reactions independently, thereby minimizing costs or improving the yield of the desired final reaction products.

This is precisely the procedure carried out by Torres and coworkers (Torres, 2007), who proposed a two-step enzymatic procedure to obtain FASEs, tocopherols, and fatty acid ethyl esters (FAEEs) from SODD. Firstly, SODD was mixed with oleic acid to reduce its melting point and to enhance the free phytosterols esterification. The first enzymatic step (using *Candida rugosa* lipase) allowed the efficient conversion of more than 90% free phytosterols within 5 h. The second one (using Novozym 435) converted more than 95% FFAs in less than 3 h. The final product obtained was used as starting material to purify FASEs, tocopherols, and FAEEs via supercritical CO_2 extraction (Torres, 2009).

Weber and coworkers (Weber *et al.*, 2002) have also reported the use of lipases for the conversion of sterols into steryl esters leading to a higher degree of purity (90%), however the methodology is more complex and involves deacidification, flash chromatography and solvent fractionation.

Another methodology was developed to focus more specifically on the conversion of FFAs into fatty acid butyl esters (FABEs) (Nagesha *et al.*, 2004). Nagesha and coworkers (Nagesha, 2004) used immobilized *Mucor miehei* lipase in supercritical carbon dioxide at high pressure and obtained a maximum recovery of 88% and a FABE purity of 95% from SODD.

5. Molecular distillation

Most of the substances that are present in soybean deodorizer distillate are molecules of high molecular weight and thermally sensitive. These properties hinder the separation or purification of these compounds through traditional methods, because they are decomposed when subjected to high temperatures.

An alternative separation/purification procedure of such products is the use of molecular or short-path distillation. It consists of transferring molecules from the surface of an evaporating liquid to the cooled surface of a condenser through a short path, which is on the order of 2-5 cm. In this process, distillation of heat-sensitive materials is accompanied by only negligible thermal decomposition (Lutisan *et al.*, 2002) because materials, by using high vacuum, are submitted to relatively reduced temperatures, and short residence times (Lutisan, 2002) inside the equipment. Furthermore, this process has advantages over other techniques that use toxic or flammable solvents as the separating agent, avoiding toxicity and environmental problems.

The combination of a small distance between the evaporator and the condenser of only a few centimetres and a high vacuum in the distillation gap, results in a specific mass transfer mechanism with evaporation outputs as high as 20–40 gm^{-2} s^{-1} (Cvengros *et al.*, 2000). Under these conditions (e.g., short residence time and low temperature), distillation of heat-sensitive materials is accomplished without or only negligible thermal decomposition. Therefore, molecular distillation shows potential in the separation, purification and/or concentration of natural products, usually constituted by complex and thermally sensitive molecules such as tocopherols.

In lipid chemistry, it has been used for the purification of monoacylglycerols (Szelag & Zwierzykowski, 1983), recovery of carotenoids from palm oil (Batistella & Wolf Maciel, 1998), fractionation of polyunsatured fatty acids from fish oils (Breivik *et al.*, 1997), recovery of squalene (Sun *et al.*, 1997), and recovery of tocopherols (Batistella *et al.*, 2002), among others.

Normally, SODD have a high content of FFA and acylglycerol. To increase the separation efficiency of the compounds of interest, esterification and/or transesterification reactions are usually carried out prior to molecular distillation. Free fatty acids and acylglycerols are converted to fatty acid methyl esters, which are more easily removed by vacuum distillation due to their higher vapor pressure than those of the corresponding free fatty acids and acylglycerols. However, this step made the whole process more complicated and labor-intensive when compared with the saponification process. Other problem is that the separation of tocopherols from phytosterols is difficult because they have similar molecular weights, boiling points and vapor pressure, and, consequently, they are distillated together (Ghosh & Bhattacharyya, 1996).

Different processes have been proposed in the literature to eliminate FFA by molecular distillation and purify tocopherols and phytosterols. Most of them include a preliminary chemical or enzymatic treatment step.

Ramamurthi and McCurdy (Ramamurthi & McCurdy, 1993) studied the pretreatment of deodorizer distillate using a lipase-catalyzed esterification reaction to convert FFA into methyl esters, followed by vacuum distillation (1-2 mm Hg) to remove them and concentrate tocopherols and sterols (recoveries were over 90%).

Hirota and coworkers (Hirota et al., 2003) isolated naturally occurring Fatty Acid Steryl Esters (FASEs) from SODD. SODD was firstly subjected to molecular distillation at 250 °C and 0.02 mm Hg to obtain a residue which was rich in DAGs and TAGs, and steryl esters. Enzymatic lipolysis was then conducted to specifically hydrolyze DAGs and TAGs at 35 °C for 24 h, resulting in a mixture from which fatty acid steryl esters were later purified using a two-stage molecular distillation (180 °C and 0.2 mm Hg, and 250 °C and 0.02 mm Hg). The recovery and purity of FASEs were about 87.7% and 97.3 wt%, respectively.

Purification of tocopherols from SODD was carried out by Shimada and coworkers (Shimada, 2000). SODD was distilled using molecular distillation at 250 °C and 0.02 mm Hg and the resulting distillate was used as a starting material. Sterols in SODD were converted to FA sterol esters and acylglycerols were completely hydrolyzed by applying lipase reactions. FASEs were recovered as residue from the reaction mixture via molecular distillation at 250 °C and 0.2 mm Hg. However, the last stage of molecular distillation failed to separate FFAs and tocopherols. A second esterification of free phytosterols was applied at 35 °C for 24 h, followed by another four-stage molecular distillation (160 °C and 0.2 mm Hg, 200 °C and 0.2 mm Hg, 230 °C and 0.04 mm Hg, and 255 °C and 0.03 mm Hg) which yielded tocopherols with purity and recovery of about 65.3 wt% and 54.6%, respectively.

Watanabe and coworkers (Watanabe, 2004) isolated tocopherols and free phytosterols as their esters from SODD tocopherol/sterol concentrate (SODDTSC). SODDTSC was obtained via molecular distillation at 240 °C and 0.02 mm Hg, resulting in a distillate rich in tocopherols and free phytosterols (SODDTSC), and a residue rich in FASEs, DAGs, and TAGs. SODDTSC, which contained also MAGs, DAGs, FFAs, and unidentified hydrocarbons, were then subjected to a two-step in situ enzymatic reaction. SODDTSC were treated with *Candida rugosa* lipase (200 U/g activity) to convert free phytosterols to FASEs, acylglycerols (MAGs and DAGs) to FFAs, and FFAs to FAMEs at 30 °C for 40 h, achieving 80% conversion of the initial sterols to FA steryl esters, complete hydrolysis of the acylglycerols, and a 78% decrease in the initial FFA content by methyl esterification. Tocopherols did not change throughout the process. To enhance degree of steryl and methyl esterification, FASEs and FAMEs enriched in the reaction product were then removed by a two-step molecular distillation. In the first step, FAMEs was removed in the distillate (160 °C and 0.2 mm Hg). In the second step (240 °C, 0.2 mm Hg), FASEs was isolated in the residue and the distillate containing tocopherols, free phytosterols, and FFAs were treated again with lipase. A three-step molecular distillation of the reaction mixture purified tocopherols to 76.4 wt% purity (89.6% recovery) and free phytosterols to 97.2 wt% purity as FASEs (86.3% recovery).

Nagao and coworkers (Nagao, 2005) carried out similar steps of isolation of tocopherols and free phytosterols as FASEs from SODDTSC. SODDTSC were first treated with *Candida rugosa* lipase (250 U/g activity) to convert free phytosterols to FASEs at 40 °C for 24 h, achieving about 95% conversion. Unreacted FFAs contained in the reaction mixture was then converted to FAMEs by *Alcaligenes* sp. lipase at 30 °C for 20 h, achieving 95% conversion. Reaction mixture was then subjected to a four-stage molecular distillation (160 °C and 0.2 mm Hg, 175 °C and 0.2 mm Hg, 230 °C and 0.02 mm Hg, and 240 °C and 0.02 mm Hg, respectively) to isolate tocopherols (72.3 wt% purity, 87.6% recovery) in the third distillate fraction and free phytosterols as FASEs (97 wt% purity, 97% recovery) in the last (fourth) residue.

Jacobs (Jacobs, 2005) proposed a method for recovering tocotrienols from fatty acid distillate FAD, which initially contained 1 wt% tocopherols and 0.3 wt% free phytosterols, by stripping FFAs (condition: 0.5–1.5 mm Hg, 180–240 °C, and 0.5–1.5 min) to form a first stripped product. Short path distillation of the first stripped product gave a first distillate. Saponifying the second stripped product resulted in a saponified product with all FFA converted to FAMEs. A second short path distillation of the saponified product generated a second distillate without FAMEs. Solvent wintering (via filtration) of the second distillate gave a stripped filtrate. The stripped filtrate from the previous step is subjected to a third short path distillation at a temperature of 180° and an absolute pressure of 0.01 mm Hg. The resulting final tocotrienol product, about 1% of the original feed, contains from about 50% tocotrienols, 1% sterols and 49% other unsaponifiables and unknowns. Additionally, the final product contains from about 15% to about 30% tocopherols.

Fizet (Fizet, 1996) esterified free phytosterols from deodorizer distillate with FFAs at either 180 °C for 2.5 h or 250 °C for 1.5 h. Esterification product was then distilled at 120–150 °C and 0.08 mm Hg to obtain a residue containing mostly tocopherols and FASEs and a distillate containing mostly fatty acids. The residue was then distilled again at 200–220 °C and 0.1 mbar to obtain a distillate containing mostly tocopherols and a residue containing mostly FASEs. Tocopherols enriched in distillate were then subjected to an ion exchange chromatography and FASEs enriched in residue were then subjected to an acid-catalyzed transesterification with methanol to produce free phytosterols.

Even so, some authors (Martins *et al.*, 2006) have been trying to achieve an efficient FFA separation from SODD with the lowest loss of tocopherols by molecular distillation, without preliminary steps. This separation is difficult to achieve although is technologically viable at least at lab scale, due to the differences between molecular weights and vapour pressures of FFA (MW 180 g·mol⁻¹, VP at 200 °C = 4 mm Hg) and tocopherols (MW 415 g·mol⁻¹, VP at 200 °C = 0.15 mm Hg). Martins and coworkers (Martins, 2006) employed molecular distillation at 160 °C, 7.5×10⁻⁴ mm Hg, and 10.4 g/min feed flow rate to remove FFAs into the distillate fraction and obtain a residue fraction, which contained 6.4 wt% FFAs and 18.3 wt% tocopherols from a SODD feed which contained 57.8 wt% FFAs and 8.97 wt% tocopherols. They succeeded in removing 96.16% FFAs and recovering 81.23% tocopherols. Martins and coworkers (Martins *et al.*, 2005) reported the isolation of tocopherols by first converting acylglycerols in SODD into FFAs through saponification at 65 °C followed by acidulation, and then submitting the unsaponifiable product to five stages of molecular distillation. They succeeded in enriching tocopherols (34.14% purity) by 5.8 times. The major disadvantage in these cases is the residual free fatty acids in the tocopherol mixture.

Among the great variety of processes that have been patented for the purification of the compounds of the SODD, only the processes of esterification of fatty acids and acylglycerols with methanol or ethanol followed by high vacuum distillation, have been developed on a commercial scale for the concentration of tocopherols (Takagi & Kai, 1984), (Su-Min et al., 1992), (Yong-Bo et al., 1994), (Rohr & Trujillo-Quijano, 2002). These processes are the most time efficient and economical methods, however high purity of sterols or tocopherols cannot be achieved due to the similar boiling points of these two substances.

In the case of separation by distillation of unsaponifiable valuable products of SODD subjected to saponification, the difference between the boiling point of volatile products, such as unsaponifiable components, and the boiling point of the sodium and potassium organic acid soaps is so great that separation is theoretically possible at a high level of efficiency. However, a problem related to this separation technique is that the soaps have a very high melting point, close to the decomposition temperature of the sodium or potassium soaps (i.e. the sodium or potassium salts of fatty acids, rosin acids etc), and, when melted, these soaps produce an extremely viscous liquids. These two factors combined make industrial handling difficult. Furthermore, while at the high temperature necessary to maintain their flow, these soaps are in permanent decomposition, compromising the separation output and the quality of the final product, as many of the unsaponifiable valuable products are heat sensitive.

6. Supercritical fluid extraction (SFE)

Although the conventional methods, vacuum and molecular distillation, have been applied to commercial production of tocopherols from SODD, there are some drawbacks such as residual solvents, high temperature, large amounts of energy consumption, high production costs and the unreliable quality of the products that require further developments. Since thermal degradation of tocopherols is commonly caused by processing at high temperatures (de Lucas et al., 2002), new alternative isolation techniques are desired.

Supercritical carbon dioxide extraction is a process where carbon dioxide passes through a mixture of interest at a certain temperature and pressure until it reaches an extractor. This process is used because supercritical carbon dioxide has a low viscosity, a high diffusivity and a low surface tension that provides selective extraction, fractionation and purification, allowing its penetration in micro- and macro-porous materials. Carbon dioxide is the most desirable supercritical fluid solvent for the separation of natural products used in foods and medicines because of its inertness, nontoxicity, low cost, and high volatility. The major advantage of this method is the easy post-reaction separation of the components by depressurization, so resultant extract does not contain solvent residue and hence natural-quality extracts can be obtained. Another advantage is the low temperatures used for the majority of the experimentations because carbon dioxide has a near-ambient critical temperature (31.1 °C), so is suitable for thermolabile natural products.

However, the use of high pressure conditions to concentrate tocopherols makes the system energetically expensive, but the industrial process can be economically viable using conditions of approximately 90 atm and 40 °C (Mendes et al., 2002). At these specific conditions, only fatty acids are separated from tocopherol (Mendes et al., 2005). An increase in pressure and temperature increases the oil extraction and tocopherol recovery, although different pressure–temperature systems need to be used in order to separate the different components (sterols, tocopherols, fatty acids and squalene). It is important to know that recycling the solvent does not endanger the viability of the process. The value of the rate of return on investment and time of return on investment for the process that does not recycle the carbon dioxide is higher than those of recycling the solvent. This is due to the compression cost that represents more than 59% of the total cost of the production (Mendes, 2002).

SODD as such will not be feasible to work with SC-CO_2 for the tocopherol enrichment, owing to its poor SC-CO_2 solubility. So, to concentrate tocopherols from SODD, pre-treatment of the raw material, including the esterification of free fatty acids and the removal of sterols with alcohol recrystallization, is needed to obtain the primary tocopherols concentrate with improve solubility in SC-CO_2 (Shishikura et al., 1988). For that, triglycerides and FFAs which constitute a major component in SODD have to be chemically modified to obtain free fatty acids and then their methyl esters by esterification.

Several researchers have tried to concentrate tocopherols from SODD by supercritical CO_2 (Lee, 1991), (Brunner et al., 1991), (Brunner, 1994b), (Zhao et al., 2000), (Nagesha, G. K. et al., 2003), (Fang et al., 2007), but the operation parameters, especially pressure, differ from author to author. Moreover, in all the cases, tocopherols content of the extract depended on the composition and properties of the natural matrix.

The interest in the tocopherol concentration using supercritical fluid extraction started with Lee and coworkers (Lee, 1991) followed by Brunner and coworkers (Brunner, 1991) and Brunner (Brunner, 1994b). The operational conditions used varied from 35 to 90 °C and from 200 to 400 bar using extractors or countercurrent columns. Lee and coworkers (Lee, 1991) attempted to modify soybean sludge chemically, to improve the solubility in SC-CO_2. A simple batch process was utilized to recover tocopherols at 40% concentration at a pressure of 400 bar from the esterified soybean sludge which initially contained 13-14% tocopherols. The solubility of the esterified soybean sludge in supercritical carbon dioxide was more than 4 to 6 times higher than that of the sterols. Brunner and coworkers (Brunner, 1991) obtained a higher enrichment of tocopherols from SODD using supercritical carbon dioxide as a solvent compared to results obtained by Lee and coworkers This group recovered tocopherols from a model mixture of squalene, tocopherols, and sterols using two continuous countercurrent fractionation columns. Squalene was separated from the model mixture in the first column. Sterols were removed from the bottom of the second column, resulting in 85-95% concentration of tocopherol being obtained at the top of the second column.

These works concluded that the fatty acids are extracted initially and the tocopherols are enriched inside the extractor. The results also indicated that the solubility of the tocopherols is intermediary when compared to the solubilities of squalene and stigmasterol.

Chang and coworkers (Chang, 2000) worked on the separation of several SODD components. Their supercritical fluid extraction apparatus had a separation and an extraction unit. Free fatty acids, squalene and tocopherols were recovered in the extract and the sterols were recovered in the raffinates. The average tocopherol concentration factor was 1.38, which means that the mixture in the extract did not separate. However, the author mentioned that with the increase of CO_2 volume, the separation factor can reach 1.7, but the poor increase in the concentration factor does not justify the raise in gas volume. The following research groups focused on the separation of the problematic pairs by supercritical fluids using synthetic mixtures: (Mendes, 2005), (Mendes et al., 2000), (Wang et al., 2004), (Nagesha, 2004) and (Nagesha, 2003).

There are some interesting relationships and conclusions that can be deduced from the results obtained by Mendes (Mendes, 2005, Mendes, 2000) and Chang (Chang, 2000) regarding the yield and concentration factor of pairs of compounds at different conditions of pressure and temperature. The binary mixture of tocopherol and squalene cannot be separated at low pressure conditions. An acceptable separation needs a raise in pressure to almost 203 bar ((Mendes, 2005) and (Mendes, 2000)). However, a recovery of 90% and a purity of 60% of α-tocopherol has been achieved using a pressure swing adsorption (PSA) device, that is a widely used process in the separation of gas mixtures for air-drying, oxygen and nitrogen separation of air, hydrogen purification, and various other separations. The PSA process is based on the regeneration of adsorber by the difference in adsorbed amounts of gas solute as a function of pressure. In the case of a two-bed process, one bed is in the adsorption step, while the other is simultaneously in the desorption step. The adsorption and desorption steps had pressure conditions of 160 and 300 bar, respectively (Wang, 2004).

The ternary mixture of tocopherol, fatty acids and squalene behaved differently from the tocopherol–fatty acid binary mixture. For the same conditions of pressure (160 and 300 bar), the binary mixture had a total separation while the ternary mixture did not achieve any separation. Squalene and stigmasterol mixtures are also very difficult to separate. At low pressure conditions, the yield is less than 10% but at higher pressure, the yield is 76%. It is important to note that low temperatures were used in these studies.

Results of these authors suggest that the supercritical-CO_2 process could be used for the separation of squalene, fatty acids and tocopherols. In order to enhance the squalene, fatty acids and tocopherols separation, deodorizer distillate mixtures should be processed several times in supercritical-CO_2 at different temperature and pressure conditions. However, there is no data reported regarding the total extraction time of the mixture and this makes it impossible to estimate the operational cost. The major advantage of this method is the total removal of free fatty acids from the mixture.

Zhao and coworkers (Zhao, 2000) concentrated tocopherols up to 75% at 120 bar using a fractionation column with a gradient of temperature from 30–80 °C in a pilot plant scale.

Nagesha and coworkers (Nagesha, 2003) performed chemical modification of SODD containing about 2.9 wt% of tocopherols, as well as triglycerides (56 wt%), free fatty acids (25.3 wt%), sterols (7.8 wt%), hydrocarbons (0.6 wt%), and unsaponifiables (6.4 wt%) apart from tocopherols. Chemical modification of SODD included saponification and esterification steps to result in fatty acid methyl esters from free fatty acids, so as to improve the solubility of SODD in SC-CO_2 extraction. Reactions were conducted in dark with continuous flushing of N2 and 1.0 wt% of pyrogallol was added to prevent the oxidation of tocopherols. After chemical modification, esterified SODD contained about 3.7 wt% of tocopherols. Tocopherols concentrates of about 36% was obtained by SC-CO_2 extraction at the pressure 180 bar and temperature 60 °C.

Fang and coworkers (Fang, 2007) carried out a pretreatment of methyl esterification and methanolysis reactions, which converted most of free fatty acids and glycerides to fatty acid methyl esters (FAMEs), respectively, to simplify the composition of SODD and improve his solubility in supercritical CO_2 extraction. The mixture was held at 3 °C in a refrigerator for 12 h, as a result most of sterols were crystallized and removed by filtering under a reduced pressure. Supercritical CO_2 fractionation was employed to concentrate tocopherols from Methyl Ester Oil Deodorizer Distillate (ME-DOD) product, mainly contained FAMEs (65–80 wt.%), tocopherols (10–15 wt.%), and impurities (such as residual sterols, glycerides, squalene, pigments, and long chain paraffins, comprising in total about 10–15 wt.%). The initial pressure, feed location, temperature gradient, and ratio of CO_2 to ME-DOD were optimized for separating FAMEs. For the following tocopherol concentration step, a final pressure of 200 bar resulted in the greatest average tocopherol content (>50%) and tocopherol recovery (about 80%).

The important step in concentrating natural tocopherols from these systems is to remove the FAMEs. FAMEs are important chemical materials in biofuel, metal-cutting oil, and cleaning agent production, as well as in the synthesis of other fatty acid products (Swern, 1986).

Some works on phase equilibrium for the realistic system of modified esterification SODD/supercritical CO_2 (Fang et al., 2005) established that the separation factor [1] between tocopherols and FAMEs was always smaller than unity in the range investigated. This indicates that when supercritical CO_2 is used as the separation solvent, tocopherols, unlike FAMEs, tend to enrich in the liquid phase. In particular, the separation factors at pressures lower than 200 bar were relatively small. At 40 °C, for instance, the separation factor remained lower than 0.2 for all pressures lower than 150 bar. As pressure increased, the separation factor increased greatly, reaching 0.35 at 200 bar. The influence of temperature was contrary to that of pressure, with an increase in temperature leading to a decrease in separation factor. Low pressure and high temperature result in high selectivity, indicated by a low separation factor, which is advantageous in the separation of FAMEs from tocopherols with supercritical CO_2.

King and coworkers (King et al., 1996) combined supercritical fluid extraction (SFE) with supercritical fluid chromatography (SFC) for concentrating tocopherols and the optimized conditions were 250 bar/80 °C for SFE and 250 bar/40 °C for SFC. Approximately 60% of the available tocopherols in soyflakes can be recovered in the SFE step, yielding enrichment factors of 1.83-4.33 for the four tocopherol species found in soybean oil. Additional enrichment of tocopherol species can be realized in the SFC stage, with enrichment factors [2] ranging from 30.8 for delta-tocopherol to 2.41 for beta-tocopherol.

Starting with a feed containing 48.3 wt% tocopherols, Gast and coworkers (Gast et al., 2005) were able to obtain tocopherols with a purity of 94.4 wt% in bottom phase by supercritical CO_2 extraction at 230 bar and 80 °C, with a solvent-to-feed ratio of 110 and a reflux ratio of 4.6. Squalene was completely recovered in top phase.

Torres and coworkers (Torres, 2007) proposed a two-step enzymatic procedure to obtain FASEs, tocopherols, and fatty acid ethyl esters (FAEEs) from SODD, together with minor amounts of squalene, free fatty acids, free sterols and triacylglycerols. The final product obtained was used as starting material to purify FASEs, tocopherols, and FAEEs via supercritical CO_2 extraction The phytosterol esters were then purified from this mixture using supercritical carbon dioxide (Torres, 2009). Experimental extractions were carried out in an isothermal countercurrent column (without reflux), with pressures ranging from 200 to 280 bar, temperatures of 45-55 °C and solvent-to-feed ratios from 15 to 35 kg/kg. Using these extraction conditions, the fatty acid esters were completely extracted and, thus, the fractionation of tocopherols and phytosterol esters was studied. At 250 bar, 55 °C and a

---

[1] The separation factor represents the process selectivity for separating methyl oleate from tocopherol. In detail, a lower value indicates higher selectivity, whereas a higher value indicates that it is more difficult to separate the two compounds under certain conditions. Furthermore, when the separation factor equals unity, the composition in the gas phase is similar to that in the liquid, and the supercritical CO_2 process cannot separate methyl oleate from tocopherol.

[2] The enrichment factors were the ratio of individual tocopherols in extracts versus the same tocopherol content initially found in the soyflakes.

solvent-to-feed ratio of 35, the phytosterol esters were concentrated in the raffinate up to 82.4 wt-% with satisfactory yield (72%).

Other supercritical fluids have been explored but unsuccessfully for the separation of different pairs of components. An attempt at using similar methodology to (Mendes, 2000) and (Mendes, 2005) but using liquid gas petroleum instead of carbon dioxide did not change the poor concentration factor between the critical pairs of components (Buczenko et al., 2003). Buczenko and coworkers (Buczenko, 2003) performed the saponification of the raw material and the extraction of unsaponifiable matter as pre-treatment of VODD.

As discussed above, there are a lot of experimental studies proving the efficiency of the supercritical extraction to concentrate the vitamin E from different raw material or in some cases, from synthetic mixtures representing the deodorizer distillate, but the extraction of sterols using supercritical fluid from the deodorizer distillate was not described in the literature.

On the other hand, due to the low content of squalene in SODD, specific extraction processes of squalene using supercritical fluid from SODD was not described in the literature. Existing studies are models of fractionation of artificial mixtures such as those mentioned above (Chang, 2000). For example, Brunner (Brunner, 1994a) studied the phase equilibrium for recovering α-tocopherol from a mixture of squalene, tocopherol, and campesterol. He concluded that the separation factor for squalene/α-tocopherol varied between a value of 4 at low squalene concentrations (0.5 wt %), to a value of 1 at high squalene concentrations (85 wt %), at pressures ranging from 200 to 300 bar and temperatures ranging from 70 to 100 °C.

Bondioli and coworkers (Bondioli et al., 1993) esterified FFAs into their corresponding glycerides and then applied a supercritical carbon dioxide extraction to produce a squalene-enriched fraction (purity 90.0%, yield 91.1%), but from olive oil deodorizer distillates.

7. Other purification techniques

There has been also limited literature reported on the following alternative methods for purification of tocopherols and sterols from SODD. Hence, Nagesha and coworkers (Nagesha, G. et al., 2003) using nonporous denser polymeric membranes to separate tocopherols from SODD by permeation. The separation in a denser membrane is generally based on a solution-diffusion mechanism. The lower polarity of tocopherols compared to that of free fatty acids appears to have facilitated the preferential permeation of tocopherols through the hydrophobic membrane. Selectivity of the membrane for tocopherols improved with esterified SODD, because the presence of FAME decreased the viscosity of the feed and thereby increased convective flow, which in turn improved permeate flux.

Alternatively, Maza and coworkers (Maza, 1992) concentrated tocopherols and sterols by addition of melted deodorizer distillates to a solution of urea and alcohol which separate fatty acids from the mixture.

In addition, Gunawan and coworkers (Gunawan et al., 2008a) proposed a facile procedure to isolate naturally occurring FASEs from SODD without degradation of FASEs. SODD was first subjected to a modified soxhlet extraction (MSE) to enrich FASEs in a non polar lipid fraction (NPLF). Modified silica gel column chromatography (MSE) was then applied to NPLF to collect FASEs in a third fraction with a purity of 79.99 wt% and a recovery of 97.38%. The third fraction was then subjected to a binary solvent (water/acetone = 20/80, v/v) extraction to purify FASEs to a purity of 86.74 wt% with a total recovery of 85.32%.

These methods do not present a significant advance regarding the most frequently utilized methods and probably their application to production scale would be little profitable. However, they can be used to isolate tocopherols and sterols from SODD at laboratory scale or complement other methods at industrial scale.

8. Degradation and oxidation of functional lipids from SODD

8.1 Influence of refining on phytosterols

During refining of edible fats and oils, the content of total sterols decreases due to degradation and formation of products through isomerization (D5 to D7-sterol), dehydration, polymerization, and formation of hydrocarbons or sterenes and sterol oxidation products (Dutta, 2006). These qualitative and quantitative changes in sterols can be traced in the refined oil and in by-products such as soapstocks and distillate fractions collected after chemical and physical refining processes (Dowd, 1998).

Acid hydrolysis of steryl esters may occur upon bleaching with an acid activated bleaching earth. The slight reduction of the total sterol content is due to the formation of steradienes and disteryl ethers. A gradual reduction in the total sterol content is observed at increasing deodorization temperature due to distillation and steradiene formation. Increasing the temperature from 220 °C to 260 °C resulted in a gradual reduction of the total sterol recovery from 90.4% to 67.7% in physical refining and from 93% to 62.7% in chemical refining. However in physical refining, an increase of 40% in the steryl ester fraction is observed due to an esterification reaction, promoted by high temperature between a sterol and a fatty acid. Due to the absence of free fatty acids in the chemical refining their esterification did not occur (Verleyen et al., 2001b). The influence of refining on free and esterified sterols has been studied by (Verleyen, 2002b).

Phytosterols are progressively lost during refining while continuously altering the ratio of free and esterified sterols (Kochhar, 1983). During chemical neutralization, the free sterol content is significantly reduced especially upon addition of weak caustic solution due to the loss in the soapstock (Gutfinger & Letan, 1974).

Bleaching effects on phytosterols are generally minor and mainly limited to the formation of some nonpolar dehydration products (Ferrari, 1996) and partial hydrolysis of sterol esters (Homberg & Bielefeld, 1982). Steradienes and disteryl ether dehydration products (Figure 1) are formed during bleaching step by the bleaching temperature and the degree of acid activation of the bleaching earth, while during the deodorization, the degree of sterol dehydration is mainly influenced by deodorization temperature giving rise to a concentration of the steradienes in the distillate   (Verleyen, 2002b, Verleyen, 2001c). The presence of steradienes can also be used as a marker for the presence of refined oils (Grob et al., 1994).

Whenever applied, hydrogenation has a tremendous effect on sterol structures, including hydrogenation of double bonds, opening of cyclopropane rings, and positional isomerization of side chain unsaturation (Strocchi & Marascio, 1993).

A part of a multinational EU research project (FOOD-CT2004-007020) was to carry out qualitative and quantitative assessment of sterols and sterol oxidation products in samples of by-products from chemical and physical refining of edible fats and oils collected from various locations in Europe. To the best of our knowledge, this is the first report on the contents of oxidized sterols in soapstock and distillate fractions from edible oil refining processes. The levels of sterol oxidation products were higher in acid oil obtained from chemical refining (AOCHE) samples than in acid oil obtained from physical refining (AOPHY) samples, with ranges 0.02–17.0 and 0.01–1.5 mg/100 g, respectively. The lower content of sterol oxidation products in AOPHY samples may be due to the high temperature applied during vacuum distillation accelerating the breakdown and transformation of the sterol oxidation products into other unidentified degradation products. Further formation of sterol oxidation products has been prevented by the high amounts of natural antioxidants in AOPHY distillate (Verleyen, 2001c). Some sterols appeared to be more liable to breakdown than others, e.g. there was a higher content of oxybrassicasterols than the other sterol oxidation products in this study, although the content of brasicasterol in the sample was lower than other sterols. Similar results have been reported previously (Dutta, 2006). This may be due to the structural arrangement in the brassicasterol molecule rendering it more easily oxidized than other sterols. However, systematic studies are required to clarify this phenomenon. Although stigmasterol has a double bond in the side-chain, similar to brassicasterol, the quantities of phytosterol oxidation products or oxyphytosterols observed in this study were quite different. Stigmasterol has an ethyl group at position C24 while brasicasterol has a methyl group, and this difference may affect in the relative rate of formation of their oxidation products (Dutta, 2006). Further studies are needed on this point.

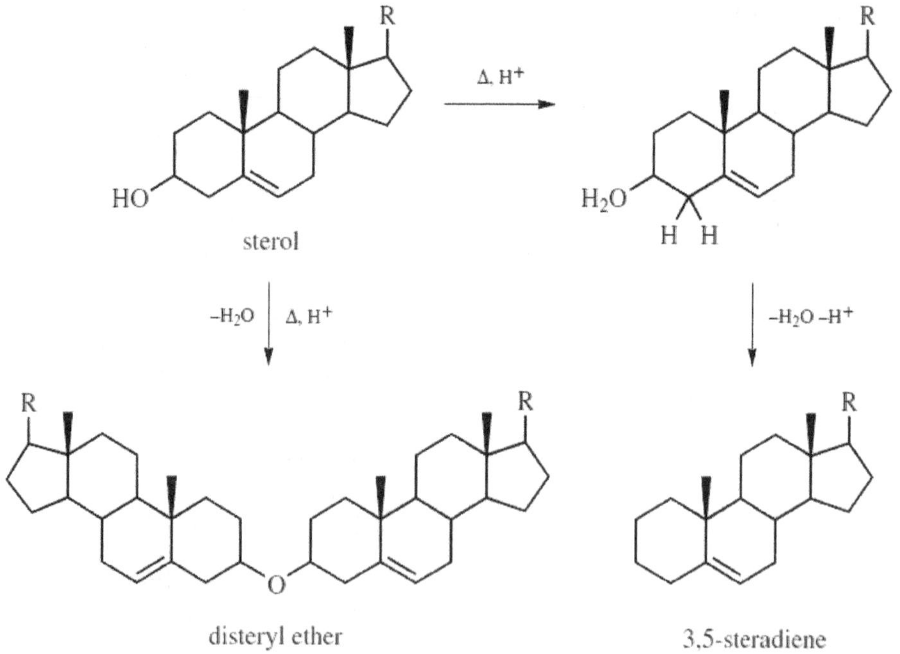

Fig. 1. Reaction products of sterols during refining.

It has been reported that the formation of sterol oxidation products is affected not only by the chemical nature of the sterols but also by their quantity (Dutta, 2006). There were positive correlations between total sterols and total phytosterol oxidation products in the by-products collected from both refining processes.

The biological effects of oxycholesterol have been extensively studied (Bjorkhem *et al.*, 2002); however, the amount of biological research on oxyphytosterols is rather scarce, mostly dated and has never been extensively reviewed before (Francesc, 2004) and (Dieter, 2004). Most reports available so far on oxyphytosterols cover the methodological aspects of their measurement in foods. The usual perception about oxyphytosterols is that these components present a concern in terms of food quality and health. This perception originates from the parallel that is made between oxycholesterol and oxyphytosterols. Whether oxyphytosterols may indeed play similar and/or different biological roles compared to oxycholesterol has not been elucidated yet.

A review (Hovenkamp *et al.*, 2008) summarise the current knowledge on the possible biological effects of oxyphytosterols and to identify future research needs, which will help in clarifying the possible impact of oxyphytosterols on human health. The review focuses on the more common oxyphytosterols which differ only in a few structural changes from the parent sterol.

Over the last thirty years a diversity of potential biological effects, including modulation of cholesterol homeostasis, anti-inflammatory and anti-tumour activities, as well as lipid-lowering and anti-diabetic properties, have been attributed to specific oxyphytosterols. Although these studies were not all carried out with oxyphytosterols also identified in the human body, these results suggest that oxyphytosterols may have systemic effects in vivo and therefore, the potential to modulate human metabolism.

Despite some putative desirable effects, oxyphytosterols may be perceived as presenting a concern in terms of food quality and health. Indeed, oxyphytosterols have been reported to exert, in vitro, cytotoxic effects comparable to those attributed to oxycholesterol. However, high, non-physiological concentrations of oxyphytosterols were needed to exert adverse effects. In addition, data from one animal study do not support a role of oxyphytosterols in atherosclerosis promotion. However, this aspect deserves more attention in future research. Altogether, the currently available observations do not suggest that oxyphytosterols, in relatively low concentrations such as those reported in human plasma, may exert in vivo deleterious effects similar to those attributed to oxycholesterol. In addition, although probably different in structure than the potentially deleterious ones, some oxyphytosterols may also have the ability to activate transcription factors involved in cholesterol metabolism. Nevertheless, more detailed investigations are needed to evaluate the biological impact of long-term exposure to physiologically relevant concentrations of oxyphytosterols in humans.

8.2 Influence of refining on tocopherols

During deodorization, all tocopherols present in the bleached oil will be partitioned either in the deodorized oil or in the deodorizer distillate. A significant loss in the tocopherol mass balance in the range of 25%-35% was observed originating from technological and/or chemical origin.

The loss of tocopherols can be caused either by a thermal breakdown at temperatures higher than 240 °C, by oxidation reaction or by chemical reaction such as the formation of tocopheryl esters (Verleyen *et al.*, 2001a). Extensive analysis of vegetable oils by HPLC and comparison with synthesized tocopheryl esters did not show any adsorption in the elution region of tocopheryl esters, indicating that esters of tocopherols with fatty acids are not present in crude oils (Verleyen, 2001c). Therefore the stability of tocopherols during deodorization has been studied under various process conditions. The presence of oxidation products has no influence on the loss of tocopherols during deodorization based on the fact that two successive deodorization steps yielded identical loss of tocopherols.

Experiments using spiked triolein with 2000 ppm of α-tocopherol showed that the addition of tertbutylhydroquinone (TBHQ) as a strong antioxidant reduces the loss of tocopherols with more than 50% in comparison with the reference procedure. α- Tocopherol (2000 ppm) was dissolved in triolein and heated to 254 °C, 5-6 mbar, for 80 min, with no steam injection. 9% of tocopherol loss was observed in the control sample and 3% for the sample with 1500 ppm TBHQ. The more active TBHQ will compete with tocopherols to scavenge radicals and consequently the tocopherol loss in the mass balance is reduced as more natural tocopherols stay in the oil or in the distillate (Verleyen et al., 2002a, Verleyen et al., 2003).

In vegetable oils, the addition of TBHQ from 0 to 1500 ppm establishes a gradual reduction in tocopherol loss from 26.7% to 17.6% while the concentration of tocopherols in the distillate rises from 1.85% to 2.35%. Performing deodorization with nitrogen as stripping agent showed an important reduction in the tocopherol loss (Verleyen, 2002a). In the model study with triolein no reduction of α-tocopherol was observed while using corn oil a reduction of 30%-50% was observed. The highest reduction was detected at severe deodorization conditions (260 °C, 3 mbar) (Verleyen, 2002a). These experiments show that tocopherols are thermally stable compounds and probably the loss of tocopherols is due to oxidation reactions, which leads to compounds such as α- tocopherol dimer quinone, 4α, 5-epoxytocopherolquinone, 7, 8-epoxy tocopherol quinone, tocopherol dimer quinone, tocopherol spirotrimer and ditocopherol ethers (Verleyen, 2001a). These compounds can be found in the finished oil and in the distillate.

In a model experiment using 3500 ppm α- tocopherol in triolein and heating at 240 °C for 90 min at a reduced pressure of 6-7 mbar 4α, 5-epoxytocopherolquinone, 7, 8-epoxy tocopherolquinone and α-tocopherol quinone were identified as oxidation products supporting that the tocopherol loss during deodorization is mainly due to oxidative degradation (Verleyen, 2002a).

9. Nonfood applications

As commented previously, soybean deodorizer distillate represent good source of valuable compounds such as phytosterols, tocopherols and squalene, which can be recovered and further used as food additives, in pharmaceutical industry and cosmetics. Alternatively, deodorizer distillates have nonfood applications, such as biodiesel or can be used mixed with the fuel oil to fire the steam boilers (Svensson, 1976).

Refined vegetable oils are the predominant feedstocks used for the production of biodiesel. However, their relatively high cost renders the resulting fuels unable to compete with petroleum-derived fuel and makes the use of side-stream refining products (soapstock, acid oil and deodorizer distillate) important alternatives as a feedstock for biodiesel production.

Biodiesel is produced from deodorizer distillates by direct esterification of the FFA or by conversion of FFA to acylglycerols prior transesterification (Figure 2). Esterification of the FFA is also performed as a preliminary step in the purification of the tocopherols and sterols in order to reduce their boiling points, thereby facilitating their separation.

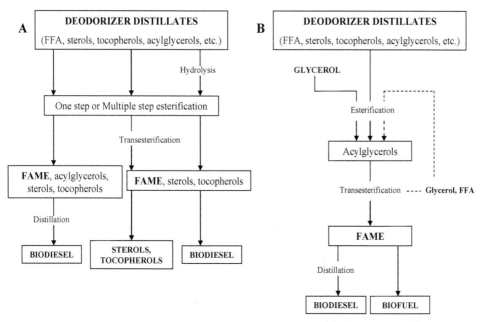

Fig. 2. Production of biodiesel from deodorizer distillates by direct conversion (A) and *via* acylglycerols route (B).

9.1 Production of biodiesel by direct conversion
9.1.1 Chemically catalyzed process

Verhé and coworkers (Verhé *et al.*, 2008) reported a process of converting the deodorizer distillates to biodiesel using methanol in a weight ratio 1:1 and 5 % w/w sulphuric acid as catalyst, at 75 °C for 5 h. Under the mentioned conditions, the FFA have undergone esterification while MAG reacted *via* transesterification, resulting in methyl esters. The crude biodiesel was further washed with 20 % water for 15 min, dried and distilled in order to increase the quality of the methyl esters. The distillation pitch was further processed for the recovery of sterols and tocopherols.

Facioli and Arellano (Facioli & Barrera-Arellano, 2002) described a process to obtain ethyl esters from SODD. SODD contained 47.5 % FFA C18:1, 26.2 % acylglycerols and 26.2 % unsaponifiable matter using concentrated sulphuric acid as catalyst. The optimum conditions found in this study were for EtOH:FFA between 6.4:1 to 11.2:1, H_2SO_4 from 0.9-1.5 % and reaction time from 1.3 h to 2.6 h. Under the described conditions a conversion of 94 % of the fatty acids to ethyl esters was achieved. Tocopherols losses were below 5.5 %. A molar excess of ethanol in relation to SODD:FFA was found to be necessary to obtain the best conversion.

Hammond and coworkers (Hammond & Tong, 2005) described a three-stage acid catalyzed esterification using a molar ratio acid oil:methanol:sulphuric acid of 1:1.3:0.03 for the first stage (25 h). The reaction mixture was centrifuged, the supernatant lipid phase was separated from the sludge (glycerol, water, acid and methanol), and further reacted with methanol and acid, keeping the previous mentioned ratios of unreacted lipid:methanol:sulphuric acid.

It was seen that the reaction proceeded rapidly during the first hour of reaction and then slowed down considerably. In contrast, the second and third stage showed a gradual increase in FAME over time. The maximum FAME conversion obtained for 12 tested acid oils averaged 81%. However, the ester phase could not be increased above 85% even after a fourth-stage reaction or if a base catalyst (sodium methoxide) was used in large excess. If higher amount of methanol was used, the initial reaction tended to go faster, but the reaction reached the plateau in a short time. Furthermore, an increase in the acid catalyst concentration above 1.2 % did not affect the initial reaction rate.

9.1.2 Enzymatically catalyzed process

Several enzymatic methods have been developed for the conversion of fatty acids into FAMEs or FAEEs with positive results. One of the main disadvantages of use of biocatalysts is the high price compared to chemical catalyst, although unfortunately, no rigorous economical viability of these enzymatic procedures has been reported.

Facioli and Arellano (Facioli & Barrera-Arellano, 2001) investigated the enzymatic esterification of the free fatty acids from SODD with ethanol using immobilized fungal lipase (Lipozyme IM) as biocatalyst. SODD contained 47.5 % FFA, 26.22 % neutral oil and 26.23 % unsaponifiable matter. The effect of three independent variables: temperature, enzyme concentration and EtOH:FFA molar ratio on the conversion rate of FFA to ethyl esters was studied. The best conversion (above 88 %) was obtained with EtOH:FFA ratio 1.7-3.2:1, temperature in the range 46.4 °C to 53.6 °C, lipase concentration from 10.7 to 23.0% and the reaction time of 2 h. All three variables had statistically significant effect on the conversion of the FFAs to ethyl esters. During the above mentioned esterification process no tocopherols losses were observed.

The esterification of SODD with butanol using *Mucor miehei* lipase as a biocatalyst and supercritical carbon dioxide (SC-CO$_2$) has been described by Nagesha and coworkers (Nagesha, 2004). The SODD contained 56.0 % neutral oil, 25.3 % FFA, 7.2 % sterols, 2.9 % tocopherols, 0.6 % hydrocarbons and 0.13 % moisture. It was preliminary filtered in order to remove sediments and sterols and enzymatic hydrolyzed to free fatty acid using immobilized lipase (*Candida rugosa*) in SC-CO$_2$ reactor unit. The operational conditions were as follows: pressure 160 bar, temperature 45 °C, moisture content 60 % (w/w) and enzyme concentration 200 U/g of SODD. Hydrolyzed SODD containing 87.8 % (w/w) FFA was further esterified for 3h in presence of butanol (1.2 M) using 15 % enzyme (w/w) (*M. miehei*), pressure 120 bar and temperature 35 °C. The maximum yield of 95 % FABE was achieved.

The high content of residual glycerides (3.10 %) present in the final FABE precluded its direct use as biodiesel. However, the process was designed as preliminary step for the purification of tocopherols, since hydrolysis/esterification helps their recovery.

Wang and coworkers (Wang et al., 2006) described a process for simultaneously conversion of FFA (28 %) and acylglycerols (60 %) from SODD to alkyl esters using a mixture of two enzymes (3 % Lipozyme TL IM and 2 % Novozym 435) in the presence of *tert*-butanol as co-solvent. It was found that the negative effects on the enzyme stability caused by the excessive methanol ratio and by-product glycerol could be minimized by using *tert*-butanol. The lipase activity remained stable after 120 cycles. The maximum yield of FAME (84 %) was achieved with an increase of *tert*-butanol content up to 80 % (based on the oil weight). However, a further increase of the solvent resulted in a decrease of the FAME yield which was explained by the dilution effect on reactants.

Du and coworkers (Du et al., 2007) investigated the enzymatic esterification of SODD containing 28 % FFA, 60 % TAG and 6 % tocopherols. The reaction was lipase mediated methanolysis using Novozym 435 as catalyst, at 40 °C in a solvent free medium. The enzyme kept its activity after being reused for 10 cycles, each cycle of 24 h. The highest biodiesel yield of 95 % was achieved by adding 10 fold of 3 Å molecular sieves (based on the maximum water produced from FFA esterification). The investigation of the lipase to methanol tolerance revealed that the lipase could maintain its stability and activity in the presence of even 3 molar concentration of methanol. This tolerance was attributed to the presence of other compounds apart from triglycerides, namely FFA, sterols and tocopherols. A linear relationship between the FFA content and the lipase tolerance to methanol was observed but the presence of sterols and tocopherols showed no effect. The correlation between the initial FFA present in the feedstock and the rate of conversion was confirmed by other authors (Hammond & Tong, 2005).

9.2 Production of biodiesel via acylglycerols route

Another approach reported in the literature consists on esterification of FFA with glycerol to form acylglycerols, followed by conventional transesterification. Synthesis of MAG from deodorizer distillate was mainly studied due to the large number of applications as additives (e.g. enhancing plasticity of fats) in the food, medicine and cosmetic industry. Among synthesized acylglycerols, the monoester has the highest surface activity and therefore, its concentration is very important for direct utilization of the reaction mixture as emulsifier.

Although the use of a large number of different heterogeneous catalysts have been reported in literature, most of the research has been done on the synthetic samples and less on the side stream refining products. Different studies summarized hereunder describe processes for synthesis of acylglycerols as an intermediate step in the production of biodiesel/biofuels. These processes are catalyzed either chemically or enzymatically, or conducted under non-catalytic conditions.

9.2.1 Enzymatically catalyzed process

Tangkam and coworkers (Tangkam et al., 2008) described the enzymatic esterification in a solvent free medium of different deodorizer distillates resulting from the refining of various vegetable oils. A direct esterification of mixed distillates (61 % FFA and 39 % acylglycerols) with glycerol using immobilized lipase B from *Candida Antarctica* (Novozym 435) led to moderate proportions (46 %) of DAG. Application of a two-stage reaction consisting of a hydrolysis step of deodorizer distillate to increase the FFA content followed by esterification with glycerol led to a higher formation (>61 %) of DAG. Furthermore, it was observed that the high initial concentration of free fatty acids in the distillate has a positive influence on the concentration of DAG in the final product (>71 %). This observation is consistent with other literature data (Yamada et al., 1999). Enrichment of DAG in the final products by short-path vacuum distillation led to concentrates containing up to 94 % DAG, ~ 5 % TAG and no unesterified fatty acids and MAG.

9.2.2 Non-catalytic process

Smet (Smet, 2008) described a process for the esterification of fatty acid distillate (93 % FFA) with technical grade glycerol. The reaction was carried out in a high pressure Parr reactor (stirred and thermostated reactor of stainless steal). The following parameters have been checked: temperature, reactor design, agitation speed, molar ratio and influence of the catalyst. The best results were obtained at 200 °C, pressure of 90 mbar and agitation speed of 60 rpm. It was seen that by using a molar ratio 1:1 FFA:glycerol, a total glycerides content of 85.3 % was obtained within 345 min reaction time. The formation of MAG was faster in the first hours and than reached the plateau, while the formation of DAG was slower at the beginning of the reaction and faster at the end. Furthermore, an increase in the molar ratio of 1:2 FFA:glycerol slow down the reaction, the total glycerides content reaching 64.9 % within 345 min reaction time. A molar ratio 2:1 FFA:glycerol gave an increase of the MAG and DAG at the beginning of the reaction, followed by an decrease of MAG after 90 min, the glycerol being completely consumed in within 345 min reaction time. The percent of DAG and TAG increased gradually during reaction, reaching a final yield of 86.2 % of total acylglycerols.

However the FFA content was still high, a distillation step of the residual FFAs and glycerol was necessary in order to increase the purity of the synthesized acylglycerols. The by-products of distillation were further re-used as reaction products in the synthesis of acylglycerols. The novelty of the process consists in synthesizing acylglycerols in a relatively short time (<6 h) in a catalyst free medium.

10. Concluding remarks

Deodorizer distillate is an excellent source of valuable compounds such as phytosterols and tocopherols. Numerous procedures have been described to isolate bioactive compounds from soybean oil deodorizer distillates to improve the value and the quality of this by-product. All these procedures can be grouped in four generic categories: classic methods such as crystallization and precipitation, chemical and enzymatic modification, molecular distillation, and supercritical fluid extraction.

Crystallization seems successful as a simple and efficient process to remove and concentrate sterols and tocopherols from SODD. However solvent based processes are expensive, unattractive and less environmentally friendly, resulting in a scarce and expensive final product.

To increase the separation efficiency of the compounds of interest from SODD, esterification and/or transesterification reactions are usually carried out prior to the purification or fractionation procedure. Hence, the utilization of enzymes, for instance, makes easier the separation of tocopherols from SODD by converting sterols to steryl esters, acylglycerols to free fatty acids and free fatty acids to fatty acid methyl or ethyl esters (FAMEs or FAEEs). Then, it is easier to separate the new product mixture by distillation or supercritical fluid extraction. The main difficulties of the enzymatic processes are the numerous parameters involved such as moisture content, enzyme concentration, time, temperature, ratio of the reactants, stability, recovery and reutilization of the enzyme preparation, among others. However, it is possible to separate the sterol esterification and ethyl esterification in time or space to optimize each of these reactions independently, thereby minimizing costs or improving the yield of the desired final reaction products.

Among the great variety of processes that have been patented for the purification of the compounds of the SODD, only the processes of esterification of fatty acids and acylglycerols with methanol or ethanol followed by high vacuum distillation, have been developed on a commercial scale for the concentration of tocopherols

Regarding supercritical fluid fractionation SODD is not adequate feed material to work with SC-CO_2 for tocopherol enrichment, owing to its poor SC-CO_2 solubility. So, to concentrate tocopherols from SODD, pre-treatment of the raw material is needed to obtain the primary tocopherols concentrate with improve solubility in SC-CO_2.

Alternatively, deodorizer distillates have also non-food applications, such as biodiesel or can be used mixed with the fuel oil to fire the steam boilers. The use of deodorizer distillate instead of refined vegetable oils is an important alternative as a feedstock for biodiesel production.

Two main degradation products from sterols can be observed in SODD, namely dehydration and oxidation products. The degree of sterol dehydration is mainly influenced by deodorization temperature giving rise to a variable concentration of steradienes in the distillate. The content of oxidized sterols in deodorization distillate fractions from edible oil refining processes fluctuates depending on both the temperature applied during vacuum distillation and the breakdown and transformation of oxidized sterols into other unidentified degradation products. Finally, formation of oxidized sterols can be partially prevented by the high amounts of natural antioxidants in acid oil obtained from physical refining distillate. The current knowledge on the possible biological effects of oxyphytosterols is limited and further research to clarify the possible impact of oxyphytosterols on human health is needed.

11. References

Andrews, J. S. (1941). Process for preparation of vitamin e concentrate and antioxidant. U. S. Patent 2.263.550.

Awad, A. B., Roy, R., & Fink, C. S. (2003). Beta-sitosterol, a plant sterol, induces apoptosis and activates key caspases in MDA-MB-231 human breast cancer cells. *Oncol Rep.* Vol. 10, No. 2, pp. 497-500.

Balazs, I. (1987). Refining and use of byproducts from various fats and oils. *Journal of the American Oil Chemists' Society.* Vol. 64, No. 8, pp. 1126-1128.

Batistella, C. B., Moraes, E. B., Maciel Filho, R., & Maciel, M. R. (2002). Molecular distillation: rigorous modeling and simulation for recovering vitamin E from vegetal oils. *Appl. Biochem. Biotechnol.* Vol. 98-100, No. pp. 1187-206.

Batistella, C. B. & Wolf Maciel, M. R. (1998). Recovery of carotenoids from palm oil by molecular distillation. *Computers & Chemical Engineering.* Vol. 22, No. Supplement 1, pp. S53-S60.

Bjorkhem, I., Meaney, S., & Diczfalusy, U. (2002). Oxysterols in human circulation: which role do they have? *Current Opinion in Lipidology.* Vol. 13, No. 3, pp. 247-253.

Block, G. & Langseth, L. (1994). Antioxidant vitamins and disease prevention. *Food Technology.* Vol. 48, No. 7, pp. 80-84.

Bockisch, M. (1998). *Fats and oils handbook.* (AOCS, Champaign, IL.

Bondioli, P., Mariani, C., Lanzani, A., Fedeli, E., & Muller, A. (1993). Squalene recovery from olive oil deodorizer distillates. *Journal of the American Oil Chemists' Society.* Vol. 70, No. 8, pp. 763-766.

Bouic, P. J. D. (2001). The role of phytosterols and phytosterolins in immune modulation: a review of the past 10 years. *Current Opinion in Clinical Nutrition & Metabolic Care.* Vol. 4, No. 6, pp. 471-475.

Breivik, H., Haraldsson, G., & Kristinsson, B. (1997). Preparation of highly purified concentrates of eicosapentaenoic acid and docosahexaenoic acid. *Journal of the American Oil Chemists' Society.* Vol. 74, No. 11, pp. 1425-1429.

Brown, W. & Meag, K. H. (1963). Process for recovery of tocopherols and sterols. U.S. Patent 3,108,120.

Brown, W. & Smith, F. E. (1964). Process for separating tocopherols and sterols from deodorizer sludge and the like. U.S. Patent 3,153,055.

Brunner, G. (1994a). *Gas Extraction.* Springer: New York.

Brunner, G. (1994b). *Gas Extraction-An Introduction to Fundamentals of Supercritical Fluid and the Application to Separation Processes.* Springer, Berlin

Brunner, G., Malchow, T., Stürken, K., & Gottschau, T. (1991). Separation of tocopherols from deodorizer condensates by countercurrent extraction with carbon dioxide. *The Journal of Supercritical Fluids.* Vol. 4, No. 1, pp. 72-80.

Buczenko, G. M., de Oliveira, J. S., & von Meien, O. F. (2003). Extraction of tocopherols from the deodorized distillate of soybean oil with liquefied petroleum gas. *European Journal of Lipid Science and Technology.* Vol. 105, No. 11, pp. 668-671.

Cvengros, J., Lutisan, J., & Micov, M. (2000). Feed temperature influence on the efficiency of a molecular evaporator. *Chemical Engineering Journal.* Vol. 78, No. 1, pp. 61-67.

Czuppon, T., Kemeny, Z., Kovari, E., & Recseg, K. (2003). Process for recovery of plant sterols from by-product of vegetable oil refining. WO2004000979.

Chang, C. J., Chang, Y.-F., Lee, H.-z., Lin, J.-q., & Yang, P.-W. (2000). Supercritical Carbon Dioxide Extraction of High-Value Substances from Soybean Oil Deodorizer Distillate. *Industrial & Engineering Chemistry Research.* Vol. 39, No. 12, pp. 4521-4525.

Chu, B. S., Baharin, B. S., & Quek, S. Y. (2002). Factors affecting pre-concentration of tocopherols and tocotrienols from palm fatty acid distillate by lipase-catalysed hydrolysis. *Food Chemistry.* Vol. 79, No. 1, pp. 55-59.

de Jong, A., Plat, J., & Mensink, R. P. (2003). Metabolic effects of plant sterols and stanols (Review). *The Journal of Nutritional Biochemistry.* Vol. 14, No. 7, pp. 362-369.

de Lucas, A., Martinez de la Ossa, E., Rincón, J., Blanco, M. A., & Gracia, I. (2002). Supercritical fluid extraction of tocopherol concentrates from olive tree leaves. *The Journal of Supercritical Fluids.* Vol. 22, No. 3, pp. 221-228.

Dieter, L. (2004). Sterol autoxidation: from phytosterols to oxyphytosterols. *British Journal of Nutrition.* Vol. 91, No. pp. 3-4.

Donova, M. (2007). Transformation of steroids by actinobacteria: A review. *Applied Biochemistry and Microbiology.* Vol. 43, No. 1, pp. 1-14.

Dowd, M. K. (1998). Gas chromatographic characterization of soapstocks from vegetable oil refining. *Journal of Chromatography A.* Vol. 816, No. 2, pp. 185-193.

Du, W., Wang, L., & Liu, D. (2007). Improved methanol tolerance during Novozym435-mediated methanolysis of SODD for biodiesel production. *Green Chemistry.* Vol. 9, No. 2, pp. 173-176.

Dumont, M.-J. & Narine, S. S. (2007). Soapstock and deodorizer distillates from North American vegetable oils: Review on their characterization, extraction and utilization. *Food Research International.* Vol. 40, No. 8, pp. 957-974.

Dutta, P. C., Przybylski, R., & Eskin, M. (2006). Formation, analysis and health effects of oxidized sterols in frying fat, In: *Deep frying: chemistry, nutrition and practical applications,* M. D. Erickson, Editor, pp. 111-164, AOCS Press, Urbana, IL.

Facioli, N. & Barrera-Arellano, D. (2002). Optimization of direct acid esterification process of soybean oil deodorizer distillate. *Grasas y Aceites.* Vol. 53, No. 2, pp. 206-212.

Facioli, N. L. & Barrera-Arellano, D. (2001). Optimisation of enzymatic esterification of soybean oil deodoriser distillate. *Journal of the Science of Food and Agriculture.* Vol. 81, No. 12, pp. 1193-1198.

Fang, T., Goto, M., Sasaki, M., & Hirose, T. (2005). Phase Equilibria for the Ternary System Methyl Oleate + Tocopherol + Supercritical CO_2. *Journal of Chemical & Engineering Data.* Vol. 50, No. 2, pp. 390-397.

Fang, T., Goto, M., Wang, X., Ding, X., Geng, J., Sasaki, M., & Hirose, T. (2007). Separation of natural tocopherols from soybean oil byproduct with supercritical carbon dioxide. *The Journal of Supercritical Fluids.* Vol. 40, No. 1, pp. 50-58.

Fernandes, P. & Cabral, J. M. S. (2007). Phytosterols: Applications and recovery methods. *Bioresource Technology.* Vol. 98, No. 12, pp. 2335-2350.

Ferrari, R., Schulte, E., Esteves, W., Brühl, L., & Mukherjee, K. (1996). Minor constituents of vegetable oils during industrial processing. *Journal of the American Oil Chemists' Society.* Vol. 73, No. 5, pp. 587-592.

Fizet, C. (1996). Process for tocopherols and sterols from natural sources. U.S. Patent 5,487,817.

Francesc, G. (2004). Phytosterol oxidation products: state of the art. *Reprod. Nutr. Dev.* Vol. 44, No. 6, pp. 597-598.

Gangopadhyay, S., Nandi, S., & Ghosh, S. (2007). Biooxidation of Fatty Acid Distillates to Dibasic Acids by a Mutant of Candida tropicalis. *Journal of Oleo Science.* Vol. 56, No. 1, pp. 13-17.

Gapor, A., Leong, L. W., & Ong, A. (1989). Production of high concentration tocopherols and tocotrienols from palm oil by-products. European Patent Application 0,3 33,472442.

Gast, K., Jungfer, M., Saure, C., & Brunner, G. (2005). Purification of tocochromanols from edible oil. *The Journal of Supercritical Fluids.* Vol. 34, No. 1, pp. 17-25.

Gavin, A. (1978). Edible oil deodorization. *Journal of the American Oil Chemists' Society.* Vol. 55, No. 11, pp. 783-791.

Ghosh, S. & Bhattacharyya, D. (1996). Isolation of tocopherol and sterol concentrate from sunflower oil deodorizer distillate. *Journal of the American Oil Chemists' Society.* Vol. 73, No. 10, pp. 1271-1274.

Grob, K., Biedermann, M., Bronz, M., & Giuffré, A. M. (1994). The Detection of Adulteration with Desterolized Oils. *Lipid / Fett.* Vol. 96, No. 9, pp. 341-345.

Gunawan, S., Ismadji, S., & Ju, Y.-H. (2008a). Design and operation of a modified silica gel column chromatography. *Journal of the Chinese Institute of Chemical Engineers.* Vol. 39, No. 6, pp. 625-633.

Gunawan, S., Kasim, N. S., & Ju, Y.-H. (2008b). Separation and purification of squalene from soybean oil deodorizer distillate. *Separation and Purification Technology*. Vol. 60, No. 2, pp. 128-135.

Gutfinger, T. & Letan, A. (1974). Quantitative changes in some unsaponifiable components of soya bean oil due to refining. *Journal of the Science of Food and Agriculture*. Vol. 25, No. 9, pp. 1143-1147.

Hammond, E. G. & Tong, W. (2005). Method of converting free fatty acids to fatty acid methyl esters with small excess of methanol. U.S. Patent 6965044.

Haraldsson, G. (1992). *The application of lipases in organic synthesis*. (John Wiley,

Hedström, G., Slotte, J. P., Backlund, M., Molander, O., & Rosenholm, J. B. (1992). Lipase-Catalyzed Synthesis and Hydrolysis of Cholesterol Oleate in Aot/Isooctane Microemulsions. *Biocatalysis and Biotransformation*. Vol. 6, No. 4, pp. 281 - 290.

Hickman, K. C. D. (1944). Purification of sludges, scums, and the like to prepare relatively purified tocopherol. U. S. Patent 2.349.270

Hirota, Y., Nagao, T., Watanabe, Y., Suenaga, M., Nakai, S., Kitano, A., Sugihara, A., & Shimada, Y. (2003). Purification of steryl esters from soybean oil deodorizer distillate. *Journal of the American Oil Chemists' Society*. Vol. 80, No. 4, pp. 341-346.

Homberg, E. & Bielefeld, B. (1982). Free and bound sterols in crude and refined palm oils. . *Fette Seifen Anstrichm*. Vol. 84, No. pp. 141-146.

Hovenkamp, E., Demonty, I., Plat, J., Lütjohann, D., Mensink, R. P., & Trautwein, E. A. (2008). Biological effects of oxidized phytosterols: A review of the current knowledge. *Progress in Lipid Research*. Vol. 47, No. 1, pp. 37-49.

Hunt, T. K., Jeromin, L., Johannisbauer, W., Gutsche, B., Jordon, V., & Wogatzki, H. (1997). Recovery of tocopherols. U.S. Patent 5,646,3 1 1.

Ito, V., Martins, P., Batistella, C., Filho, R., & Wolf Maciel, M. (2006). Natural compounds obtained through centrifugal molecular distillation. *Applied Biochemistry and Biotechnology*. Vol. 131, No. 1, pp. 716-726.

Jacobs, L. (2005). Process for the Production of Tocotrienols. U.S. Patent 6,838,104.

Jonzo, M. D., Hiol, A., Druet, D., & Comeau, L. C. (1997). Application of Immobilized Lipase from *Candida rugosa* to Synthesis of Cholesterol Oleate. *Journal of Chemical Technology & Biotechnology*. Vol. 69, No. 4, pp. 463-469.

Jung, M., Yoon, S., & Min, D. (1989). Effects of processing steps on the contents of minor compounds and oxidation of soybean oil. *Journal of the American Oil Chemists' Society*. Vol. 66, No. 1, pp. 118-120.

Jung, M. Y. & Min, D. B. (1990). Effects of α-, γ-, and δ-Tocopherols on Oxidative Stability of Soybean Oil. *Journal of Food Science*. Vol. 55, No. 5, pp. 1464-1465.

Kasim, N. S., Gunawan, S., & Ju, Y.-H. (2009). Isolation and identification of steroidal hydrocarbons in soybean oil deodorizer distillate. *Food Chemistry*. Vol. 117, No. 1, pp. 15-19.

Khatoon, S., Raja Rajan, R., & Gopala Krishna, A. (2010). Physicochemical Characteristics and Composition of Indian Soybean Oil Deodorizer Distillate and the Recovery of Phytosterols. *Journal of the American Oil Chemists' Society*. Vol. 87, No. 3, pp. 321-326.

Kijima, S., Ichikana, N., & Naito, K. (1964). Purification process of tocopherol containing materials. U.S. Patent 3,122,565.

Kim, S. K. & Rhee, J. S. (1982). Isolation and purification of tocopherols and sterols from distillates of soy oil deodorization. *Kor. J. Food Sci. Technol.* Vol. 14, No. pp. 174-178.

King, J. W., Favati, F., & Taylor, S. L. (1996). Production of Tocopherol Concentrates by Supercritical Fluid Extraction and Chromatography. *Separation Science and Technology.* Vol. 31, No. 13, pp. 1843 - 1857.

Kline, K., Lawson, K. A., Yu, W., & Sanders, B. G. (2007). Vitamin E and Cancer, in *Vitamins and Hormones.* pp. 435-461.

Kochhar, S. P. (1983). Influence of processing on sterols of edible vegetable oils. *Progress in Lipid Research.* Vol. 22, No. 3, pp. 161-188.

Kritchevsky, D. & Chen, S. C. (2005). Phytosterols--health benefits and potential concerns: a review. *Nutrition Research.* Vol. 25, No. 5, pp. 413-428.

Lee, H., Chung, B., & Park, Y. (1991). Concentration of tocopherols from soybean sludge by supercritical carbon dioxide. *Journal of the American Oil Chemists' Society.* Vol. 68, No. 8, pp. 571-573.

Lin, K.-M. & Koseoglu, S. S. (2003). Separation of sterols from deodorizer distillate by crystallization. *Journal of Food Lipids.* Vol. 10, No. 2, pp. 107-127.

Lin, K.-M., Zhang, X., & Koseoglu, S. S. (2004). Separation of tocopherol succinates from deodorizer distillate. *Journal of Food Lipids.* Vol. 11, No. 1, pp. 29-43.

Lin, K. M. (2002). National Chung Hsing University, M.S., Texas A& M University. No.

Lutisan, J., Cvengros, J., & Micov, M. (2002). Heat and mass transfer in the evaporating film of a molecular evaporator. *Chemical Engineering Journal.* Vol. 85, No. 2-3, pp. 225-234.

Marangoni, A. G. & Rousseau, D. (1995). Engineering triacylglycerols: The role of interesterification. *Trends in Food Science & Technology.* Vol. 6, No. 10, pp. 329-335.

Martins, P. F., Batistella, C. s. B., Maciel-Filho, R., & Wolf-Maciel, M. R. (2005). Comparison of Two Different Strategies for Tocopherols Enrichment Using a Molecular Distillation Process. *Industrial & Engineering Chemistry Research.* Vol. 45, No. 2, pp. 753-758.

Martins, P. F., Ito, V. M., Batistella, C. B., & Maciel, M. R. W. (2006). Free fatty acid separation from vegetable oil deodorizer distillate using molecular distillation process. *Separation and Purification Technology.* Vol. 48, No. 1, pp. 78-84.

Maza, A. (1992). Process for separating mixed fatty acids from deodorant distillate using urea. U.S. Patent 5078920.

Mendes, M. F., Pessoa, F. L. P., Coelho, G. V., & Uller, A. M. C. (2005). Recovery of the high aggregated compounds present in the deodorizer distillate of the vegetable oils using supercritical fluids. *The Journal of Supercritical Fluids.* Vol. 34, No. 2, pp. 157-162.

Mendes, M. F., Pessoa, F. L. P., & Uller, A. M. C. (2002). An economic evaluation based on an experimental study of the vitamin E concentration present in deodorizer distillate of soybean oil using supercritical CO_2. *The Journal of Supercritical Fluids*. Vol. 23, No. 3, pp. 257-265.

Mendes, M. F., Uller, A. M. C., & Pessoa, F. L. P. (2000). Simulation and thermodynamic modeling of the extraction of tocopherol from a synthetic mixture of tocopherol, squalene and CO_2. *Brazilian Journal of Chemical Engineering*. Vol. 17, No. pp. 761-770.

Moghadasian, M. H., McManus, B. M., Godin, D. V., Rodrigues, B., & Frohlich, J. J. (1999). Proatherogenic and Antiatherogenic Effects of Probucol and Phytosterols in Apolipoprotein E–Deficient Mice : Possible Mechanisms of Action. *Circulation*. Vol. 99, No. 13, pp. 1733-1739.

Moghadasian, M. H., McManus, B. M., Pritchard, P. H., & Frohlich, J. J. (1997). "Tall Oil"–Derived Phytosterols Reduce Atherosclerosis in ApoE-Deficient Mice. *Arterioscler Thromb Vasc Biol*. Vol. 17, No. 1, pp. 119-126.

Moreira, E. & Baltanás, M. (2004). Recovery of phytosterols from sunflower oil deodorizer distillates. *Journal of the American Oil Chemists' Society*. Vol. 81, No. 2, pp. 161-167.

Mulligan, J. D., Flowers, M. T., Tebon, A., Bitgood, J. J., Wellington, C., Hayden, M. R., & Attie, A. D. (2003). ABCA1 Is Essential for Efficient Basolateral Cholesterol Efflux during the Absorption of Dietary Cholesterol in Chickens. *Journal of Biological Chemistry*. Vol. 278, No. 15, pp. 13356-13366.

Munteanu, A. & Zingg, J. M. (2007). Cellular, molecular and clinical aspects of vitamin E on atherosclerosis prevention. *Molecular Aspects of Medicine*. Vol. 28, No. 5-6, pp. 538-590.

Nagao, T., Kobayashi, T., Hirota, Y., Kitano, M., Kishimoto, N., Fujita, T., Watanabe, Y., & Shimada, Y. (2005). Improvement of a process for purification of tocopherols and sterols from soybean oil deodorizer distillate. *Journal of Molecular Catalysis B: Enzymatic*. Vol. 37, No. 1-6, pp. 56-62.

Nagesha, G., Subramanian, R., & Sankar, K. (2003). Processing of tocopherol and FA systems using a nonporous denser polymeric membrane. *Journal of the American Oil Chemists' Society*. Vol. 80, No. 4, pp. 397-402.

Nagesha, G. K., Manohar, B., & Udaya Sankar, K. (2003). Enrichment of tocopherols in modified soy deodorizer distillate using supercritical carbon dioxide extraction. *European Food Research and Technology*. Vol. 217, No. 5, pp. 427-433.

Nagesha, G. K., Manohar, B., & Udaya Sankar, K. (2004). Enzymatic esterification of free fatty acids of hydrolyzed soy deodorizer distillate in supercritical carbon dioxide. *The Journal of Supercritical Fluids*. Vol. 32, No. 1-3, pp. 137-145.

Nergiz, C. & Çelikkale, D. (2010). The effect of consecutive steps of refining on squalene content of vegetable oils. *Journal of Food Science and Technology*. Vol. No. pp. 1-4.

Norris, F. A. (1979). Handling, Storage and Grading of Oils and Oil Bearing Materials, In: *Bailey's Industrial Oil and Fat Products*, D. Swern, Editor, pp. 601-635, Interscience Publishers, New York.

Plat, J. & Mensink, R. P. (2005). Plant Stanol and Sterol Esters in the Control of Blood Cholesterol Levels: Mechanism and Safety Aspects. *The American Journal of Cardiology.* Vol. 96, No. 1, Supplement 1, pp. 15-22.

Quílez, J., García-Lorda, P., & Salas-Salvado, J. (2003). Potential uses and benefits of phytosterols in diet: present situation and future directions. *Clinical nutrition (Edinburgh, Scotland).* Vol. 22, No. 4, pp. 343-351.

Ramamurthi, S., Bhirud, P. R., & McCurdy, A. R. (1991). Enzymatic methylation of canola oil deodorizer distillate. *JAOCS, Journal of the American Oil Chemists' Society.* Vol. 68, No. (12), pp. 970-975.

Ramamurthi, S. & McCurdy, A. (1993). Enzymatic pretreatment of deodorizer distillate for concentration of sterols and tocopherols. *Journal of the American Oil Chemists' Society.* Vol. 70, No. 3, pp. 287-295.

Rimm, E. B., Stampfer, M. J., Ascherio, A., Giovannucci, E., Colditz, G. A., & Willett, W. C. (1993). Vitamin E Consumption and the Risk of Coronary Heart Disease in Men. *New England Journal of Medicine.* Vol. 328, No. 20, pp. 1450-1456.

Rohr, R. & Trujillo-Quijano, J. A. (2002). Process for Extraction and Concentration of Liposoluble Vitamins, Growth Factors and Animal and Vegetable Hormones from Residues and By-products of Industrialized Animal and Vegetable Products. U.S. Patent No 6,344,573.

Rohr, R. & Trujillo-Quijano, J. A. (2005). Process for separating unsaponifiable valuable products from raw materials. U.S. Patent No. 846941.

Rozner, S., Garti, N. (2006). The activity and absorption relationship of cholesterol and phytosterols. *Colloids and Surfaces A: Physicochemical and Engineering Aspects* Vol. No. pp. 435–456.

Sheabar, F. Z. & Neeman, I. (1987). Concentration of tocopherols from soy oil deodorizer scum. *La Rivista Italiana Delle Sostanze Grasse ZXZV.* Vol. 2, No. pp. 19-222.

Shimada, Y., Hirota, Y., Baba, T., Sugihara, A., Moriyama, S., Tominaga, Y., & Terai, T. (1999). Enzymatic synthesis of steryl esters of polyunsaturated fatty acids. *JAOCS, Journal of the American Oil Chemists' Society.* Vol. 76, No. 6, pp. 713-716.

Shimada, Y., Nakai, S., Suenaga, M., Sugihara, A., Kitano, M., & Tominaga, Y. (2000). Facile purification of tocopherols from soybean oil deodorizer distillate in high yield using lipase. *JAOCS, Journal of the American Oil Chemists' Society.* Vol. 77, No. 10, pp. 1009-1013.

Shishikura, A., Fujimoto, K., Kaneda, T., Arai, K., & Saito, S. (1988). Concentration of Tocopherols from Soybean Sludge by Supercritical Fluid Extraction. *J. Jpn. Oil Chem. Soc.* Vol. 37, No. pp. 8-12.

Smet, P. (2008). Valorisatie van vetzuurdestillaten als biobrandstof door herverestering met glycerol. Master thesis: 1-68.

Smith Frank, E. (1967). Separation of tocopherols and sterols from deodorizer sludge and the like. U. S. Patent 3.335.154.

Strocchi, A. & Marascio, G. (1993). Structural Modifications of 4,4'-Dimethyl Sterols during the Hydrogenation of Edible Vegetable Oils. *Lipid / Fett.* Vol. 95, No. 8, pp. 293-299.

Su-Min, J., Yong-Hun, P., Bong-Hyon, J., Hyon-Ho, S., & Jong-Dok, S. (1992). Method of separation and concentration of sterol and tocopherol from distillated soybean powder. KR Patent No 9,205,695.

Sun, H., Wiesenborn, D., Tostenson, K., Gillespie, J., & Rayas-Duarte, P. (1997). Fractionation of squalene from amaranth seed oil. *Journal of the American Oil Chemists' Society.* Vol. 74, No. 4, pp. 413-418.

Svensson, C. (1976). Use or disposal of by-products and spent material from the vegetable oil processing industry in Europe. *Journal of the American Oil Chemists' Society.* Vol. 53, No. 6, pp. 443-445.

Swern, D. (1986). *Bailey's Industrial Oils and Fats.* (John Wiley & Sons, New York.

Szelag, H. & Zwierzykowski, W. (1983). The Application of Molecular Distillation to Obtain High Concentration of Monoglycerides. *Fette, Seifen, Anstrichmittel.* Vol. 85, No. 11, pp. 443-446.

Takagi, Y. & Kai, Y. (1984). Process for preparation of tocopherol concentrates. U.S. Patent 4454329.

Tangkam, K., Weber, N., & Wiege, B. (2008). Solvent-free lipase-catalyzed preparation of diglycerides from co-products of vegetable oil refining. *Grasas y Aceites.* Vol. 59, No. 3, pp. 245-253.

Torres, C. F., Fornari, T., Torrelo, G., Señoráns, F. J., & Reglero, G. (2009). Production of phytosterol esters from soybean oil deodorizer distillates. *European Journal of Lipid Science and Technology.* Vol. 111, No. 5, pp. 459-463.

Torres, C. F., Torrelo, G., Señorans, F. J., & Reglero, G. (2007). A two steps enzymatic procedure to obtain sterol esters, tocopherols and fatty acid ethyl esters from soybean oil deodorizer distillate. *Process Biochemistry.* Vol. 42, No. 9, pp. 1335-1341.

Trautwein, E. A., Duchateau, G. S. M. J. E., Lin, Y., Mel'nikov, S. M., Molhuizen, H. O. F., & Ntanios, F. Y. (2003). Proposed mechanisms of cholesterol-lowering action of plant sterols. *European Journal of Lipid Science and Technology.* Vol. 105, No. 3-4, pp. 171-185.

Ubhayasekera, S. & Dutta, P. (2009). Sterols and Oxidized Sterols in Feed Ingredients Obtained from Chemical and Physical Refining Processes of Fats and Oils. *Journal of the American Oil Chemists' Society.* Vol. 86, No. 6, pp. 595-604.

van Rensburg, S. J., Daniels, W. M. U., van Zyl, J. M., & Taljaard, J. J. F. (2000). A Comparative Study of the Effects of Cholesterol, Beta-Sitosterol, Beta-Sitosterol Glucoside, Dehydro-epiandrosterone Sulphate and Melatonin on In Vitro Lipid Peroxidation. *Metabolic Brain Disease.* Vol. 15, No. 4, pp. 257-265.

Verhé, R., Van Hoed, V., Echim, C., Stevens, C., De Greyt, W., & Kellens, M. (2008). *Production of Biofuel from Lipids and Alternative Resources.* (John Wiley & Sons, Inc., 9780470385869)

Verleyen, T., Kamal-eldin, A., Dobarganes, C., Verhé, R., Dewettinck, K., & Huyghebaert, A. (2001a). Modelling of alphatocopherol loss and oxidation products formed during thermoxidation in triolein and tripalmitin mixtures. *Lipids in Health and Disease.* Vol. 36, No. 7, pp. 719-726.

Verleyen, T., Kamal-Eldin, A., Mozuraityte, R., Verhé, R., Dewettinck, K., Huyghebaert, A., & De Greyt, W. (2002a). Oxidation at elevated temperatures: competition between α-tocopherol and unsaturated triacylglycerols. *European Journal of Lipid Science and Technology.* Vol. 104, No. 4, pp. 228-233.

Verleyen, T., Sosinska, U., Ioannidou, S., Verhe, R., Dewettinck, K., Huyghebaert, A., & De Greyt, W. (2002b). Influence of the vegetable oil refining process on free and esterified sterols. *Journal of the American Oil Chemists' Society.* Vol. 79, No. 10, pp. 947-953.

Verleyen, T., Verhe, R., Cano, A., Huyghebaert, A., & De Greyt, W. (2001b). Influence of triacylglycerol characteristics on the determination of free fatty acids in vegetable oils by Fourier transform infrared spectroscopy. *J. Amer. Oil Chem. Soc.* Vol. 78, No. 10, pp. 981-984.

Verleyen, T., Verhe, R., Garcia, L., Dewettinck, K., Huyghebaert, A., & De Greyt, W. (2001c). Gas chromatographic characterization of vegetable oil deodorization distillate. *Journal of Chromatography A.* Vol. 921, No. 2, pp. 277-285.

Verleyen, T., Verhé, R., & Kamal-eldin, A. (2003). Competitive oxidation between alfa tocopherol and unsaturated fatty acids under thermoxidation conditions. *Lipid Oxidation Pathways.* Vol. No. pp. 70-84.

Wang, H., Goto, M., Sasaki, M., & Hirose, T. (2004). Separation of α-Tocopherol and Squalene by Pressure Swing Adsorption in Supercritical Carbon Dioxide. *Industrial & Engineering Chemistry Research.* Vol. 43, No. 11, pp. 2753-2758.

Wang, L., Du, W., Liu, D., Li, L., & Dai, N. (2006). Lipase-catalyzed biodiesel production from soybean oil deodorizer distillate with absorbent present in tert-butanol system. *Journal of Molecular Catalysis B: Enzymatic.* Vol. 43, No. 1-4, pp. 29-32.

Warner, K. (2005). Effects on the Flavor and Oxidative Stability of Stripped Soybean and Sunflower Oils with Added Pure Tocopherols. *Journal of Agricultural and Food Chemistry.* Vol. 53, No. 26, pp. 9906-9910.

Watanabe, Y., Nagao, T., Hirota, Y., Kitano, M., & Shimada, Y. (2004). Purification of tocopherols and phytosterols by a two-step in situ enzymatic reaction. *JAOCS, Journal of the American Oil Chemists' Society.* Vol. 81, No. 4, pp. 339-345.

Weber, N., Weitkamp, P., & Mukherjee, K. D. (2001). Fatty acid steryl, stanyl, and steroid esters by esterification and transesterification in vacuo using Candida rugosa lipase as catalyst. *Journal of Agricultural and Food Chemistry.* Vol. 49, No. 1, pp. 67-71.

Weber, N., Weitkamp, P., & Mukherjee, K. D. (2002). Cholesterol-lowering food additives: Lipase-catalysed preparation of phytosterol and phytostanol esters. *Food Research International.* Vol. 35, No. 2-3, pp. 177-181.

Xu, Y., Shi, X., Du, X., Xing, M., Xu, T., Meng, J., & Feng, Z. (2005). A method to extract natural vitamin E from by-product of refined vegetable. CN Patent ZL 200510114851.X.

Yamada, Y., Shimizu, M., Sugiura, M., & Yamada, N. (1999). Process for producing diglycerides. WO 1999/09119.

Yang, H., Yan, F., Wu, D., Huo, M., Li, J., Cao, Y., & Jiang, Y. (2009). Recovery of phytosterols from waste residue of soybean oil deodorizer distillate. *Bioresource Technology.* Vol. 101, No. 5, pp. 1471-1476.

Yong-Bo, H., Kim, N. H., Park, Y. H., Jong, B. H., Shin, H. H., & Son, J. D. (1994). Method of Separation and Purification of Natural Tocopherol for Deodorized Sludge of Soybean oil. KR Patent No 9,402,715.

Zhao, Y., Sheng, G., & Wang, D. (2000). Pilot-scale isolation of tocopherols and phytosterols from soybean sludge in a packed column using supercritical carbon dioxide, *Proceedings of the Fifth International Symposium on Supercritical Fluids*, Atlanta, Georgia, U.S.A.

12

Characterization of Enzymes Associated with Degradation of Insoluble Fiber of Soybean Curd Residue by *Bacillus subtilis*

Makoto Shoda and Shinji Mizumoto
Chemical Resources Laboratory, Tokyo Institute of Technology,
Nagatsuta, Midori-ku, Yokohama,
Japan

1. Introduction

Soybean curd residue is a residue of soy milk processing in which most soluble nutrients of soybean are extracted to liquid phase, and thus major carbon sources of the residue are insoluble fibers (O'tool, 1999) which amount to 40.2- 43.6 % on a dry matter basis (Van der Riet et al.,1989). Approximately 700,000 tons of the soybean curd residue were produced annually as a byproduct of *tofu* manufacturing in Japan and most of them is incinerated as an industrial waste. We re-utilized the soybean curd residue as a solid substrate of solid-state fermentation (SSF) using *Bacillus subtilis* (Mizumoto et al., 2006).

The insoluble fibers of soybean consist of cellulose, hemicellulose and lignin. Cellulose is the most abundant biological polymer on earth and is the major constituent of the plant cell wall. This lineal polymer is composed of D-glucose subunits linked by β-1,4 glycosidic bonds forming cellobiose molecules and the long chains are linked together by hydrogen bonds and van der Waals forces (Perez et al., 2002). Hemicellulose is a complex of polymeric carbohydrates which contains xylan, xyloglucan, (heteropolymer of D-xylose and D-glucose), glucomannan (heteropolymer of D-glucose and D-mannose), galactoglucomannan (heteropolymer of D-galactose, D-glucose and D-mannose) and arabinogalactan (heteropolymer of D-galactose, D-glucose and arabinose). Among them, xylan, a complex polysaccharide comprising a backbone of xylose residues linked by β-1,4-glycosidic bonds, is the major component. Xylan is the second most abundant polysaccharide in nature, accounting for approximately one-third of all renewable organic carbon on earth (Collins et al., 2005). Lignin is an amorphous non-water soluble and optically inactive heteropolymer. It consists of phenylpropane units joined together by different types of linkages (Perez et al., 2002) Although lignin is the most abundant polymer in wood fiber along with cellulose, its content in non-wood fiber such as straw, grass and seed hull is low (Sun & Cheng, 2002). The lignin content in the soybean seed coat is reported to be low (Krzyzanowski et al., 2001), and thus it is speculated that the soybean curd residue contains relatively small amount of lignin.

B. subtilis has ability to produce several antibiotics with a variety of structures, especially peptides that are either ribosomally or non-ribosomally synthesized (Leclere et al., 2005; Ongena et al., 2005; Stein, 2005). We previously isolated several strains of *B. subtilis* and the wild strains and their derivatives suppressed 26 types of plant pathogen *in vitro* (Phae et al., 1990) and a fungal disease *in vivo* (Asaka & Shoda, 1996) by producing three lipopeptide antibiotics, iturin A, surfactin and plipastatin (Asaka & Shoda, 1996; Hiraoka et al., 1992; Tsuge et al., 1996, 1999). The suppressive effect of one of the isolates, *B. subtilis* RB14, was mainly associated with the cyclolipopeptide antibiotic iturin A, which contains seven α-amino acids and one β-amino acid. *B. subtilis* RB14-CS, a derivative of the original strain RB14 and a sole producer of iturin A, produced iturin A in SSF using the soybean curd residue 3-fold higher than in submerged fermentation (SmF) (Mizumoto et al., 2006). This suggests that RB14-CS could degrade some kinds of insoluble fibers in soybean curd residue and utilize them as carbon sources during SSF. In this chapter, insoluble fibers in soybean curd residue that RB14-CS could degrade during SSF were clarified and the fiber-degrading enzymes were purified and characterized.

2. Materials and methods

2.1 Strain
B. subtilis RB14-CS which is a spontaneous mutant derived from RB14-C is a single iturin A producer. *B. subtilis* RB14-C is a streptomycin-resistant mutant from a parent strain RB14 and is a co-producer of the antibiotics iturin A and surfactin (Asaka & Shoda, 1996).

2.2 Solid-state fermentation (SSF)
The detail of SSF was described in the previous paper (Mizumoto et al., 2006). The L medium used for the growth of the bacterium contained 10 g of Polypepton (Nippon Pharmaceutical Co., Tokyo, Japan), 5 g of yeast extract and 5 g of NaCl (per liter). One ml of L medium culture broth after 24 h cultivation at 30°C was inoculated into 100 ml of number 3S (no. 3S) medium consisting of 30 g of Polypepton S (Nippon Pharmaceutical Co., Tokyo), 10 g of glucose, 1 g of KH_2PO_4, and 0.5 g of $MgSO_4\cdot7H_2O$ (per liter) (pH 6.8), and the culture was incubated at 120 strokes per minute (spm) at 30°C for 24 h in a shaking flask and used as a seed for SSF.

The soybean curd residue was supplied from a *tofu* company in Tokyo and stored at - 20°C. Each of fifteen grams of thawed soybean curd residue was placed in a 100-ml conical flask and autoclaved twice at 120°C for 20 min at an interval of 8-12 h to kill spore-forming microorganisms inhabiting the material. After cooling to room temperature, the following solutions were added as nutrient supplements for every 15 g of soybean curd residue and moisture content was adjusted to 79%: 833 μL of 0.45 g glucose /ml, 75 μL of 1 M KH_2PO_4, 150 μL of 1 M $MgSO_4\cdot7H_2O$ and 367 μL of deionized distilled water. Then, 3 mL of an RB14-CS culture grown in no. 3S medium was added to 15 g of soybean curd residue and mixed with a stainless steel spatula. All flasks were incubated statically in a water incubator at 25°C, and at a specified time, one flask was taken and the whole soybean curd residue in a flask was used as a sample for analysis.

2.3 Preparation of samples for acid and neutral detergent fiber analysis
After 5 days of SSF by *B. subtilis* RB14-CS, the whole solid culture was dried by microwave and ground by using a pestle and a mortar. Raw soybean curd residue was used as a control.

2.4 Acid and neutral detergent fiber analysis
2.4.1 Acid detergent fiber
The content of acid detergent fiber, which contains mainly cellulose and lignin, was analyzed in the following manner (Van Soest, 1963). In a 150 mL-flat bottom flask, 0.45 – 0.55 g of ground sample was weighed using micro-balance and 50 mL of acid detergent solution (20 g/L cetyl trimethylammonium bromide in 0.5 M sulfuric acid) was mixed. The flask was placed in an oil bath under the cold water condenser and boiled within 5-10 min. Sample was refluxed for 60 min from onset of boil. After approximately 30 min, the inside of flask was washed with minimal amount of acid detergent solution. After refluxed, sample was filtrated under reduced pressure with a tared Gooch crucible. The crucible was washed twice with hot water, then twice with acetone and was dried at 105°C overnight. After cooled to room temperature in a desiccator, the weight of the crucible was measured.

2.4.2 Neutral detergent fiber
The content of neutral detergent fiber which contained mainly cellulose, lignin, and hemicellulose was analyzed in the following manner (Van Soest, et al., 1991). In a 300 mL-round bottom flask, 0.45 – 0.55 g of ground sample, 50 mL of neutral detergent solution (13.5 g of sodium dodecyl sulfate, 8.38 g of EDTA disodium salt, 3.07 g of $NaB_4O_7 \cdot 10H_2O$, 5.18 g of $Na_2HPO_4 \cdot 12H_2O$ and 4.5 mL of tryethylene glycol per 450 mL) and 0.5 g of sodium sulfite were mixed. The flask was placed in an oil bath under the cold water condenser and boiled for 5 min. After 5 min of boiling, 2 mL of α-amylase solution, which consists of heat-stable α-amylase (Kleistase T10S; Daiwa Kasei, Shiga, Japan) and 50 mM sodium phosphate buffer (pH 6.0) (1:39 [vol/vol]), were mixed. Then, the sample was refluxed for 60 min. After approximately 30 min, the inside of flask was washed down with minimal amount of neutral detergent solution. After refluxed, the sample was filtrated under reduced pressure with a tared Gooch crucible. The crucible was filled with 2 mL of α-amylase solution and hot water, and incubated for at least 2 min. Then, the crucible was washed twice with hot water, and then twice with acetone. The crucible was dried at 105°C overnight. After cooled to room temperature in a desiccator, the weight of the crucible was measured.

2.4.3 Calculation of content of insoluble fibers
As the amount of acid detergent fiber was regarded as total amount of cellulose and lignin, the amount of the neutral detergent fiber minus the amount of acid detergent fiber was regarded as the content of hemicellulose.

2.5 Iturin A production in liquid culture using insoluble fibers
In a 200-mL conical flask, 40 mL of liquid medium consisting of 10 g of fibrous carbon sources, 10 g of Polypepton S, 1 g of KH_2PO_4 and 0.5 g of $MgSO_4 \cdot 7H_2O$ (per liter) (pH 6.8) was prepared. As fibrous carbon sources, xylan (Tokyo Chemical Industry, Tokyo, Japan), avicel, carboxymethyl cellulose, and pectin were used. As a control carbon source, glucose was used. Four hundreds μL of a seeding culture was inoculated into the medium and the flasks were incubated at 30°C at 120 spm.

For measurement of iturin A concentration, 1 mL of culture broth was acidified to pH 2.0 with 12 N HCl. Iturin A was collected by centrifugation at 18,000 $\times g$, at 4°C for 10 min, and extracted with 1 mL of methanol. The extract was injected into a high-performance liquid chromatography (HPLC) with a column (Chromolith Performance RP-18eb 4.6 mm diameter× 100 mm height, Merck, Germany) to determine iturin A concentrations. The HPCL system was operated at a flow rate of 2.0 mL/min with acetonitrile-10 mM ammonium acetate (65:35 [vol/vol]) at a column temperature of 40°C. The elution was monitored at 205 nm by a UV detector (880-UV, Intelligent UV/VIS Detector, Jasco, Tokyo, Japan).

Although iturin A has 8 homologues with different side-chain structures (Asaka & Shoda, 1996), the concentration of iturin A was defined as the total amount of five major homologues. The correlation between the peak heights and the concentration of pure iturin A (Sigma-Aldorich, Tokyo, Japan) was used for quantification. Iturin A concentration was expressed as μg/ g initial wet soybean curd residue.

2.6 Xylanase activity assay

Dinitrosalicylic acid (DNS) solution was prepared in the following manner. Solution A was prepared by mixing 300 mL of 4.5 % NaOH, 880 mL of 1 % 3,5- DNS and 225 g of potassium sodium (+)-tartrate tetrahydrate. For the preparation of solution B, 22 mL of 10 % NaOH and 10 g of phenol was mixed and filled up to 100 mL. To 69 mL of the mixture, 6.9 g of $NaHCO_3$ was added. Solutions A and B were mixed thoroughly and placed at room temperature for 2 days. After filtration, the mixture was used as DNS solution.

Xylanase activity was determined by measuring the amount of reducing sugar released from xylan. One hundred μL of enzyme sample was added to 1 mL of 1 % xylan in 100 mM sodium phosphate buffer (pH 6.5) in a test tube (15 mmΦ × 10.5 cm) and incubated statically at 50°C for 5 min. Two mL of DNS solution was added and cooled immediately in an ice bath. Then the test tubes were boiled for 5 min and cooled in an ice bath. After centrifugation at 18,000×g at 4°C for 5 min, absorbance of the supernatant at 540 nm was measured by spectrophotometer (UV2400; Shimadzu, Kyoto, Japan). Xylose was used as the standard. One unit (U) of xylanase activity was defined as the amount of enzyme that liberates 1 μmol of reducing sugars (xylose equivalent) per min.

2.7 Measurement of xylanase activity during SSF

SSF was carried out as described in Section 2.2 without addition of glucose. One gram of solid culture sample and 9 ml of sterile distilled water were mixed in a sterile 18-mm-diameter test tube, the test tube was vortexed thoroughly and shaken at 150 spm for 5 min at room temperature. The suspension was centrifuged at 18,000 ×g at 4 °C for 10 min and the supernatant obtained was used for xylanase assay.

2.8 Measurement of concentration of protein

Protein concentrations were determined by the Bradford method (Bradford, 1976) with the Protein Assay Kit II (Bio-Rad, Tokyo, Japan) with bovine serum albumin as the standard protein.

2.9 Purification of xylanase

Solid cultures (90 g) incubated for 5 days in SSF were mixed with 900 mL of distilled water and stirred for 10 min. The suspension was centrifuged at 6,500×g at 4°C for 20 min and the supernatant was frozen at -20°C and then thawed. The sample was centrifuged for removal of polysaccharides under the same condition. Ammonium sulfate was added to the supernatant to 30 % saturation, and the precipitate was removed by centrifugation. Then, ammonium sulfate was added to 70 % saturation. The precipitate was recovered by centrifugation, suspended in 50 mM MES buffer (pH 6.0) and dialyzed overnight against the same buffer. Then the sample was concentrated by ultrafiltration with YM10 (molecular mass cut-off 10 kDa; Advantec, Tokyo, Japan).

The concentrate was applied to a CM-Toyopearl column (1.3 cm Φ×8.3 cm; Tosoh, Tokyo, Japan) pre-equilibrated with buffer A (50 mM MES buffer, pH 6.0), and fractions were eluted with a continuous linear gradient of 0-0.5 M NaCl in buffer A (total volume 120 mL). The flow speed and the volume of one fraction were 4 mL/min and 8 mL, respectively. In this process, xylanase activity was detected in two fractions, one of which was trapped in the column (Fraction I) and the other was not trapped in the column but passed through (Fraction II). These fractions were subjected to further purification processes.

Fraction I was concentrated using Centriprep YM-10 (molecular mass cut-off 10 kDa; Millipore, Tokyo, Japan), diluted with buffer A and applied to a RESOURCES column (0.6 cm Φ×3.0 cm; Pharmacia Biotech, Uppsala, Sweden) pre-equilibrated with buffer A. Fractions were eluted with a continuous linear gradient of 0-0.15 M NaCl in buffer A (total volume 30 mL). The flow speed and the volume of one fraction were 1 mL/min and 1 mL, respectively. The xylanase active fractions were concentrated with Centriprep YM-10 and applied to a Superdex 75 column (1.6Φ×60 cm; Amersham Bioscience, Tokyo, Japan) pre-equilibrated with buffer A containing 0.2 M NaCl. The elution was carried at a flow rate of 1 mL/min and a volume of one fraction was 2 mL.

The pH of the Fraction II was adjusted to 9.5 by adding NaOH and applied to a QAE-Toyopearl (1.6Φ×3.7 cm; Tosoh) pre-equilibrated with buffer B (25 mM piperazine buffer, pH 9.5), and fractions were eluted with a continuous linear gradient of 0-0.5 M NaCl in buffer B (total volume 120 mL). The flow speed and the volume of one fraction were 4 mL/min and 8 mL, respectively. The xylanase active fractions were concentrated with Centriprep YM-10, and fractions were diluted with buffer B and applied to a QAE column. Step elution was performed with 0.07 M NaCl (total elution volume 96 mL). The flow speed and the volume of one fraction were 4 mL/min and 8 mL, respectively.

The xylanase active fractions were supplied to the subsequent Butyl-Toyopearl chromatography. A column of Butyl-Toyopearl (1.6Φ×4.5 cm; Tosoh) pre-equilibrated with 25 mM piperazine buffer containing 1 M ammonium sulfate was used. Ammonium sulfate was added to the active fractions and its concentration was adjusted to 1 M. This solution was then applied to the column and the elution was carried out with a linear gradient of 1-0 M ammonium sulfate in 25 mM Piperazine buffer (total volume 180 mL). The flow speed and the volume of one fraction were 4.5 mL/min and 9 mL, respectively.

2.10 Molecular mass determination
Sodium dodecyl sulfate-polyacrylamide gel electrophoresis (SDS-PAGE) was performed with a 12.5 % gel in accordance with the Laemmli method (Laemmli, 1970). M. W. Marker "Daiichi" II (Daiichi Pure Chemicals, Tokyo, Japan) was used as a molecular mass marker. After electrophoresis, the gel was stained with Coomassie brilliant blue R-250 (CBB).

2.11 N-terminal sequence analysis
SDS-PAGE of xylanases was performed according to the above-described method and then the xylanases on the gel were electroblotted to a commercial membrane (Immobilon-P;

Millipore, Tokyo, Japan) with a horizontal blotting apparatus (ATTO, Tokyo, Japan). For the blotting of pure enzyme of Fraction II, 0.01 % of SDS was added to transfer buffer to improve protein transfer efficacy. Parts of the membrane blotted with xylanases were cut out and then amino acid sequencing analysis was performed with an amino acid sequencing apparatus (PPSQ-21; Shimadzu, Kyoto, Japan) according to the standard method (Edman, 1949).

Searches for homologous amino acid sequences were performed by a *B. subtilis* database BSORF (http://bacillus.genome.jp/) and the nonredundant database at The National Center for Biotechnology Information (http://www.ncbi.nlm.nih.gov/) with the BLASTP.

2.12 pH and temperature profiles and thermostability of xylanases
Xylanase activity was examined in pH range of 3.0 to 11.0. For pH from 3.0 to 4.0, 100 mM sodium citrate buffer was used. For pH from 4.0 to 6.0, 100 mM sodium acetate buffer was used. For pH from 6.0 to 8.0, 100 mM sodium phosphate buffer was used. For pH from 8.0 to 9.0, 100 mM Tris-HCl buffer was used. For pH from 9.0 to 11.0, 100 mM glycine-NaOH buffer was used. To investigate the effect of temperature, the xylanase activity was measured at 20-70°C at pH 6.5. Xylanase thermostability was measured at 50, 55 and 60°C.

2.13 Thin layer chromatography (TLC) analysis of the digestion products
The digestion products of xylan and xylooligosaccharides (Wako Pure Chemical Industries, Osaka, Japan) by xylanase were analyzed by thin layer chromatography (TLC) according to the method previously reported (Kiyohara et al., 2005) with some modifications.

As a substrate solution, 0.5 % xylan or 0.5 % xylooligosaccharides in 100 mM sodium phosphate buffer (pH 6.5) was used. In a test tube (15 mmΦ × 10.5 cm), 0.5 mL of substrate solution and 0.5 mL of enzyme solution containing 0.5 U of xylanase in 100 mM sodium phosphate buffer (pH 6.5) were mixed and the reaction mixture was incubated at 120 spm at 37°C. After 1, 3, and 16 h of incubation, 100 µL of reaction mixture was sampled to microtube, and mixed with 200 µL of ethanol. Then, the mixture was centrifuged at 18,000×g for 10 min and the supernatant obtained was evaporated with a centrifugal concentrator (VC-36N; Taitec, Saitama, Japan). The dried material was dissolved in distilled water and spotted on a Silica Gel 60 TLC plate (Merck, Tokyo, Japan), which was then developed with *n*-butanol/acetic acid/ water (10:5:1, by vol.). After development, the TLC plate was sprayed with aniline hydrogen phthalate reagent. The reagent consisted of 0.93 g of aniline, 1.48 g of phthalic anhydride, 84.5 mL of *n*-butanol and 15.5 mL of distilled water (Partridge, 1949), and heated at 100°C to visualize the digestion products.

3. Results

3.1 Degradation of insoluble fibers in soybean curd residue by *B. subtilis* RB14-CS in SSF
To evaluate the ability of *B. subtilis* RB14-CS to degrade insoluble fibers in soybean curd residue, residual fibers after SSF were analyzed by acid and neutral detergent fiber methods. The same analyses were repeated three times. The average values of three samples are shown in Figure 1. After SSF of RB14-CS, no change in content of cellulose and lignin was observed. On the other hand, the content of hemicellulose decreased to 15 % of initial one, indicating that RB14-CS degraded hemicellulose in soybean curd residue.

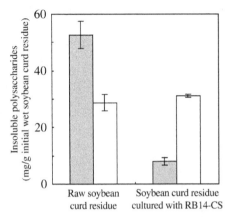

Fig. 1. Analysis of insoluble fiber contents in raw soybean curd residue and soybean curd
residue cultured with B. subtilis RB14-CS (N=3). Gray bars, hemicellulose; Open bars,
cellulose and lignin.

3.2 Iturin A production by *B. subtilis* RB14-CS using insoluble fibers in submerged fermentation

To investigate the effect of insoluble fibers on iturin A production of RB14-CS, each of
insoluble fibers was added to a liquid medium as a carbon source and RB14-CS was
cultivated in the medium. Results are shown in Figure 2. Xylan exhibited iturin A
production at the same level with glucose which has been used as a carbon source for iturin
A production in the previous reports (Asaka & Shoda, 1996; Tsuge et al., 2001). Other
insoluble fibers, avicel and carboxymethyl cellulose, showed the similar level of iturin A
production with control where no additional carbon was added. Pectin, a hardly-soluble or
sometimes insoluble fiber which is contained in soybean curd residue (Kasai et al., 2004) did
not enhance the iturin A production.

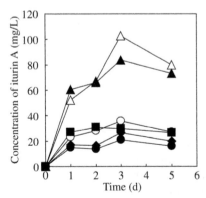

Fig. 2. Iturin A production during submerged fermentation in liquid medium containing
fibers (N=3). Symbols: open circles, no additional carbon sources (control 1); open triangles,
glucose (control 2); solid circles, pectin; solid triangles, xylan; solid squares, avicel; solid
diamonds, carboxymethyl cellulose.

3.3 Xylanase activity of *B. subtilis* RB14-CS during SSF

As RB14-CS degraded xylan, a major hemicellulose in plant cell wall (Beg et al., 2001), in submerged fermentation, xylanase activity was measured during SSF, in which glucose was not added as medium component. Results are shown in Figure 3. The culture of RB14-CS exhibited xylanase activity in SSF. The activity increased after 12 h of incubation, reached the maximum value of approximately 50 U/g wet soybean curd residue at 3 d, and maintained the level during fermentation. When xylanase activity was detected, almost no reducing sugars were detected (data not shown), indicating that RB14-CS immediately utilized the saccharides released from hemicellulose as carbon sources. Changes in cell number and pH were similar to those in SSF of RB14-CS using soybean curd residue previously reported (Mizumoto et al., 2006).

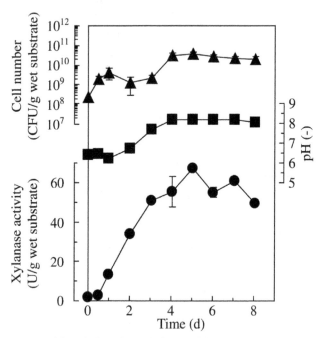

Fig. 3. Xylanase activity of *B. subtilis* RB14-CS during SSF.
Symbols: circles, xylanase activity; squares, pH; triangles, viable cell number.

3.4 Purification of xylanases produced by *B. subtilis* RB14-CS in SSF

Xylanases were purified as described in materials and methods. When the crude enzyme solution was applied to a cation exchange CM-Toyopearl column, xylanase activity was found in both the trapped fraction (Fraction I) and non-trapped fraction (Fraction II). From these fractions, two enzymes were purified and the two enzymes are homogeneous and have different sizes because each single protein band on SDS-PAGE was observed (Figure 4). This indicates that RB14-CS produces two different xylanases. Purified enzymes of Fraction I and II were designated as Xyl-I and Xyl-II, respectively. The molecular masses of the Xyl-I and Xyl-II estimated from SDS-PAGE were 24 and 58 kDa, respectively.

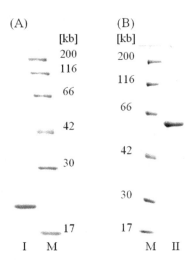

Fig. 4. SDS-PAGE of purified xylanases.(A) Xyl-I, (B) Xyl-II
Lanes: M, molecular mass standards; I, Xyl-I; II, Xyl-II.

3.5 Physicochemical properties of xylanases

Effects of temperature and pH on xylanase activity and thermal stability of the two enzymes
are shown in Figure 5. The optimal temperature and optimal pH of Xyl-I were 50-60°C and
6-7, respectively. At 50°C, approximately 30 % of the initial activity of Xyl-I remained after 3
h. At 55 and 60°C, Xyl-I was completely inactivated within 2 and 3 h and the half lives were
approximately 18 and 8 min, respectively. The optimal temperature of Xyl-II was 70°C or
higher and the optimum pH was 5.5-6. At 50°C, approximately 80 % of the initial activity of
Xyl-II remained after 3 h. At 60°C, Xyl-II was inactivated within 3 h and the half life was
approximately 40 min.

3.6 Analysis of hydrolytic products

The hydrolysis products released from xylan or xylooligosaccharides by Xyl-I and Xyl-II
were analyzed by TLC. From hydrolysis of xylan by both Xyl-I and Xyl-II xylotriose was
liberated, but neither xylose nor xylobiose was released. This indicats that these xylanases
were not β-D-xylosidase.

3.7 Identification of xylanases by N-terminal sequencing and database matching

The N-terminal sequences of Xyl-I and Xyl-II were determined by automated Edman
degradation and compared with databases. Results are summarized in Table 1. Xyl-I
displayed 90 % amino acid identity with endo-1,4-β-xylanase (XynA) of B. subtilis 168, a
standard strain whose complete genome has been sequenced (Kunst et al., 1997). The
molecular mass estimated by SDS-PAGE was similar to the database value. Moreover, pI
value (9.64) of database was identical to that of purified Xyl-I.

Xyl-II has exactly the same N-terminal sequence as α-amylase (AmyE) secreted by B. subtilis
X-23 (Ohdan et al., 1999). Actually, Xyl-II exhibited α-amylase activity because reducing
sugar was increased when soluble starch was treated with Xyl-II (data not shown). It is

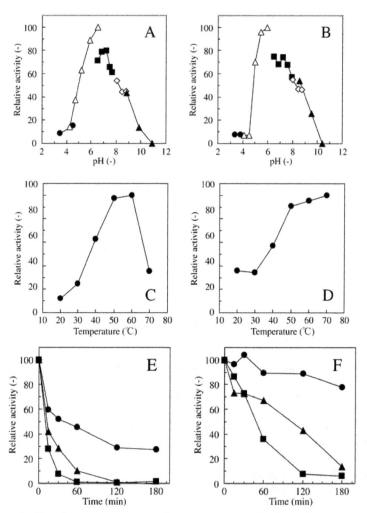

Fig. 5. Effects of pH and temperature on xylanase activities of Xyl-I and Xyl-II. Effects of pH on Xyl-I (A) and -II (B), respectively; Effects of temperature on Xyl-I (C) and -II (D), respectively; Thermal stability of Xyl-I (E) and -II (F), respectively. Symbols in (E) and (F): circles, 50°C; triangles, 55°C; squares, 60°C.

assumed that 45 amino acid residues prior to these sequenced residues deduced from the nucleotide sequence of the *B. subtilis* X-23 are the signal peptide that is removed during the secretion process. Xyl-II also displayed 80 % amino acid identity with α-amylase of *B. subtilis* 168 (Kunst et al., 1997). Although the molecular mass of Xyl-II estimated from SDS-PAGE was different from those in the previous reports, the C-terminal structures of α-amylase of *B. subtilis* were reported to be variable (Ohdan, et al., 1999). The pI value of α-amylase of *B. subtilis* 168 (5.85) was identical with the value of purified Xyl-II. This also reflected in that Xyl-II was trapped in anion exchange chromatography when piperazine buffer of pH 9.5 was used for elution.

This work			Database					
Sequence	Size (kDa)		Sequence	Gene	Protein identity	Size (kDa)	pI	References
Xyl-I AGTDYWQNWT	24		ASTDYWQNWT	*xynA*	endo-1,4-β-xylanase	23	9.64	Kunst et al.
Xyl-II SVKNGTILHA	58		SVKNGTILHA	*amyE*	α-amylase	47, 67	-	Ohdan et al.
			SIKSGTILHA	*amyE*	α-amylase	73	5.85	Kunst et al.

Table 1. N-terminal amino acid sequences of purified xylanases.

4. Discussion

B. subtilis RB14-CS degraded xylan in soybean curd residue and utilized it as a carbon source during SSF by producing xylanases. Xylanases are produced from xylan by fungi, yeast and bacteria, including *Bacillus* sp. (Beg et al., 2001; Blanco et al.,1995; Gallardo et al., 2004;Heck et al., 2005; Sa-Pereira et al., 2003) and physicochemical properties, structures and specific activities of these xylanases were diverse.

In this study, two xylanase-active enzymes were isolated. One of them (Xyl-I)was endo-1,4-β-xylanase (XynA), which has been found in many strains of *Bacillus* sp.(Blanco et al., 1995; Gallardo et al., 2004; Nishomoto et al., 2002). Characteristics of the Xyl-I obtained in this work are similar to those previously reported in that there is β-D-glucosidase activity and the values of optimum pH and temperature of Xyl-I are similar to those in other xylanases (Table 2). Another xylanase-active enzyme obtained (Xyl-II) was identified as α-amylase. As shown in Table 2, physicochemical properties of Xyl-II except for molecular mass were similar to those reported previously. Distribution of α-amylase is wide from common mesophilic bacteria to hyperthermophilic archaeon *Pyrococcus furiosus* (Jorgensen et al., 1997). Alpha-amylase of *B. subtilis* is used commercially in various categories such as starch hydrolysis in starch liquefaction process and additives to detergents for both washing machines and automated dish-washers because of its high thermo-stable activity (Nielsen & Borchert, 2000). As α-amylase, which catalyzes the hydrolysis and transglycosylation at α-1,4- and α-1,6-glycosidic linkages, it doesn't seem to be responsible for degradation of xylan. However, it has been shown that, due to the heterogeneity and structural complexity of xylan, the complete hydrolysis of xylan requires a large variety of cooperatively acting enzymes; such as endo-1,4-β-D-xylanases, β-D-xylosidase, α-L-arabinofuranosidases, α-D-glucuronidases, acetylxylan esterases, ferulic acid esterases and *p*-coumaric acid esterases (Collins et al., 2005). Thus, α-amylase of RB14-CS which hydrolyzed α-1,4- or 1,6-glucoside linkage in the reagent grade xylan used in this study may act as the cooperatively acting enzymes to release reducing sugars from xylan.

Two enzymes isolated in this work liberated xylooligosaccharides but not xylose from xylan. However, almost no reducing sugars were detected when xylanase activity was detected in SSF. This indicates that RB14-CS degraded xylooligosaccharides into xylose and utilized it as a carbon source. RB14-CS may produce other enzymes such as β-D-xylosidase for this reaction.

In recent years, biomass containing hemicellulose, such as agricultural and forestry residues, waste paper, and industrial wastes, has been recognized as inexpensive and abundantly available sources of sugar (Katahira et al., 2004). Since the production of iturin A by RB14-

CS in soybean curd residue was almost equivalent to that when glucose was used as carbon source, the utilization of soybean curd residue will be one possible nutrient in peptide production.

| | This work | References | |
		Gallardo et al.	Blanco et al.
Molecular mass	24 kDa	24 kDa	32 kDa
Optimum pH	6-7	6	5.5
Optimum temperature	50-60°C	60°C	50°C
Thermal stability	Decreased to 30 % at 50°C after 3 h. Deactivated within 1 h at 60°C.	Remained stable at 50°C for at least 3 h. Deactivated within 1 h at 60°C.	-

(A) Xyl-I

| | This work | Reference | (Ohdan et al.) |
		Ba-S	Ba-L
Molecular mass	58 kDa	47 kDa	67 kDa
Optimum pH	5.5-6.0	5.5	5.5
Optimum temperature	70°C	65°C	65°C
Thermal stability	80 % was retained after 3 h at 50°C. Deactivated within 2 h at 60°C.	60 % was retained after 10 min at 65°C	30 % was retained after 10 min at 65°C

(B) Xyl-II

Table 2. Comparison of characteristics of purified xylanases with previous reports.

5. Conclusion

Soybean curd residue which is the residue of *Tofu* production was used for nutrients for production of a lipopeptide antibiotic, iturin A in solid state fermentation(SSF) using *Bacillus subtilis*. As the main carbon sources of soybean curd residue were insoluble fiber, we expected that *B. subtilis* produced the soybean curd residue-degrading enzymes. Among insoluble fibers in soybean curd residue, hemicellulose was mainly degraded by *B. subtilis* during SSF. Xylan, a major hemicellulose in plant cell wall was degraded by *B. subtilis*, and two enzymes which showed xylanase activity were purified and identified as endo-1,4-β-xylanase and α-amylase. As productivity of iturin A in soybean curd residue was almost equivalent to that in glucose medium, this study gave a possible way to use soybean curd residue in higher and economical production of lipopeptides.

6. References

Asaka, O. & Shoda, M. (1996). Biocontrol of *Rhizoctonia solani* damping-off of tomato with *Bacillus subtilis* RB14. *Applied and Environmental Microbiology*, Vol.62, pp. 4081-4085

Beg, Q. K.; Kapoor, M.; Mahajan, L. & Hoondal, G. S. (2001). Microbial xylanases and their industrial applications: a review. *Applied Microbiology and Biotechnology*, Vol.56, pp. 326-338

Characterization of Enzymes Associated with Degradation of Insoluble Fiber of Soybean Curd
Residue by Bacillus subtilis
271

Blanco, A.; Vidal, T.; Colom, J. F. & Pastor, F.J. I. (1995). Purification and properties of xylanase A from alkali-tolerant Bacillus sp. strain BP-23. Applied and Environmental Microbiology, Vol.61, pp.4468-4470

Bradford, M. M. (1976). A rapid and sensitive method for the quantification of microgram quantities of protein utilizing the principle of protein-dye binding. Analytical Biochemistry, Vol.72, pp. 248-254

Collins, T.; Gerday, C. & Feller, G. (2005). Xylanases, xylanase families and extremophilic xylanases. FEMS Microbiological Review, Vol.29, pp. 3-23

Edman, P. (1949). A method for the determination of the amino acid sequence in peptides. Archives of Biochemistry and Biophysics, Vo.11, pp. 475-476

Gallardo, O.; Diaz, P. & Javier Pastor, F. I. (2004). Cloning and characterization of xylanase A from the strain Bacillus sp. BP-7: Comparison with alkaline pI-low molecular weight xylanases of family 11. Current Microbiology, Vol.48, pp.276-279

Heck, J. X.; Flores, S. H.; Hertz, P. F. & Ayub, M. A. Z. (2005). Optimization of cellulase-free xylanase activity produced by Bacillus coagulans BL69 in solid-state cultivation. Process Biochemistry, Vol.40, pp.107-112

Hiraoka, H.; Ano, T. & Shoda, M.(1992). Characterization of Bacillus subtilis RB14, coproducer of peptide antibiotics iturin A and surfactin. Journal of General and Applied Microbiology, Vol.38, pp.635-640

Jorgensen, S.; Vorgias, C. E. & Antranikian, G. (1997). Cloning, sequencing, characterization, and expression of an extracellular α-amylase from the hyperthermophilic archaeon Pyrococcus furiosus in Escherichia coli and Bacillus subtilis. Journal of Biological Chemistry, Vol.272, pp.16335-16342

Kasai, N.; Murata, A.; Inui, H.; Sakamoto, T. & Kahn, R. I. (2004).Enzymatic high digestion of soybean milk residue (Soybean curd residue). Journal of Agricultural and Food Chemistry, Vol.52, pp. 5709-5716

Katahira, S.; Fujita, Y.; Mizuike, A.; Fukuda, H. & Kondo, A. (2004). Construction of a xylan-fermenting yeast strain through codisplay of xylanolytic enzymes on the surface of xylose-utilizing Saccharomyces cerevisiae cells. Applied and Environmental Microbiology, Vol.70, pp.5407-5414

Kiyohara, M.; Sakaguchi, K.; Yamaguchi, K.; Araki, T.; Nakamura, T. & Ito, M. (2005). Molecular cloning and characterization of a novel β-1,3-xylanase possessing two putative carbohydrate-binding modules from a marine bacterium Vibrio sp. strain AX-4. Biochemical Journal, Vol.388, pp. 949-957

Krzyzanowski F. C.; Franca-Neto, J. B.; Mandarino, J. M. G. & Kaster, M. (2001). Comparison between two gravimetric methods to determine the lignin content in soybean seed coat. Seed Science and Technology, Vol.29, pp. 619-624

Kunst, F.; Ogasawara, N. & 149 other authors (1997). The complete genome sequence of the Gram-positive bacterium Bacillus subtilis. Nature, Vol.390, pp.249-256

Laemmli, U. K. (1970). Cleavage of structural proteins during the assembly of the head bacteriophage T4. Nature, Vol.277, pp. 680-685

Leclere, V.; Bechet, M.; Adam, A.; Guez, J.S.; Wathelet, B.; Ongena, M.; Thonart, P.; Gancel, F.; Chollet-Imbert, M. & Jacques, P. (2005). Mycosubtilin overproduction by Bacillus subtilis BBG100 enhances the organism's antagonistic and biocontrol activities. Applied and Environmental Microbiology, Vol. 71, pp.4577-4584

Mizumoto, S.; Hirai M. & Shoda, M. (2006). Production of lipopeptide antibiotic iturin A using soybean curd residue cultivated with Bacillus subtilis in solid-state fermentation, Applied Microbiology and Biotechnology, Vol.72, pp.869-875

Nielsen, J. E. & Borchert, T. V. (2000). Protein engineering of bacterial α-amylases. *Biochimica et Biophysica Acta*, Vol.1543, pp.253-274

Nishimoto, M.; Honda, Y.; Kitaoka, M. & Hayashi, K. (2002). A kinetic study on pH-activity relation ship of XynA from alkaliphilic *Bacillus halodurans* C-125. *Journal of Bioscience and Bioengineering*, Vol.93, pp.428-430

Ohdan, K.; Kuriki, T.; Kaneko, H.; Shimada, J.; Takada, T.; Fujimoto, Z.; Mizuno, H. & Okada, S. (1999). Characteristics of two forms of α-amylases and structural implication. *Applied Environmental Microbiology*, Vol.65, pp. 4652-4658

Ongena, M.; Jacques, P.; Toure, Y.; Destain, J.; Jabrane, A. & Thonart, P. (2005). Involvement of fengycin-type lipopeptides in the multifaceted biocontrol potential of *Bacillus subtilis*. *Applied Microbiology and Biotechnology*, Vol.69, pp.29-38

O'tool, D. K. (1999). Characteristics and use of soybean curd residue, the soybean residue from soy milk production – A review. *Journal of Agricultural and Food Chemistry*, Vol.47, pp.363-371

Partridge, S. M. (1949). Aniline hydrogen phthalate as a spraying reagent for chromatography of sugars. *Nature*, Vol.164, pp. 443

Perez, J.; Munoz-Durado, J.; de la Rubia, T. & Martinez, J. (2002). Biodegradation and biological treatments of cellulose, hemicellulose and lignin: an overview. *International Microbiology*, Vol.5, pp. 53-63

Phae, C. G.; Shoda, M. & Kubota, H. (1990). Suppressive effect of *Bacillus subtilis* and its products to phytopathogenic microorganisms. *Journal of Fermentation and Bioengineering*, Vol. 69, pp.1-7

Sa-Pereira, P.; Paveia, H.; Costa-Ferreira, M. & Aires-Barros, M. R. (2003). A new look at xylanases: an overview of purification strategies. *Molecular Biotechnology*, Vol.24, pp.257-281

Stein, T. (2005). *Bacillus subtilis* antibiotics: structures, syntheses and specific functions. *Molecular Microbiology*, Vol. 56, pp.845-857

Sun, Y. & Cheng, J. (2002). Hydrolysis of lignocellulosic materials for ethanol production: a review. *Bioresources Technology*, Vol.83, pp. 1-11

Tsuge, K.; Ano T. & Shoda, M. (1996). Isolation of a gene essential for biosynthesis of the lipopeptide antibiotics plipastatin B1 and surfactin in *Bacillus subtilis* YB8. *Archives of Microbiology*, Vol. 165, pp. 43-251

Tsuge, K.; Ano, T.; Hirai, M.; Nakamura Y. & Shoda, M. (1999). The genes, *degQ, pps* and *lpa-8* (*sfp*) are responsible for conversion of *Bacillus subtilis* strain 168 to plipastatin production. *Antimicrobial Agents and Chemotherapy*, Vol. 43, pp. 2183-2192

Tsuge, K.; Akiyama, T. & Shoda, M. (2001). Cloning, sequencing, and characterization of the iturin A operon. *Journal of Bacteriology*, Vol.183, pp. 6265-6273

Van der Riet, W. B.; Wight, A. W.; Cilliers, J. J. L. & Datel, J. M. (1989). Food chemical investigation of tofu and its byproduct soybean curd residue. *Food Chemistry*, Vol.34, pp.193-202

Van Soest, P. J. (1963) .Use of detergents in the analysis of fibrous feeds. II. A rapid method for the determination of fiber and lignin. *Journal- Association of Official Analytical Chemistry*, Vol.46, pp. 825-835

Van Soest, P. J.; Robertson, J. B. & Lewis, B. A. (1991). Methods for dietary fiber, neutral detergent fiber, and nonstarch polysaccharides in relation to animal nutrition. *Journal of Dairy Science*, Vol. 74, pp. 3583-3597

Permissions

The contributors of this book come from diverse backgrounds, making this book a truly international effort. This book will bring forth new frontiers with its revolutionizing research information and detailed analysis of the nascent developments around the world.

We would like to thank Prof. Dora Krezhova, for lending her expertise to make the book truly unique. She has played a crucial role in the development of this book. Without her invaluable contribution this book wouldn't have been possible. She has made vital efforts to compile up to date information on the varied aspects of this subject to make this book a valuable addition to the collection of many professionals and students.

This book was conceptualized with the vision of imparting up-to-date information and advanced data in this field. To ensure the same, a matchless editorial board was set up. Every individual on the board went through rigorous rounds of assessment to prove their worth. After which they invested a large part of their time researching and compiling the most relevant data for our readers. Conferences and sessions were held from time to time between the editorial board and the contributing authors to present the data in the most comprehensible form. The editorial team has worked tirelessly to provide valuable and valid information to help people across the globe.

Every chapter published in this book has been scrutinized by our experts. Their significance has been extensively debated. The topics covered herein carry significant findings which will fuel the growth of the discipline. They may even be implemented as practical applications or may be referred to as a beginning point for another development. Chapters in this book were first published by InTech; hereby published with permission under the Creative Commons Attribution License or equivalent.

The editorial board has been involved in producing this book since its inception. They have spent rigorous hours researching and exploring the diverse topics which have resulted in the successful publishing of this book. They have passed on their knowledge of decades through this book. To expedite this challenging task, the publisher supported the team at every step. A small team of assistant editors was also appointed to further simplify the editing procedure and attain best results for the readers.

Our editorial team has been hand-picked from every corner of the world. Their multi-ethnicity adds dynamic inputs to the discussions which result in innovative outcomes. These outcomes are then further discussed with the researchers and contributors who give their valuable feedback and opinion regarding the same. The feedback is then collaborated with the researches and they are edited in a comprehensive manner to aid the understanding of the subject.

Apart from the editorial board, the designing team has also invested a significant amount of their time in understanding the subject and creating the most relevant covers. They scrutinized every image to scout for the most suitable representation of the subject and create an appropriate cover for the book.

The publishing team has been involved in this book since its early stages. They were actively engaged in every process, be it collecting the data, connecting with the contributors or procuring relevant information. The team has been an ardent support to the editorial, designing and production team. Their endless efforts to recruit the best for this project, has resulted in the accomplishment of this book. They are a veteran in the field of academics and their pool of knowledge is as vast as their experience in printing. Their expertise and guidance has proved useful at every step. Their uncompromising quality standards have made this book an exceptional effort. Their encouragement from time to time has been an inspiration for everyone.

The publisher and the editorial board hope that this book will prove to be a valuable piece of knowledge for researchers, students, practitioners and scholars across the globe.

List of Contributors

C. Pirola, D.C. Boffito, G. Carvoli, A. Di Fronzo, V. Ragaini and C.L. Bianchi
Università degli Studi di Milano – Dipartimento di Chimica, Fisica ed Elettrochimica, Milano, Italy

Joanna McFarlane
Oak Ridge National Laboratory, USA

Vanderléa de Souza, Marcos Paulo Vicentim, Lenise V. Gonçalves, Maurício Guimarães da Fonseca and Viviane Fernandes da Silva
INMETRO- National Institute of Metrology, Standardization and Industrial Quality, Directorate of Industrial and Scientific Metrology, Division of Chemical Metrology, Brazil

Rogério de Paula Lana
Universidade Federal de Viçosa, Brazil

Miguel Araujo Medeiros
Universidade Federal do Tocantins, Brasil

Carla M. Macedo Leite and Rochel Montero Lago
Universidade Federal de Minas Gerais, Brasil

Roberto Guimarães Pereira
Fluminense Federal University/TEM/PGMEC/MSG, Niterói, RJ, Brazil

Oscar Edwin Piamba Tulcan
National University of Colombia, Bogota, Colombia

Valdir de Jesus Lameira
INESC, Coimbra, Portugal

Dalni Malta do Espirito Santo Filho
LAFLU/DIMEC/INMETRO, RJ, Brazil

Ednilton Tavares de Andrade
Fluminense Federal University/TER/PGMEC, Brazil

George. E. Meyer
University of Nebraska, Department of Biological Systems Engineering, USA

Daicheng Liu and Fucui Ma
College of Life Science, Shandong Normal University, China

Tatjana Rijavec and Živa Zupin
University of Ljubljana, Slovenia

J.A. Campbell and S.C. Goheen
Battelle, Pacific Northwest National Laboratory, Chemical and Biological, Signature Sciences
Richland, WA, U.S.A.

P. Donald
USDA/ARS, Crop Genetics Research Unit, Jackson, TN, U.S.A.

Carlos F. Torres, Guzmán Torrelo and Guillermo Reglero
Departamento de Producción y Caracterización de Nuevos Alimentos, Instituto de Investigación en Ciencias de la Alimentación (CIAL), CSIC-UAM, Madrid, Spain

Makoto Shoda and Shinji Mizumoto
Chemical Resources Laboratory, Tokyo Institute of Technology, Nagatsuta, Midori-ku, Yokohama, Japan

9 781632 393012